Adaptive Optics

for **Biological Imaging**

ADAPTIVE
OPTICS
for Biological Imaging

Edited by
Joel A. Kubby

CRC Press
Taylor & Francis Group
Boca Raton London New York

CRC Press is an imprint of the
Taylor & Francis Group, an **informa** business

Cover Image: Photograph courtesy of Bob Jagendorf.

CRC Press
Taylor & Francis Group
6000 Broken Sound Parkway NW, Suite 300
Boca Raton, FL 33487-2742

First issued in paperback 2020

Version Date: 20130311

ISBN 13: 978-0-367-57646-2 (pbk)
ISBN 13: 978-1-4395-5018-3 (hbk)

Library of Congress Cataloging-in-Publication Data

Adaptive optics for biological imaging / editor, Joel A Kubby.
  pages cm
  Includes bibliographical references and index.
  ISBN 978-1-4395-5018-3 (hardback)
  1. Microscopy. 2. Optics, Adaptive. I. Kubby, Joel A.

QH205.2.A33 2013
570.28'2--dc23   2013004189

Visit the Taylor & Francis Web site at
http://www.taylorandfrancis.com

and the CRC Press Web site at
http://www.crcpress.com

If we have seen deeper and more clearly it is by standing on the shoulders of giants. One of those giants was Horace Welcome Babcock (1912–2003).

Contents

SECTION I Principles

SECTION II Methods

SECTION III   Applications

PART 1   Indirect Wavefront Sensing

PART 2   Direct Wavefront Sensing

Foreword

This book is a broad and comprehensive introduction to the use of adaptive optics (AO) in biological microscopy. It provides a much-needed entrée to the field and includes not only the basics and general principles but also discussion of practical implementations and key application areas.

Adaptive optics is a technique to correct aberrations in optical systems. It is no surprise that AO found its first use on ground-based telescopes, where it was initially suggested for astronomy in the 1950s and implemented for imaging space objects in the 1980s. The problem was relatively simple: the light from astronomical sources was distorted by the turbulence in the Earth's atmosphere, resulting in blurring of the image. The aberrations caused by this turbulence could be measured either from the astronomical source itself, if it was small enough, or from a nearby star. Distortions were measured using fast wavefront sensors such as Hartmann–Shack and curvature-sensing devices. In the simplest cases—for example, searching for faint companions near bright stars—the aberrated object itself was the source for the turbulence measurement.

Once the aberrations could be measured, it was only a matter of time before wavefront technology, fast computers, and deformable mirrors enabled their real-time correction. The maturation of laser guide stars freed astronomers from the necessity of finding a nearby bright natural star to measure the aberrated wavefront and enabled AO correction in almost any direction on the sky. Starting from only a few peer-reviewed science papers per year in 1996, it took a decade to surpass 100 papers per year from observations on virtually all of the largest state-of-the-art telescopes.

Also, in 1996 the field turned in an entirely new direction when AO was used in an ophthalmoscope to image the living human retina. For retinal imaging, atmospheric turbulence is replaced by static and dynamic aberrations in the eye's optical system. Such aberrations impose an upper limit on the resolution of any type of ophthalmoscope. This application introduced a new challenge: the retina is an extended object and not a source of light. But this was not a huge barrier, as external sources of light have been used to illuminate the retina for over a century. The use of a simple laser beacon enabled direct wavefront sensing, and other light sources were used as illuminators for imaging once the correction had been made. The first AO ophthalmoscope was used to take images of the cone photoreceptor mosaic in 1996. And, as in astronomy, it took about a decade to go from a few peer-reviewed science publications per year to more than 100.

In the last two decades of the twentieth century, AO was used for a number of other applications as well; examples include free-space optical communication, control of beams in resonant laser cavities, and shaping laser beams for high-energy irradiation. In astronomy and vision science, AO is enabling a growing list of new discoveries, from direct observations of the motion of stars circling the black hole in the center of the Milky Way galaxy to the first-ever in vivo maps of the three classes of cone photoreceptor that enable color vision.

Why was it that in the past few decades the use of AO for microscopy—surely the largest field for advanced optical imaging—was largely absent? Was it because the optical design of the microscope and the use of a carefully designed objective lens obviated its use? Far from it! Microscope images are rife with aberrations—from the simple aberrations that arise when the cone of focused light passes through the cover plate to the more-complex aberrations caused by the myriad refractive index changes within the sample itself. In fact, it is the aberrations and scattering originating from the sample that have made the direct use of AO so challenging for microscopy. Application of AO for ophthalmoscopy required little more than borrowing ideas from astronomers to make the first AO ophthalmoscope run. But in microscopy, it is not easy to measure the wave aberrations. The light returning from a simple laser beacon comes from scattering throughout the entire depth of the tissue, both in front of and behind the point of interest. How does one extract the relevant light for wavefront sensing?

This challenge has not deterred the field: the demand for sharper optical images and the fact that conventional wavefront sensing concepts cannot be borrowed directly from vision or astronomy are driving a whole new wave of innovation in how AO is applied to microscopy. The innovative approaches to date are reviewed in this book: sensorless optimization and phase diversity, pupil segmentation, coherence-gated wavefront sensing, and direct wavefront sensing.

Modeled on the historical precedents of AO-based research in astronomy and in vision, the application to biological microscopy is now poised for an extremely exciting decade of growth ahead.

Austin Roorda
Professor of Optometry and Vision Science
University of California at Berkeley

Claire Max
Professor of Astronomy and Astrophysics
University of California at Santa Cruz

Preface

This book on adaptive optics for biological imaging builds on prior works on the use of adaptive optics in astronomy and vision science. Adaptive optics was first applied to biological imaging in the 2000 time frame, and much of the work has been published in various applied optics conference proceedings and journals, none of which are likely to be read by the intended users in the areas of biological and biomedical science and engineering. The aim of this book is to compile the results of the first decade of research in a format that will be useful to new biological and biomedical investigators who want to use these techniques in their research.

The book is split into three sections, including principles, methods, and applications of adaptive optics for biological imaging. In the section on principles, the intent is to provide a general overview for readers who do not have a background in applied optics. From my experience in interdisciplinary research, the specialized language and acronyms that are used in one field can be a source of confusion for people in other fields. The section on principles should provide the nonspecialist with the definitions and vocabulary that will be helpful to understanding the materials in the following sections.

In the section on methods, the goal was to provide the user with background on what sort of optical aberrations arise in imaging through various biological tissues, and the technology that is being used to make corrections for these aberrations. After the close of the first decade of research in this area, commercial microscopes with adaptive optics are just now being introduced. Since there is no standard approach, as can be seen in the last section on applications, new users have a lot of choices to make in deciding which approach will work best for their own applications. Because the sample is an integral part of the optical system in biological imaging, the field will benefit from participation by biologists and biomedical researchers with expertise in this area. Hopefully, this section will help lower the barriers to entry for these users.

The final section is on applications that include research that has been done with different biological samples and different imaging instruments, including wide-field, confocal, and two-photon microscopes. One of the major differentiators in the early applications is how the aberrations were measured and corrected. A similar differentiation arose in the early days of astronomical adaptive optics. One approach used parallel correction of the wavefront across the entire optical aperture. The wavefront was measured and corrected in a single step using multiple control paths between a wavefront sensor and a wavefront corrector. A second approach was serial correction using sequential measurement and correction of the wavefront for each zone or mode of the aperture. Serial correction simplified the hardware but limited the time available for collecting photons during each iteration. Astronomers have settled on parallel correction, since they are usually more starved for photons than for funding (although that may be changing), and they must make rapid corrections to compensate the dynamics of the Earth's turbulent atmosphere.

Here, the approaches for the application of adaptive optics in biological imaging have been divided into indirect and direct wavefront sensing. Indirect wavefront sensing uses an iterative approach, without the need for a wavefront sensor, which simplifies the hardware but can take longer to make the correction. Direct wavefront sensing uses a parallel approach, with some form of wavefront sensing that complicates the hardware, but it can increase the speed for correction. Which approach is best for different applications may depend on how much light is available from the sample and how quickly the sample is changing.

Finally, what can be expected in the next decade for applications of adaptive optics in biological imaging? Now that refractive aberrations have been overcome using adaptive optics, we can reach "the diffraction limit." Like the test pilots who broke through the sound barrier, microscopy researchers are pushing the envelope for resolution beyond the diffraction limit. New forms of super-resolution microscopy are being developed for "diffraction-unlimited imaging." Most likely, these microscopes will require adaptive optics to reach their ultimate level of performance for in vivo imaging through thick tissue. Beyond that, it is a hazy crystal ball—limited, no doubt, by scattering. Again, adaptive optics is being used to overcome scattering losses, so we may envision optical systems that can image deeply into dynamic live samples with molecular resolution.

Joel A. Kubby
University of California at Santa Cruz

Editor

Joel A. Kubby is the chair and an associate professor in the Department of Electrical Engineering in the Baskin School of Engineering at the University of California at Santa Cruz. His research is in the area of microelectromechanical systems (MEMS) with applications in optics, fluidics, and bio-MEMS. Before joining the University of California at Santa Cruz in 2005, he was an area manager with the Wilson Center for Research and Technology and a member of the technical staff in the Xerox Research Center Webster in Rochester, New York (1987–2005). While with Xerox, he received a Xerox Excellence in Science and Technology award. Prior to Xerox, he was at the Bell Telephone Laboratories in Murray Hill, New Jersey, working in the area of scanning tunneling microscopy. While at Bell Laboratories, he received an Exceptional Contribution award. He has led a six-company industrial research consortium under the National Institute of Standards and Technology's Advanced Technology Program to develop a new process for optical MEMS and has over 78 patents and 40 journal publications. He is the cochair of the SPIE Silicon Photonics and the MEMS Adaptive Optics conferences.

Contributors

David A. Agard
Department of Biochemistry
   and Biophysics
Howard Hughes Medical
   Institute
University of California at
   San Francisco
San Francisco, California

Oscar Azucena
W. M. Keck Center for
   Adaptive Optical
   Microscopy
Baskin School of
   Engineering
University of California at
   Santa Cruz
Santa Cruz, California

Jerome Ballesta
Imagine Optic
Fair Oaks, California

Eric Betzig
Janelia Farm Research
   Campus
Howard Hughes Medical
   Institute
Ashburn, Virginia

Jonas Binding
Biomedical Optics
Max Planck Institute for
   Medical Research
Heidelberg, Germany

Martin J. Booth
Department of Engineering
   Science
Centre for Neural Circuits and
   Behaviour
University of Oxford
Oxford, United Kingdom

Jae Won Cha
Department of Mechanical
   Engineering
Massachusetts Institute of
   Technology
Cambridge, Massachusetts

Diana C. Chen
Lawrence Livermore National
   Laboratory
Livermore, California

Donald T. Gavel
Laboratory for Adaptive Optics
University of California at
   Santa Cruz
Santa Cruz, California

John M. Girkin
Department of Physics
Biophysical Sciences Institute
Durham University
Durham, United Kingdom

Alexander Jesacher
Division of Biomedical Physics
Innsbruck Medical University
Innsbruck, Austria

Na Ji
Janelia Farm Research Campus
Howard Hughes Medical
   Institute
Ashburn, Virginia

Zvi Kam
Department of Molecular Cell
   Biology
Faculty of Biology
Weizmann Institute of Science
Rehovot, Israel

Peter Kner
Faculty of Engineering
University of Georgia
Athens, Georgia

Joel A. Kubby
W. M. Keck Center for Adaptive
   Optical Microscopy
Baskin School of
   Engineering
University of California at
   Santa Cruz
Santa Cruz, California

Markus Rückel
BASF SE
Material Physics
Ludwigshafen, Germany

Michael Schwertner
confovis GmbH
Jena, Germany

John Sedat
Department of Biochemistry
  and Biophysics
University of California at San
  Francisco
San Francisco, California

Peter T. C. So
Department of Mechanical
  Engineering

and

Department of Biological
  Engineering

Massachusetts Institute of
  Technology
Cambridge, Massachusetts

Xiaodong Tao
W. M. Keck Center for Adaptive
  Optical Microscopy
Baskin School of Engineering
University of California at
  Santa Cruz
Santa Cruz, California

Elijah Y. S. Yew
Singapore-MIT Alliance for
  Research and Technology
  (SMART)
Singapore

Yaopeng Zhou
Abbott Laboratories
Princeton, New Jersey

I

Principles

1

Principles of Wave Optics

Donald T. Gavel
University of California
at Santa Cruz

1.1 Introduction

Adaptive optics system development requires a basic understanding of the behavior of light both in terms of classical electromagnetic waves and in terms of the quanta, or photon, phenomena. Key issues relevant to system performance, such as limits to sensitivity, diffraction, transmission, interference, and so on, are explained and calculable given a set of fundamental principles that incorporate the wave theory of optics combined with the quantum mechanical interpretation of the wave fields. The wave behavior of light is perhaps best explained using the Huygens wavelet theory, which accurately predicts how the waves propagate and add coherently in space and time. Ultimately, however, our understanding of the detection of light (and the fundamental signal-to-noise limits) requires interpreting the wave amplitudes as probability amplitudes for photon arrivals.

The objective of this chapter is to provide the reader with a firm grounding in the physical theory that best explains the behavior of light. As with all theories, it is not complete. Many questions remain, which are the subject of current research. However, armed with a few fundamental ideas, the practitioner can go a long way. We try to minimize the often cumbersome mathematical detail here to get right to the key physical ideas. (And too much math can sometimes yield an undue sense of perfection to a theory, distracting from the weaknesses or limitations in the assumptions!) There is some necessary mathematical detail, but our emphasis is on the fundamental physical understanding and not on the mathematical theory. The reader is referred to the excellent texts in the field for the rigorous mathematical development.

The wave optic/quantum optic physical model, and the interaction with the electron via quantum electrodynamics, has been experimentally verified to great accuracy. The speed of light in a vacuum is known to nine digits' accuracy.* The theory that light has a wave nature with a definite wavelength very accurately explains observed interference phenomena. However, the fact that individual photons are observed in photo detection experiments is not explained by a purely wave hypothesis and so a quantum interpretation is used. Planck's constant (ratio of photon energy to wave frequency) is known to approximately 10 digits. This dual, wave/particle, nature of light leaves many open questions. One question to ponder is what is the nature of the light before it is detected? It certainly rigorously obeys a wave theory and does not behave like a particle. How exactly does the photon interact with matter and with other photons? What exactly determines a "detection" which forces the photon to stop acting like a wave and manifest itself as a particle? These questions are the subject of active study (and much speculation) today. Although fascinating open questions remain, the present theory is quite practical and very accurate for application in instrument building and performance analysis.

1.2 Physical Nature of Light

1.2.1 Original Concepts of Light

The ancient Greeks had the hypothesis (Empedocles, fifth century BC) that human vision was an active event initiated by the eye. From this hypothesis they concluded that the speed of light was infinite, as one could see arbitrarily distant objects the moment the eyelids are opened. In the 1600s, both Galileo and Newton became interested in light. Galileo attempted to measure the speed of light with timing experiments but was unsuccessful. It was successfully measured (to within 30%) by Olaf Roemer in 1676 using calculations involving the eclipse of Jupiter's moons. Isaac Newton posited (*Hypothesis of Light*, 1670) that light was composed of "corpuscles," or particles. This became the popularly accepted theory for over a century, but he had not worked out the corpuscle theory to a convincingly quantitative level and had trouble explaining interference and diffraction. Christian Huygens developed a wave theory of light (*Treatise on Light*, 1690) that accounted for refraction, diffraction, and interference. Huygens wavelet principle, with the appropriate intensity normalizing constants subsequently derived, forms a basic wave theory that now accurately predicts all the wave phenomena of light. It is fascinating to note that Huygens' construction of wave optics well predates the electromagnetic theory (Maxwell, *Dynamical Theory of the Electromagnetic Field*, 1865) where the equations of electricity and magnetism combine to predict free-space propagation of electromagnetic waves. The Maxwell theory united the whole spectrum of radiation phenomena, from low frequency (radio) to visible light to hard X-rays and gamma rays, as manifestations of electromagnetic waves differentiated only by wavelength.

A particle explanation for light had to be revived (Einstein 1905) when it was discovered that light interacts with matter in discrete energy packets (Planck 1901). As it is not easy to describe light's behavior in terms of particles (as Newton had found), an uneasy truce was reached in the interpretation of the wave function—the electromagnetic wave in the case of the photon—as probability amplitude. In this interpretation, the squared amplitude of the wave amplitude, expressed as a function of position, is proportional to the probability that the particle is at that position. The Copenhagen interpretation (Bohr and Heisenberg 1924–1929), as this came to be called, incorporated the wave behavior through the wave function and the particle behavior through the interpretation in terms of probabilities.

1.2.2 Maxwell's Equations and the Plane Wave Solutions

We start at Maxwell's equations, a basis from which we can generate the classical wave concepts including Huygens theory. The electric and magnetic fields obey

---

* Since 1983, the speed of light has established the definition of a meter through the definition $c = 299{,}792{,}458$ m/s.

$$\nabla \cdot \mathbf{E} = 4\pi\rho$$

$$\nabla \cdot \mathbf{B} = 0$$

$$\nabla \times \mathbf{E} = -\frac{1}{c}\frac{\partial \mathbf{B}}{\partial t}$$

$$\nabla \times \mathbf{B} = \frac{1}{c}\frac{\partial \mathbf{E}}{\partial t} + \frac{4\pi}{c}\mathbf{J}$$

where **E** and **B** are the electric and magnetic vector fields, respectively, c is the speed of light, and ρ and **J** are the source terms' charge and current density, respectively. In the absence of source terms, the equations have a free-space propagation solution, where E and B components are normal to the direction of propagation, as depicted in Figure 1.1. The waves obey the Helmholtz equation,

$$\nabla^2 u(r,t) = \frac{1}{c^2}\frac{\partial^2}{\partial t^2} u(r,t)$$

where $u(r,t)$ represents any of the transverse components of E or B.

We detect the presence of the electromagnetic field (photon) because it interacts with the electron. Classically, this is through the force equation:

$$\mathbf{F} = q\left[\mathbf{E} + \left(\frac{v}{c}\right) \times \mathbf{B}\right]$$

where **F** is the vector force on the electron, q is the charge of the electron, and v is its velocity.

One free-space traveling wave solution to the Helmholtz equation is the plane wave depicted in Figure 1.1 and expressed mathematically as

$$E(\mathrm{x},t) = \tilde{E}(\mathrm{k})e^{i(\omega t - \mathrm{k}\cdot\mathrm{x})}$$

where $k = 2\pi/\lambda$ is the spatial wavenumber (as a vector, **k** has amplitude k and points in the direction of propagation), ω is the spatial frequency in radians per second, $\omega = 2\pi\nu$, and ν is the frequency expressed in hertz.

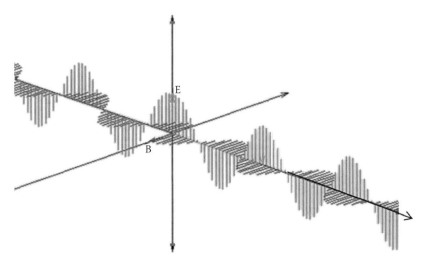

FIGURE 1.1 Electromagnetic wave propagation.

Substituting this into the Helmholtz equation, we derive the relation

$$k^2 \tilde{E}(\mathbf{k}) = \left(\frac{\omega}{c}\right)^2 \tilde{E}(\mathbf{k})$$

from which we can conclude a linear free-space dispersion relation $k = \omega/c$. The waveform, a sinusoid with frequency ω and wavelength λ, remains intact as it propagates through space with time. An arbitrary solution to the Helmholtz equation can be composed of linearly superposed sinusoidal plane waves since they each solve the equation and the equation is linear. The linearly superposed waves can have different wavelengths and directions. Thus, any waveform traveling in a given direction propagates intact at the speed of light, and multiple waveforms traveling in different directions travel through each other without interaction.

1.3 Optical Paths through Materials and Free Space

1.3.1 Dispersion and Refraction

In a propagation medium, such as glass, the light wave's resonant interaction with the atoms in the glass causes the phase front to effectively slow down. The slow-down factor is known as the index of refraction, n, which is different for different materials and generally depends on the frequency of the light. In a refracting medium, a plane wave of a given frequency travels at a phase velocity, v_p, given by

$$v_p = \frac{c}{n(\upsilon)}$$

Since each frequency ν travels at a different speed, the medium is dispersive, meaning that waveforms composed of multiple frequencies will change in form as they travel through the medium. There is no dispersion when the light travels in a vacuum, where the phase speed is c, hence $n = 1$ for all frequencies in vacuum.

We can get an idea of the practical consequences of dispersion by comparing refractive index for various materials. Figure 1.2 plots index versus wavelength for some glasses, crystals, and water. All these materials have "normal" dispersion, where the index gets smaller at the longer wavelengths. That means the waves at longer wavelengths travel faster than those at shorter wavelengths. "Anomalous" dispersion is just the opposite: shorter wavelength waves travel faster than the longer wavelength waves, at least for part of the range of wavelengths.

The normal dispersion of glass accounts for refraction and the separation of colors seen in a prism (Figure 1.3). Light waves that enter the glass at a direction vector off normal to the surface slow down on their leading edge, pulling the propagation direction, what we might call the light ray, closer to normal, resulting in Snell's law:

$$\frac{\sin\theta_1}{\sin\theta_2} = \frac{n_1}{n_2}$$

where θ_1 is the angle from normal of the incident wave, θ_2 is the angle off normal of the refracted wave, n_1 is the index (= 1 in the case of free space) of the initial medium, and n_2 is the index of the final medium. Since blue light (short wavelength) slows down more than red (long wavelength), the blue rays deflect the most when passing through a prism, creating a spectrum on exit.

Another consequence of dispersion is that a localized wave packet will travel at a different velocity than the phase velocity, generally slower in normally dispersive materials and faster in anomalously dispersive materials, but never more than the vacuum speed of light, c. First, let us define what we mean by a wave packet. We consider a waveform that is, initially, localized in a region Δx, as shown in

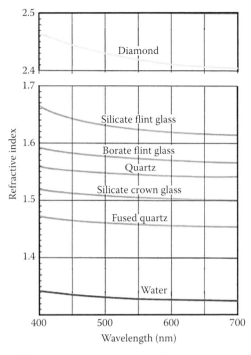

FIGURE 1.2 Dispersion curves for various transparent materials.

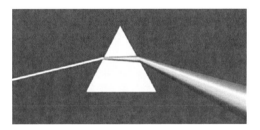

FIGURE 1.3 A glass prism separates light into its constituent colors because of the dispersion of the glass.

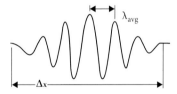

FIGURE 1.4 Wave packet.

Figure 1.4. If Δx is large compared to a wavelength, then the wave packet can be composed of waves within a narrow range of wavenumbers k around a mean wavenumber k_0. The waveform can then be expressed as

$$E(x,t) = \int \tilde{E}_k e^{i(\omega(k)t - kx)} \, dk$$
$$= e^{i(\omega_0 t - k_0 x)} \int \tilde{E}_k e^{i(\Delta\omega t - \Delta k x)} \, d\Delta k$$

where the second form is composed of two factors, the sinusoidal wave at the mean wavenumber, varying rapidly in space at k_0, and an envelope function, varying slowly in space at Δk. The rapid sinusoids travel at the phase velocity

$$v_p = \frac{\omega_0}{k_0} = \frac{c}{n}$$

whereas the envelope travels at the group velocity

$$v_g = \frac{\Delta\omega}{\Delta k} = \frac{\Delta(ck/n)}{\Delta k} = \frac{c}{n} - \frac{ck}{n^2}\frac{\Delta n}{\Delta k}$$

$$= \frac{c}{n}\left[1 - \frac{\Delta n/n}{\Delta k/k}\right] = \frac{c}{n}\left[1 + \frac{\Delta n/n}{\Delta\lambda/\lambda}\right]$$

Referring to Figure 1.2 for the case of silica flint glass, we see that

$$\lambda = 400 - 500 \text{ nm}, \quad \frac{\Delta\lambda}{\lambda} = \frac{100}{450} = 0.222$$

$$n = 1.63 - 1.66, \quad \frac{\Delta n}{n} = \frac{-0.03}{1.645} = -0.207$$

and thus

$$v_g = 0.918 v_p$$

$$v_p = \frac{c}{n} = 0.61c$$

So in this case of normal dispersion, the group velocity of the packet is slower than the phase front velocity, and both are less than the speed of light.

1.3.2 Polarization

So far our picture of the electromagnetic plane wave is like that shown in Figure 1.2, where the **E** vector increases and decreases along a single direction orthogonal to the direction of propagation. This wave is considered linearly polarized, with the line of polarization defined by the direction of the **E** vector. If we add a second wave to this one, with the same wavenumber and propagation direction but with the **E** vector in the orthogonal direction (orthogonal to the original **E** and also to the direction of propagation), we can generate a family of wave functions depending on the relative phases of the first and second oscillations. If these waves are in phase, then the resulting wave is also linearly polarized, but in the direction that is the vector sum of the two wave's polarization vectors. If the waves are not matching in phase, then the **E** vector will spin around the propagation axis, tracing out a helical path. Circular polarization is the particular case when the waves are $\pi/2$ out of phase and of equal amplitude. There is a right circular polarization and a left circular polarization case, depending on if the phase difference is plus or minus $\pi/2$ (Figure 1.5).

One method of describing the polarization state is through the Stokes parameters. These parameters are convenient combinations of the **E** vector components. Let

$$\mathbf{E} = (\hat{x}E_x \; \hat{y}E_y)e^{i(\omega t - kz)}$$

$$E_y = |E_y|e^{i\delta}$$

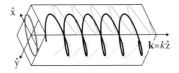

FIGURE 1.5   Electromagnetic plane wave in a general state of polarization.

and $E_x = |E_x|$ is a real number. δ represents the relative phase between the x and y components. The four Stokes parameters are defined:

$$I = \left\langle E_x E_x^* + E_y E_Y^* \right\rangle, \quad Q = \left\langle E_x E_x^* - E_y E_y^* \right\rangle$$

$$U = \left\langle E_x E_y^* + E_y E_x^* \right\rangle, \quad V = i \left\langle E_x E_y^* - E_y E_x^* \right\rangle$$

The advantage of using the Stokes parameters is they are physically measureable with a combination of waveplates, polarized film, and photometers in the laboratory, and they are a complete representation from which the helical picture of the **E** vector can be reconstructed.

The polarization state represented in terms of Stoke parameters can be traced through a series of polarization-dependent optical elements using the Mueller matrix calculus [Mueller 1948]. The polarization state expressed in terms of complex E components directly can be traced through optical elements using the Jones calculus [Jones 1941].

Some light-transmitting materials have the property that the index of refraction is different for the two axes of linear polarization states. The property is called *birefringence*. Calcite crystal is a commonly used material with this property. If the wave polarization is aligned along the crystal's *ordinary* axis, it encounters a relatively low index. If aligned along the *extraordinary* axis, it encounters a relatively high index. These crystals are useful for analyzing or changing the polarization state, per the following examples:

1. Wollaston prism: An incoming wave at nonnormal incidence to the face of the birefringent crystal is separated into its two polarization component waves since the different indices impart different refraction angles.
2. Quarter wave plate: An incident wave is converted from linear to circular polarization or vice-versa. One aligns the linear polarized input wave normal to the crystal face but with the polarization axis at 45° with respect to the ordinary axis of the crystal. The crystal is just thick enough such that slow component, the extraordinary wave, ends up ¼ wavelength retarded from the faster ordinary wave. The resulting output wave thus has the two components shifted $\pi/2$ radians out of phase and so is circularly polarized. Similarly, a circularly polarized input wave is converted to linear polarized.
3. Half wave plate: The half wave plate flips the polarization state of linear polarized waves by 90° and converts right-hand circular waves to left hand circular waves. The birefringent plate is thick enough that, at the output, the extraordinary wave lags ½ of a wavelength behind the ordinary wave. The resulting output wave has the two components shifted π radians out of phase.

1.3.3 Optical Path Length, Geometric Rays, and Diffraction

Fermat postulated that light follows paths of least-time, accounting for changes in velocity within refractive media. For example, light appears to go along in the straightest path from point A to point B in free space (verified by the fact that an opaque screen will block the light only if it is placed directly on the straight line path between A and B), but if the index of refraction is not uniform, then the light paths are curved going from A to B (verified again by blocking the path by an opaque screen, watching when the light does not arrive at point B). The reflected path of a mirror seems to be a minimum distance path in a local sense. The path of a ray through a refractive interface (e.g., air to water) is bent and appears to follow the path of least time, accounting for the slowing down of the light in the higher-index medium.

These observed phenomena are ultimately explained by a wave theory of light. It turns out that waves add constructively and destructively in such a way as to favor traveling through a region immediately surrounding the least-time path. An added benefit of the wave theory is that it also accurately predicts the size of the conducting region (e.g., how far light rays can actually stray from straight lines!).

We start by defining the optical path distance (OPD). This is the distance a phase front would have traveled at the speed of light during the time it spent traveling through the refractive medium:

$$d\,\text{OPD} = c\,dt = c\left(\frac{dx}{v_{\text{phase}}}\right) = c\left(\frac{dx}{(c/n)}\right) = n(x)\,dx$$

$$\text{OPD}(A,B) = \int_A^B n(x)\,dx$$

Why does light appear to preselect only the path of least time (or equivalently, least OPD) before it heads out on its journey from A to B? To answer this, let us imagine that light emitted from source A actually travels along *all the possible* paths in space from A to B. The light that travels on the shortest optical path from A to B takes a total time equal to the minimum OPD between A and B divided by the speed of light. All other paths require longer travel time, and these waves have a phase further advanced than the wave taking the shortest path when they arrive at B. A cluster of paths around the shortest path, differing by no more than one-half wavelength in total optical path length, will add constructively in phase, contributing to intensity at B. All the other paths (an infinite number of them) have various path lengths, which, modulo wavelength, span the full range of phases; and thus, these light waves tend to cancel each other and not contribute much intensity at B.

Now, let us put a screen with a hole in it at an intermediate plane between A and B. For simplicity, we now consider A and B and the surrounding space to be of constant index of refraction, so the minimum OPD path is a straight line from A to B. For the most part, unless the screen's hole is located on the line from A to B, the light from A will not reach B. Assuming the hole is much larger than a wavelength, we see that the possible paths from A, going through this hole, then on to B have a wide variety of OPDs, thus many of the paths tend to cancel and contribute nothing to the detector at B. There is one path, however, through the hole and on a straight line from A to B, which has a cluster of nearby paths around it with very similar path lengths, all within one-half wavelength of each other. The wavefronts along those paths constructively add and light is detected at B.

A little geometry shows how big this cluster of paths is (Figure 1.6). If the straight line distance from A to B is L, then the cluster of contributing paths (OPD difference less than $\lambda/2$) cover an area with diameter approximately $\sqrt{L\lambda}$. These paths deviate from a straight line by an angle up to $\pm\sqrt{\lambda/L}$. This region of allowable deviation from a straight line is known as the *Fresnel zone*.

The Fresnel zone size is significant at laboratory scales with visible light. For example, with visible $\lambda = 0.5$ μm light and a 2 m propagation distance, the allowed deviation is ~1 mm or 0.25 mrad angle. This means that if the edge of an opaque screen is blocking the path from A to B but is within ½ mm of

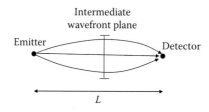

FIGURE 1.6 The light paths from an emitter to a detector.

the straight line between the two, some of the light will "get around" the screen and still illuminate B. The presence of a finite size Fresnel zone is strong evidence of a wave explanation to light propagation.

If the hole's diameter is less than or equal to that of a Fresnel zone, then a different wave optic effect takes place, *diffraction* from the hole. If the hole is, once again, in line with the path from A to B, then light is still detected at B, but at a lower intensity since some of the possible contributing paths are choked off.

But now imagine that the hole is placed off-center from the line connecting A to B. Optical paths from A, through the hole, and ending at B differ in length over the range $\pm yr/L$, where y is the perpendicular distance of the hole center from the line connecting A to B and r is the radius of the hole. If the hole size is chosen such that the range of possible path lengths of paths through it is less than one-half the wavelength, then all the optical paths through the hole contribute positively at B, and light is detected. It is as if the light travels from A to the hole, changes direction, and then travels to B!

If the hole size radius is on the order of a quarter wavelength, there is no path through it that is not within a half wave of optical path length. Thus, any quarter wavelength sized hole, on axis or off axis, and located as far as off axis as we like, will diffract light from source A into a detector at B.

By making the hole small, we choked off all the paths that would have contributed to cancelations at B and allowed only a small bundle of paths that have nearly equal (within one-half wavelength) path lengths to B. This argument applies no matter where we put the detector. So light emerging from a small hole acts like it came from a point source at the hole, spreading out in all directions.

The presence of a finite size Fresnel zone is strong evidence of a wave explanation to light propagation. Another is the presence of fringes when there are two emitters. Young's two-slit experiment is the most famous example. Figure 1.7 shows the concept of the experiment and the observed results.

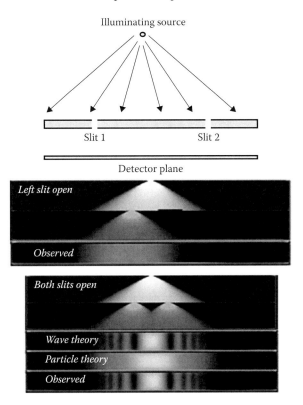

FIGURE 1.7 Young's two-slit experiment. *Top*: Young's two slit experimental set-up. *Middle*: Observed pattern if only the left slit is open. *Bottom*: Both the right and left slits are open. The patterns predicted by wave and particle theories of light. The observed pattern confirms the wave theory of light.

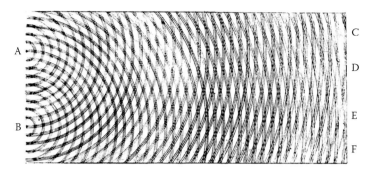

FIGURE 1.8 Huygen's drawing of the two-slit interference phenomenon. (Picture taken from Wiki commons. This image is in the public domain because its copyright has expired.)

If only one slit is open, then the light must travel from the illuminating source through this slit and on to the detector plane beyond it. A straight-line theory would have all lines that pass from the source to the detector plane build an illumination pattern on the plane. With Fresnel's theory, this area is larger since the slit screen is an intermediate wavefront surface and the very narrow slits act themselves like emitters. If both slits are open, then one might expect the sum of the two intensity patterns at the detector, but this is not what is observed. Instead, one sees (in monochromatic light) a pattern of interference fringes, the fringe spacing being on the order of $L\lambda/d$, where d is the separation distance between the slits and L is the distance from the slit plane to the detector plane. This implies that waves traversing through slit 1 are combining with waves traversing through slit 2 and that the illuminated areas are where the waves add constructively, that is, the OPDs to the emitter differ by integer multiples of a wavelength with a tolerance of within one-half wave of each other. All areas are, therefore, illuminated because of the plus or minus half-wave tolerance, but the brightest illumination is where the optical paths match exactly, modulo a wavelength and the darkest part is where they exactly cancel at a half-wave different from each other.

Huygens drew this picture depicting the waves emanating from two slits with waves adding and cancelling coherently depending on position in space, as shown in Figure 1.8.

1.3.4 Photons, Scattering, Energy, and Momentum

The early evidence for light being a wave-like phenomenon contrasts with later evidence that showed that light is composed of discrete energy particles. Furthermore, particles, such as the electron, were demonstrated to exhibit wave like behavior. For example, a Young's two-slit experiment performed with electrons shows fringes.

The hybrid theory that tries to consistently explain these disparate observed properties is based on the concept that the electromagnetic wave/photon has a wave–particle dual nature. Light acts as a wave when propagating but as a particle when detected. Formally, the light is composed of discrete packets known as photons *which do not interfere with each other* (expect in extreme cases of photon–photon collisions) but only with themselves individually behaving as full-wave fields until they are detected as discrete events. In summary,

- light originates as photons, each emitted by the oscillation of a single atom,
- light travels as a wave via all possible paths (with most of the energy picking paths of "least time"), and
- light, when detected, is realized as random single-photon events, probabilistically distributed according to the intensity of the wave at the detector.

Dirac points out that in the quantum theory of light, the wave amplitude when squared is an indicator of the probability of detecting the photon at that location and not a measure of the density of photons at that location. This subtle distinction emphasizes the point that interference is a single photon phenomenon and is not caused by interactions of photons.

Photons interact exclusively with charged particles, most often with electrons. In fact, almost everything we know about the world is transmitted to us through the photon–electron interaction. For example, heat in the sun or a lightbulb excites electrons, which when accelerated emit electromagnetic waves (photons). The photons either reflect off of surfaces or interact with electrons on the surface which absorbs and reemits photons. Finally, the photons are detected in the eye. The photon moves an electron on the retina's detectors to a higher energy state. This produces an electrical signal, which is then transmitted to the brain.

When a photon interacts with a charged particle, the process can be explained as a scattering phenomenon known as Compton scattering (Figure 1.9). This is consistent with a classical explanation where the electromagnetic wave exerts force on the electron, making the electron accelerate, which then results in a new electromagnetic wave (a moving charge is a source term in Maxwell's equations) superposing intensity with the first. Quantum mechanically, a photon collides with an electron and exchanges energy and momentum with it. The photon scatters with a different wavelength (different energy and momentum), which mixes with all the other photons that missed the electron.

Electromagnetic radiation carries energy and momentum density in its fields. Classically, the field energy density is a function of the magnitude of the **E** and **B** fields:

$$E_{EM} = \frac{1}{8\pi} \int (E^2 + B^2)\, dV \quad \mathbf{p}_{EM} = \frac{E_{EM}}{c}\hat{\mathbf{k}} = \frac{1}{4\pi c} \int (\mathbf{E} \times \mathbf{B})\, dV$$

Under the quantum mechanical explanation, each photon individually has energy and momentum:

$$E_{ph} = h\nu \quad \mathbf{p}_{ph} = \frac{E_{ph}}{c}\hat{\mathbf{k}} = \hbar\mathbf{k}$$

where h is Plank's constant, $\hbar = 2\pi/h$, $\nu = c/\lambda n$ is the frequency of the light, and **k** is the vector wavenumber with magnitude $2\pi/\lambda$ and direction pointing in the direction of photon motion.

We explored earlier how diffraction is a consequence of wave phenomena. When an incident wave is confined to pass through an opening, there are additional components in the resulting wave that diffract away from the nominal direction of the original wave. Although diffraction is not an obvious consequence of particle motions, diffraction of particles can be explained quantum mechanically, through the Heisenberg uncertainty principle, with the same resulting diffraction as the wave theory. The uncertainty principle of Heisenberg states that the energy and momentum cannot be simultaneously determined to infinite accuracy. Instead, the product of the uncertainties has a minimum value. In the case of the photon,

$$\Delta x\, \Delta p \cong h$$

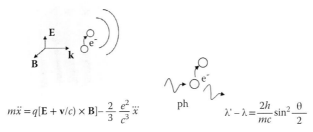

$$m\ddot{x} = q[\mathbf{E} + \mathbf{v}/c) \times \mathbf{B}] - \frac{2}{3}\frac{e^2}{c^3}\dddot{x} \qquad \qquad \lambda' - \lambda = \frac{2h}{mc}\sin^2\frac{\theta}{2}$$

FIGURE 1.9 Compton scattering explained by classical electromagnetism (EM) (*left*) or by quantum mechanics (*right*).

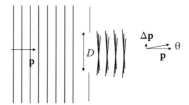

FIGURE 1.10   Diffraction as a consequence of the Heisenberg uncertainty principle.

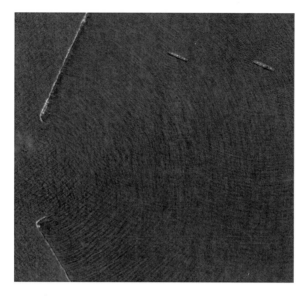

FIGURE 1.11   Ocean waves at the entrance to the Panama canal.

Now a plane wave constrained to pass through an aperture of size D has its position uncertainty constrained (in that direction) to the region D. Thus, the momentum in the direction orthogonal to the original propagation direction is uncertain to the order of h/D. With that in mind, one can see that the uncertainty in the direction of the emerging wave, $\Delta p/p$, is on the order λ/D, which corresponds exactly to what is observed in diffraction by such an aperture (Figure 1.10).

So the fact that diffraction is observed is not, after all, a discriminator between a wave theory and a particle theory of light! With this in mind, we are driven to use the Copenhagen interpretation with roles played by both waves and particles at the appropriate time, as our best theory of light to date.

We conclude this section with a photograph that illustrates the classic wave phenomena with ocean waves. Taken from above, the photo shows waves incident on the opening of a harbor (in this case, the Atlantic side entrance to the Panama canal). The planar open ocean waves diffract as they pass through the "slit" formed by the two breakwaters. Looking closely, one can also see that each ship is a source of scattered waves (Figure 1.11).

1.4 Analytic Methods for Light Wave Propagation and Analysis

1.4.1 Huygens Wavelets

In 1678, Huygens proposed an analytical theory to describe light wave propagation. The theory starts with a basic idea that the wave field in 3-space is formed by a sum of point source emitters each radiating spherical waves symmetrically into 3-space. The phase of each spherical wave wiggles once per wavelength and the amplitude drops linearly with distance so that the intensity, the squared amplitude, from each point source obeys an inverse square law.

We now imagine a surface that divides the space between emitters and detectors. The unique (and controversial) aspect of this theory is that we can replace the original emitters by spherical wave emitters of appropriate strength at each point on the surface and achieve the same solution on the detector side.

Huygens was able to explain reflection and refraction with this approach, and Fresnel was subsequently able to use this model to explain diffractive phenomena.

The Huygens–Fresnel integral is given by

$$u(\mathbf{x}) = \frac{i}{\lambda} \iint_S u(x') \frac{\exp\{ik\rho(x-x')\}}{\rho(x-x')} \cos\varphi(x,x') \mathrm{d}^2 x'$$

where u is any component of the electromagnetic wave (with the "rapidly varying" term $\exp\{i\omega t\}$ ignored), $\rho(x-x')$ is the distance from x' on surface S to x (a point in the "detector" half-space), and $\phi(x,x')$ is the angle of the line from x' to x relative the normal to S at x'. The cosine term is known as the *obliquity factor*. The Huygens–Fresnel integral is a differential description of light propagation. The field at points $\mathbf{x}'$ on surface S is propagated to points $\mathbf{x}$ on another surface further into the detector half-space. As such, the field is propagated from surface to surface in a general direction from sources to detectors.

The factors i and λ have interesting interpretations. These are normalizing factors that are necessary to, among other things, preserve energy in the wave as it propagates. The i factor is known as the Gouy phase shift and says that the Huygens emitter must be out of phase with the original wave at that point by $\pi/2$. Imagine for the moment that the surface S was completely an opaque screen except for a tiny (wavelength size) pinhole at x'; this pinhole seems to act like it instantaneously shifts the phase of the impinging original wave as it passes through the pinhole. This strange but true phenomenon has actually been observed in experiments (Ruffin et al. 1999). The λ factor is also very interesting. It hints at the size of a photon—from a theory that predates the photon theory by more than two centuries! As plane wave electromagnetic energy passes through our hypothetical small hole in surface S, it reduces the amplitude by the factor $1/\lambda$ but diffracts it to all angles in an expanding spherical wave. If the hole is made bigger ($D > \lambda$), then the diffraction is less, now concentrated over a cone angle of λ/D, but has total transmitted amplitude of D^2/λ (the hole area is D^2, and the Huygens normalizing factor is $1/\lambda$), resulting in a forward amplitude proportional to D. If the hole is smaller ($D < \lambda$), then the forward amplitude is proportional to D^2. That is, the forward energy (amplitude squared) getting through a pinhole is proportional to D^2 until the hole gets smaller than λ, at which point the forward energy drops drastically with hole size. This is exactly the behavior one would expect of particles with size roughly λ!

The Huygens integral seems to imply that each point in free space is a reemitter of waves, masking the original emitters. Even if this is not really the case physically, from a mathematical standpoint we can heuristically "prove" the Huygens integral from Maxwell's equations by realizing that the Helmholtz wave equation is linear in the field amplitude and so can be formed by a linear superposition of basic solutions. The basic solution in this case is the wave that passes through a screen pinhole. Adding up the results of screens with pinholes (each of "size" λ and experiencing the Guoy phase shift) results in the superposed solution.

Substituting a plane wave for $u(\mathbf{x}')$ in the Huygens–Fresnel integral results in an identical amplitude but appropriately phase-shifted plane wave $u(\mathbf{x})$. The normalizing factor i/λ was actually discovered in this manner (it is required to preserve the amplitude and give the correct phase), but the equation correctly maintains the planar iso-phase surface solution.

We now return to the idea that the Huygens integral is a differential explanation of wave propagation and we find an interesting connection to the minimum time principle. Imagine substituting any number of intermediate Huygens-wavelet-generating surfaces between surface A and surface B. The Huygens wavelets propagate from one surface and add up coherently at the next surface. Then each point on this subsequent surface reemits spherical wavelets that propagate on to the next surface and so on. As we consider an increasing density of these intermediate surfaces, we realize that the spherical emitters and their Huygens–Fresnel integrals are tracing out *all possible paths of light* from points on A to points on B, both straight and meandering. Each path has a complex phase factor, cycling once per wavelength of optical path, and an amplitude factor that is inversely proportional to distance traveled. Coherent addition of terms from all the paths reveals that light energy prefers those paths confined to a Fresnel zone around the minimum time path through all the surfaces, where the total integrated phase factors are nearly fixed, so the terms can add constructively.

1.4.2 Computer Simulation of Wave Propagation

Numerical evaluation of the Huygens–Fresnel integral can provide useful insight into the behavior of an optical system, especially if diffractive effects could be important or dominant. To assess whether diffractive calculations are necessary or if a simpler ray trace will suffice, one can use the back-of-the-envelope calculation in Section 1.5. Of particular importance is to determine if beams are going to propagate a significant fraction of their Rayleigh range (Section 1.5.6), in which case the ray tracing would not predict the diffraction rings that will grow near the beam edge as the beam propagates. This *Fresnel ringing* is a result of interference of the beam's plane wave with the diffracted waves scattered off the hard edges of apertures.

Assuming a beam is propagating in a direction mostly normal to the surface of Huygens emitters (the *Paraxial* condition), we can use the approximation

$$\rho(x) = L\sqrt{1 + \left(\frac{x}{L}\right)^2} \cong L + \frac{1}{2}\frac{x}{I}$$

in the phase factor, $\rho(x) \cong L$ in the amplitude factor, and $\cos\varphi(x,x') \cong 1$ for the obliquity factor. It is actually not necessary for $x \ll L$ for this approximation to work, as only the cluster of paths within the Fresnel zone will actually add substantial contribution to the integral. Thus, it is only necessary that $L \gg \sqrt{\lambda L}$, that is $L \gg \lambda$, to use this approximation. The Huygens–Fresnel integral becomes

$$u(\mathbf{x}) = \frac{i}{L\lambda}\iint_S u(\mathbf{x}')e^{-i\pi\frac{|\mathbf{x}-\mathbf{x}'|^2}{L\lambda}}\,\mathrm{d}^2x'\,e^{-ikL}$$

The integral can be recognized as a convolution, and so we can exploit Fourier transform techniques to evaluate it numerically. The Fourier transforms of the wave field at each surface are related through

$$\tilde{u}(\mathbf{k}_\perp) = \frac{4\pi^3}{L\lambda}\tilde{u}(\mathbf{k}'_\perp)e^{ik_\perp^2 L/2k}\,e^{-ikL}\quad\text{(Fresnel Method 1)}$$

where $k_\perp$ is the spatial frequency of the wave field projected on the planar Huygens emitter surface (the x–y plane, where z is the axis of propagation). The two-dimensional Fourier transform pair is here defined by

$$\tilde{u}(\mathbf{k}_\perp) = \frac{1}{2\pi}\iint_S u(\mathbf{x})e^{-i\mathbf{k}_\perp\cdot\mathbf{x}}\mathrm{d}^2x$$

$$u(\mathbf{x}) = \frac{1}{2\pi}\iint \tilde{u}(\mathbf{k}_\perp)e^{i\mathbf{k}_\perp\cdot\mathbf{x}}\mathrm{d}^2k_\perp$$

where $k_\perp$, the spatial frequency of the wave field on the planar Huygens emitter surface, which consists of the k_x and k_y components of the wavenumber vector $\mathbf{k}$. In a paraxial beam, $k_z \gg k_\perp = |\mathbf{k}_\perp|$ and $k_z \cong 2\pi/\lambda$.

An alternative approach is to expand the square and recognize the cross term as the exponent of a Fourier transform kernel:

$$u(\mathbf{x}) = \frac{i}{L\lambda} e^{-i\pi\frac{x^2}{L\lambda}} \iint_S u(\mathbf{x}') e^{-i\pi\frac{x'^2}{L\lambda}} e^{i2\pi\frac{\mathbf{x}\cdot\mathbf{x}'}{L\lambda}} d^2x' e^{-ikL} \quad (\text{Fresnel Method 2})$$

The alternative method starts by multiplying the wave field at the starting surface by the phase factor $e^{-i\pi\frac{x'^2}{L\lambda}}$, taking the Fourier transform of the result, then multiplying by the phase factor $e^{-i\pi\frac{x^2}{L\lambda}}$.

For numerical accuracy of either method, it is important that the phase factors not vary too rapidly over the sample grid of the Huygens surface, as stored in the computer. Obviously, Method 1 works well for small $k_\perp$ and small L (low spatial frequencies and short propagation distances). Method 2 works for large L and x and x' inside the Fresnel zone, that is, $x < \sqrt{L\lambda}$ and $x' < \sqrt{L\lambda}$. The rule of thumb from experience is that Method 1 should be used for propagation distances less than 10% of the Rayleigh range $= D^2/\lambda$ (see Section 1.5.6) and that Method 2 should be used for propagation distances larger than this.

Figure 1.12 shows a slice through a circular beam after it has propagated 20% of the Rayleigh range from a hard aperture. Note the ringing caused by diffraction.

At very long propagation distance, diffraction dominates and the beam width forms a *far-field* pattern that expands linearly with distance. At these distances, $x \ll L$ and $x' \ll L$, so the phase factors are very close to unity. Defining $\theta = \mathbf{x}/L$, the integral becomes

$$u(\theta) = \frac{i}{L\lambda} \iint_S u(\mathbf{x}') e^{ik\theta\cdot x'} d^2x' e^{-ikL} \quad (\text{Fraunhofer method})$$

The far field diffraction pattern of a circular aperture illuminated by a plane wave is the *Airy pattern* shown in Figure 1.13.

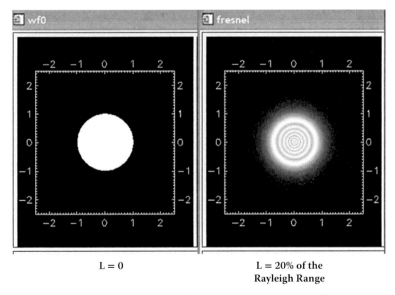

FIGURE 1.12 Results of a numerical propagation of a circular beam using Fresnel's approximation to Huygens integral.

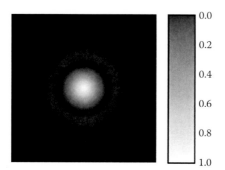

FIGURE 1.13 Diffraction pattern of a circular beam propagated to the far field. (Courtesy of http://en.wikipedia.org/wiki/File:Airy-pattern.svg.)

1.5 Application of Wave Optics to Optical System Design

Now with a firm adherence to the theoretical ideas just presented that light waves add coherently and that the phase of the wave is sum of the OPD along the shortest path from source to detector, we can present some basically useful properties for optical design application. Lenses or powered mirrors modify the path of the light to bring it into focus. It is useful to know just how sharp a focus is physically allowable in the wave theory. This can then be compared with that achieved by a real system. In the case of a pristine free-space environment, the lenses or mirrors must be very precise, figured to a fraction of a wavelength to achieve coherent addition.

1.5.1 Ray Tracing

Typically, designers use ray tracing to characterize and help tune an optical design. A ray enters a lens glass at an off-normal incident angle and is bent according to Snell's law of refraction as shown in Figure 1.14. This law is actually a consequence of the Fermat principle and, more fundamentally, Huygens wavelet theory. As a ray enters the lens, the light is slowed down. This bends the ray since the portion of the iso-phase wavefront that has not yet contacted the lens surface moves ahead of that portion that is now in the glass.

Note that as the speed of light is reduced so is the wavelength, but the frequency remains unchanged, according to

$$\nu = \frac{c}{n_1 \lambda_1} = \frac{c}{n_2 \lambda_2}$$

A focusing lens is thicker in the middle, which slows down the center rays with respect to the edge rays. Incident plane wave rays emerge from the lens with the center rays' phases, lagging the phases of those rays emerging from the lens edges. A correctly figured lens will refract an incident plane wave into a converging spherical wavefront (Figure 1.15). Similarly, the correctly figured mirror will reflect a plane wave into a converging spherical wave front.

The total optical path length through the center of a focusing lens is longer than the path going through the edges. Standing at the focus, the bent-edge ray has traveled a longer distance in the air than the central ray; however, since the central ray spent more time (more optical path) in the glass lens, the two OPDs are actually equal as measured starting from the incident plane wave.

We can exploit this idea to derive useful properties about the optical focus, including transverse resolution and depth of field.

To continue further, it is necessary to define a few more practical terms.

The *focal length* is the distance from the center of the lens to the focus. The *power* of the lens, in *diopters*, is the reciprocal of the focal length of the lens in meters.

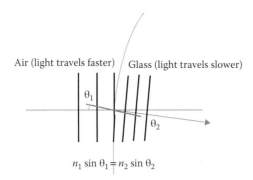

FIGURE 1.14 Snell's law is a consequence of Fermat's least-time principle and is predicted by Huygens wavelet theory.

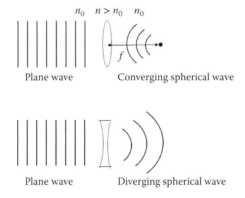

FIGURE 1.15 Lenses shape the wavefront by selectively slowing the rays.

The *aperture* is the hole size that sets a transverse spatial limit on the bundle of incoming light rays. The aperture could be defined by the edge of the glass lens, but often the aperture is made smaller than the objective lens, because the figure of the lens near the edge is not well controlled in the manufacturing process. In optical telescopes, often the primary mirror sets the aperture (all rays hitting the mirror surface are gathered). In a camera, and in the human eye, the adjustable *iris* sets the aperture.

The aperture and the lens power set the *f-number*(f/#) or *focal ratio* of the focused beam, which is defined as the ratio of focal length to the aperture diameter:

$$f/\# = \frac{f}{D}$$

Note that the focal ratio is a property of the beam, not the lens. For a given power lens, the *f*-number can be adjusted by opening or closing the entrance aperture.

Numerical aperture is another term used to describe beam convergence. It is the sine of the half angle of the converging cone beam multiplied by the index of refraction. For small angles, numerical aperture is approximately one-half the reciprocal of the *f*-number:

$$NA = n\sin(\theta) \cong \frac{1}{2f/\#}$$

f-Number is most often used for describing astronomical telescopes and instrument systems. Numerical aperture is used in microscopy and fiber optics application.

1.5.2 Lagrange Invariant

By combining various optical elements, light can be manipulated to produce magnified images. Simple two-lens systems designed for plane wave input and output is shown in Figure 1.16. Incoming light is gathered with a large aperture input lens and then presented to the eye or subsequent camera system with a small aperture output lens. An example of an astronomical telescope is shown on the left image of Figure 1.16, although modern optical telescopes use mirrors that perform the same function as the lenses depicted here. Incoming star light is essentially composed of plane waves since the source is at a great distance away. A microscope optical system is shown in the right figure. The microscope is designed to image an object that is very close to the objective lens—a biological specimen, for example. The eyepiece lens presents a collimated output beam matched to the eye's pupil and the eye acts to focus light on the retina. Again, the system magnifies angles in the sense that small displacements at the object plane, which give small input ray angles, are magnified to larger angles in the output beam presented to the eye.

The *Lagrange invariant* is a statement about the light-gathering power of an optical system. If the optical system reduces the beam diameter, then angles are proportionally magnified. Conversely, angle magnification must be accompanied by a compression of the beam size.

Figure 1.17 shows the geometry. From a wave-optic perspective, the optical path length along each ray is what matters. For a normal plane wave beam, the optical system adds the same optical path length to

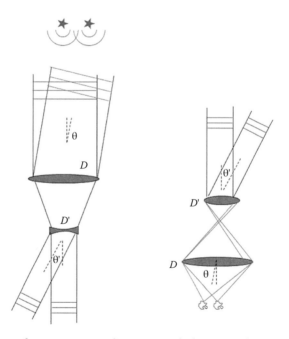

FIGURE 1.16 Optical magnification systems. *Left*: astronomical telescope. *Right*: microscope.

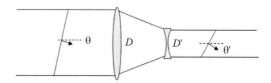

FIGURE 1.17 Lagrange invariant: compressing the beam diameter magnifies the angles.

each ray to produce a plane wave output beam, but it compresses the bundle of rays. The tilted incoming wave has a more advanced phase front on one side of the aperture compared to the other side. Again, the lenses act to bundle the rays, but since the optical system adds the same optical path length to each ray, the relative phase advance is the same on the output beam as it was on the input. But since the beam diameter is compressed, this must effectively increase the tilt of the phase front.

The light-gathering power is the product of the maximum field angle and the maximum aperture. This number is preserved throughout the optical system

$$\Xi = \theta D = \theta' D$$

The Lagrange invariant is fundamental physical concept in that it is a statement of the conservation of energy. The Lagrange invariant squared, multiplied by the radiance of the sources over the input field, is the total light energy entering the optical system per unit time. In an idealization where no light is lost within the system, the amount of energy emerging at the output must match the total energy entering.

1.5.3 Maréchal's Criterion

As mentioned earlier, the surface of a lens or mirror must be polished to great precision when we are dealing with the visible spectrum of light (0.4–0.7 μm). In the 1940s, the French optical researcher Andre Maréchal developed a criterion for determining how accurately an optic surface must be polished to achieve the diffraction limit, that is, the physical limit of focusing performance as set by the wave optics rather than by imperfections in the lens or mirror surfaces themselves. The basic idea is that in order for light rays to focus into a bright, sharp spot, they must mostly arrive at the focus in phase, that is, within one-half of a wavelength in optical path of each other. Figure 1.18 shows an example of how a nominally converging wavefront can depart from a perfect spherical one but still meet this criterion.

If the wavefront (iso-phase surface) is completely within two "Maréchal plates," hypothetical confocal surfaces within half a wavelength of each other, then all the rays are close enough in phase to add coherently at focus. If some portions of the wavefront are outside this boundary, then the rays from these regions would tend to, in coherent addition, cancel rays from other regions, reducing the overall brightness at focus. When the brightness is reduced in this manner, the energy (number of photons) is not "lost" in the cancellation process, but instead falls in a region surrounding the focal point, making the focus appear blurred and dim compared to a perfectly coherent focus.

1.5.4 Diffraction Limit

The diffraction-limited focus is not an infinitesimal point but instead is set to a finite size by Maréchal's considerations. A nominal concentric spherical wave has a single point center. If the spherical wavefront

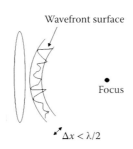

FIGURE 1.18 Maréchal's criterion for light arriving in phase at focus.

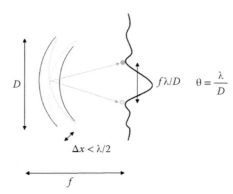

FIGURE 1.19   Consideration of Maréchal's condition sets the diffraction limit for focused waves.

is then tilted back and forth just so much that the wavefront at the edge of the entrance aperture touches one or the other plate (Figure 1.19), then light from these wavefronts would also arrive at the center in phase. Now invert the argument. Imagine standing at a point that is slightly off center of the converging perfectly spherical wavefront. If this is a center of a spherical surface that still remains within the Maréchal plates (as defined by the actual wavefront), then this point must also be bright. Therefore, *all* the points within a transverse dimension

$$\Delta x \cong f\frac{\lambda}{D} = f\#\lambda$$

must be bright, where f is the focal length of the lens and D is the aperture diameter. In terms of OPD, all straight line paths from any point in between the Maréchal plates to any point in the diffraction-limited region are with half a wavelength of optical path. This is consistent with the idea of Huygens wavelets and light energy being concentrated within a Fresnel zone around nominal minimum time paths.

Note that the focal ratio and the wavelength set the diffraction limit. An $f/10$ beam of $\lambda = 0.5\ \mu m$ light will focus to on the order of a 5 μm spot. The shortest practical $f/\#$ is about 1 (rays entering in from 1 radian of solid angle), where the focused spot is on the order of a wavelength of the light. A little thought about the Maréchal geometry reveals that even if rays concentrate from a full hemisphere (angles up to $\pi/2$ off center), an area on the order of a wavelength in dimension will be bright.

This is an interesting limit: it says that objects smaller than a wavelength are not resolved by direct imaging (there are cleaver techniques that use interference properties or Moiré patterns that can slightly exceed this limitation—see structured illumination). Resolving smaller items requires the use of shorter wavelength illumination. Electrons have a very short wavelength (10 keV electrons, used in scanning electron microscopes, have a DeBroglie wavelength about a hundred thousand times smaller than a visible light photon).

Another interesting fact about this limit is that, again, it is consistent with a photon "size" on the order of the wavelength.

1.5.5 Depth of Field

Imagine again standing at the nominal focus of a converging spherical wave but this time moving toward or away from the light, along the centerline (Figure 1.20). These various points correspond to spherical surfaces with shorter or longer radii of curvature. If these surfaces still lie within

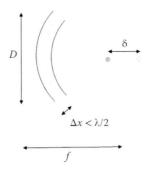

FIGURE 1.20 Consideration of Maréchal's criterion sets the depth of focus of a beam.

the Maréchal plates of the actual wave, they must be bright. Basic geometry and the formula for a circle reveal

$$\Delta z \cong 2\left(\frac{f}{D}\right)^2 \lambda = 2\left(f/\#\right)^2 \lambda$$

that is, the depth of focus is also set by the *f/#* and wavelength of the beam, this time with a square law dependence on *f/#*.

Once again, the physical limit of resolution is about one wavelength. This combined with the diffraction limit sets the resolution limit in all three spatial dimensions to about a wavelength.

1.5.6 Rayleigh Range

In applications where beams must travel significant distances, it is important to understand when simple ray tracing will no longer predict where the light will go. For short distances, aperture plane waves generally stay intact. At longer distances, however, they will be dominated by wave diffraction and the light will diverge away from the nominal beam line. Focused spherical waves exhibit a similar behavior where they remain spherical until they get to near the depth of focus region around the center of curvature.

For a plane wave, the *Rayleigh range* is the propagation distance, *L*, at which the aperture *D* is one Fresnel zone $\sqrt{\lambda/D}$ in size. That is,

$$L = \frac{D^2}{\lambda}$$

There is no lens in this definition, just an aperture limiting a plane wave. But notice that at the Rayleigh range it makes no difference if we use a lens—the diffraction of the aperture alone is as large as the diffraction limit of a lens with focal length *L*, that is (rearranging the above equation), $L\frac{\lambda}{D} = D$.

The Fresnel number is a measure of progress of a beam toward the Rayleigh range. It is defined as the number of Fresnel zones in the aperture plane wave, as a function of distance from the aperture:

$$N = \frac{D}{L\lambda/D} = \frac{D^2}{L\lambda}$$

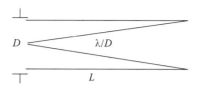

FIGURE 1.21   The Rayleigh range is the limiting distance for a geometric ray approximation to light propagation.

Recall from Section 1.3.1 that the Fresnel zone has a diameter of approximately $\sqrt{\lambda L}$. One can think of Fresnel number as the progress of diffracted waves from one edge of the aperture, opening at angle λ/D, reaching the other edge (Figure 1.21). The Fresnel number is equal to one at the Rayleigh range and is infinite at the aperture.

1.6 Image Formation and Analysis

In the Fraunhofer analysis, we showed that the field at the focal plane is the complex Fourier transform of the field at the aperture. With this in mind, we can describe the optical system performance in terms of linear system theory. The optical system is described in terms of an optical transfer function (OTF) in the spatial frequency domain, that is, what it does to the various spatial frequency components of the incoming wavefronts. The image is composed of linear superpositions of point spread functions (PSF). Assuming paraxial beams, this is a very accurate way to predict detailed properties of images produced by complex optical systems.

The Fourier transform relationship of aperture plane ("spatial domain") and image plane ("spatial frequency domain") is summarized in Figure 1.22.

The pupil function (a simple mask of ones and zeros) multiplies the electric field incident at the aperture.

An imaging system (as shown in Figure 1.23) gathers light waves from the sources in the object plane and produces an image at an image plane. The points in the image plane correspond one to one with points in the object plane. However, since the entrance aperture is finite, the resolution of the image is limited by diffraction.

Using the fact that multiplication in the spatial domain maps to convolution in the spatial frequency domain, the complex field in the image plane is the complex PSF (Fourier transform of the aperture) convolved with the complex field of the object:

$$I(u) = P(u)O(u)$$
$$i(x) = \int p(x-x')o(x')\mathrm{d}x' = p(x) \otimes o(x)$$

where $O(u)$ is the incident wave field in front of the aperture stop, which is, by Fraunhofer theory, the Fourier domain representation of the object field. $I(u)$ is the wave field just beyond the aperture stop, which is, by Fraunhofer theory, equal to the inverse Fourier transform of the image field.

The distribution of light intensity at the image plane is proportional to the squared modulus of the complex field incident on it:

$$|i(x)|^2 = \iint p(x-x')p^*(x-x'')o(x')o^*(x'')\mathrm{d}x'\mathrm{d}x'$$

In the case of an incoherent distribution of sources in the object plane, the time average of cross correlation of point sources is zero. For a scene illuminated by incandescent light or sunlight, the reemitting atoms are independent oscillators. For a laser emitter or a scene illuminated by laser light, this is

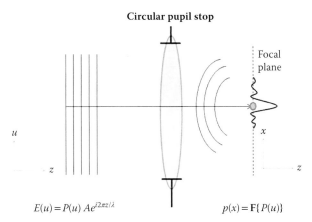

$$E(u) = P(u) \, Ae^{i2\pi z/\lambda} \qquad\qquad p(x) = \mathbf{F}\{P(u)\}$$

FIGURE 1.22 Fourier transform relation of incoming wave to the field at the image plane. Here u is the position in the aperture plane, $E(u)$ is the electric field at the aperture, $P(u)$ is the pupil function (one inside the aperture, zero outside the aperture), and $p(x)$ is the electric field at the focal plane.

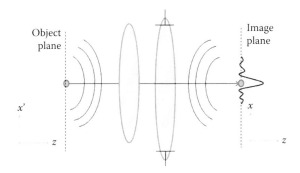

FIGURE 1.23 Two-lens system that produces an image of sources in the object plane.

not the case. A laser-illuminated scene will exhibit "speckle" due to the rough surface, causing random reinforcement and cancellation of coherent waves.

Assuming the incoherent source field

$$\left\langle o(x')o^*(x'')\right\rangle_t = \left|o(x')\right|^2 \delta(x'-x'')$$

$$\left\langle \left|i(x)\right|^2 \right\rangle_t = \int \left|p(x-x')\right|^2 \left|o(x')\right|^2 \mathrm{d}x' = \mathrm{PSF}(x) \otimes \left|o(x)\right|^2$$

that is, the distribution of intensity in the image plane is the convolution of intensity in the object plane with the PSF. The PSF is

$$\mathrm{PSF}(x) = \left|p(x)\right|^2 = F\left\{\int P(u')P(u-u')\mathrm{d}u'\right\} = F\{\mathrm{OTF}(u)\}$$

The OTF, defined as

$$\mathrm{OTF}(u) := \int P(u')P(u-u')\mathrm{d}u'$$

is useful for evaluating the spatial frequency performance of the imaging system since

$$F\left\{\left\langle\left|i(x)\right|^2\right\rangle_t\right\} = \mathrm{OTF}(u)F\left\{\left|o(x')\right|^2\right\}$$

that is, the imaging system's ability to reproduce spatial frequencies in the object is determined by the OTF. In practical use, the OTF is normalized to one at $u = 0$ so as to make it independent of overall source brightness. The OTF is typically complex valued, but from the definition it can be seen that a symmetric pupil will produce a real OTF.

The optics within the system and the propagating medium in between optical elements can introduce aberration on the propagating beam. These are accounted for by introducing phase factors on the pupil function, as in

$$P(u) = A(u)e^{i\varphi(u)}$$

Transmissivity variations can likewise be incorporated in the amplitude factor.

The plot in Figure 1.24 shows a radial slice of the OTF for a circular aperture. For a circularly symmetric aperture, the OTF is also circularly symmetric. Some apertures have asymmetries, such as the structural supports obscuring portions of the aperture of modern reflective telescopes, and these asymmetries will produce asymmetries in the OTF so that detailed evaluation will need to consider the entire 2D function. For simplification, most common optical designs have circularly symmetric or nearly symmetric apertures so we can glean information from the radial slice of the OTF.

The horizontal axis is the spatial frequency, in inverse meters. A diffraction-limited system has the highest spatial frequency transfer coefficients, but as can be seen, these fall to zero at the spatial frequency D/λ. This corresponds to spatial variations in the object intensity that are finer than the optical system's diffraction limit. These are ultimately blurred together by the diffraction-limited PSF.

Aberrations within the system suppress the response to spatial frequency, most often severe aberrations tend to suppress the high spatial frequency response—effectively reducing the optical system resolution to that of a smaller aperture.

The integral of the OTF relative to the integral of the diffraction-limited OTF is the *Strehl ratio*, named after the German physicist Karl Strehl. The Strehl ratio has a value from zero to one and is often used as a single number score of the quality of an optical system. From the definition of the OTF, one

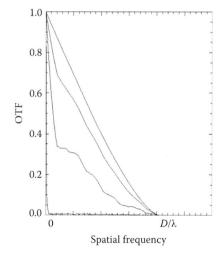

FIGURE 1.24 Examples of optical transfer functions (OTF, radial average). The upper curve is the OTF of a circular aperture of diameter D. The lower curves are those of systems with increasingly severe phase aberrations.

can confirm that the Strehl ratio is the ratio of the on-axis intensity of the aberrated PSF to that of the diffraction-limited PSF, which is its original definition.

1.7 Conclusion

In this chapter, we have given a brief introduction to the extensive field of wave optics and its practical application in optics system design. We have covered the basic physical principles of light waves viewed as an electromagnetic phenomenon (starting from Maxwell's laws) and as a traveling wave phenomenon (starting from the Huygens theory) and briefly touched on the nature of the light quanta, the photon. The polarization properties, propagation and scattering in a material medium, dispersion, refraction, and diffraction naturally follow from these basics. The limit of when light wave interference begins to dominate geometric ray behavior is understood by calculating two key scales, the Fresnel zone size and the Rayleigh range distance.

The References section can be consulted for rigorous development of the theories. In particular, consult Siegman (1986) for the Huygens–Fresnel beam propagation theory and numerical methods, Goodman (2005) for imaging systems and optical transfer functions, and Born and Wolf (2002) for geometric optics and wave optics theory, aberrations analysis, and scattering. We also highly recommend the entertaining work by Feynman (2006) for an illuminating discussion of the Fermat principle of least optical path.

References

N. Bohr, The Quantum Postulate and the Recent Development of Atomic Theory, *Nature* 121, 580–590 (1928). doi:10.1038/121580a0

M. Born and E. Wolf, *Principles of Optics Electromagnetic Theory of Propagation, Interference and Diffraction of Light*, 7th Ed., Cambridge: Cambridge University Press (2002). http://en.wikipedia .org/wiki/Wave_optics

A. Einstein, Über einen die Erzeugung und Verwandlung des Lichtes betreffenden heuristischen Gesichtspunkt, *Annalen der Physik* **17** (6): 132–148 (1905).

R. Feynman, *QED: The Strange Theory of Light and Matter*, New Jersey: Princeton University Press (2006).

J. W. Goodman, *Introduction to Fourier Optics*, Colorado: Roberts and Company Publishers (2005).

C. Huygens, *Traité de la Lumiere*, Chapter 1, Leiden, Netherlands: Pieter van der Aa (1690).

R. C. Jones, New calculus for the treatment of optical systems, *J. Opt. Soc. Am.* **31**, 488–493 (1941).

J. C. Maxwell, A dynamical theory of the electromagnetic field, Phil. Trans. R. Soc. Lond. **155**: 459–512 (1865).

H. Mueller, The foundations of optics, *J. Opt. Soc. Am.* **38**: 661 (1948).

M. Planck, Ueber das Gesetz der Energieverteilung im Normalspectrum, *Annalen der Physik* **309** (3): 553–563 (1901).

A. B. Ruffin, J. V. Rudd, J. F. Whitaker, S. Feng, and H. G. Winful, Direct observation of the Gouy phase shift with single-cycle terahertz pulses, *Phys. Rev. Lett.* **83**, 3410–3413 (1999).

A. E. Siegman, Lasers, Sausilito California: University Science Books (1986).

2

Principles of Geometric Optics

Joel A. Kubby
*University of California
at Santa Cruz*

2.1 Introduction

In this chapter, we consider geometric optics, which is an approximation to wave optics that can be used when considering an optical system composed of elements that are much larger than the wavelength of light going through the system. Then, we can ignore the wave nature of light, aside from its color, and assume that it will travel in a straight line, which is often called a ray. Although geometric optics is only an approximation to wave optics, it is technologically useful for the design and modeling of the adaptive optical systems that will be considered here. It greatly simplifies the calculations to a point that allows intuition to guide the design. This is not usually the case when considering the Huygens integral!

To determine the direction in which a light ray will pass through an optical system, we can apply Fermat's principle of least time or shortest optical path length that was discussed in Chapter 1. Fermat's principle is that the optical path distance *OPD* between points *A* and *B* given by

$$OPD(A,B) = \int_A^B n(x)\mathrm{d}x \qquad (2.1)$$

is shorter than the optical path length of any other curve that joins these points and lies in its certain regular neighborhood (Born and Wolf, 2006).

2.2 Reflection

We consider the application of Fermat's principle to two simple optical surfaces: a mirror that reflects light and an interface between two media that refracts light. With these two simple optical surfaces, we can understand the most important aspects of geometrical optics for the design of optical systems. In Figure 2.1, we show a mirror surface where a light ray starting from point *A* is reflected off the mirror surface to reach point *B*. The question is what path will the light ray take? If we assume that the

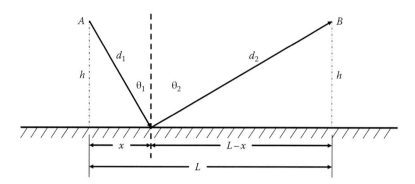

FIGURE 2.1 Reflection of the light from a mirror surface. The light travels from point A to point B by being reflected off the mirror surface at a horizontal distance x from point A. Point B is a horizontal distance L from point A. The distance that the light travels from point A to the mirror is d_1 and the distance that it travels from the mirror to point B is d_2. To find the angle of reflection, θ_2, given the angle of incidence, θ_1, we minimized the travel time along this path according to the Fermat's principle of least time.

horizontal distance between points A and B is L, then at what point x on the mirror will the light beam be reflected? We see that the light will take a path that is a distance d_1 from point A to the mirror and a path that is a distance d_2 from the mirror to point B. We assume that the light is traveling in a vacuum at a speed c.

The time t required for the light to travel from point A to B along this path is given by the following equation:

$$t(x) = \frac{d_1(x)}{c} + \frac{d_2(L-x)}{c} = \frac{1}{c}(d_1(x) + d_2(L-x))$$

$$= \frac{1}{c}\left(\sqrt{h^2 + x^2} + \sqrt{h^2 + (L-x)^2}\right)$$

To find the minimum time, we take the derivative of t with respect to x and set it equal to zero, which gives

$$\frac{dt(x)}{dx} = \frac{1}{c}\left(\frac{1}{2}(h^2 + x^2)^{-\frac{1}{2}}(2x) + \frac{1}{2}\left((h^2 + (L-x)^2)^{-\frac{1}{2}}(-2(L-x))\right)\right) = 0$$

$$= \frac{x}{\sqrt{h^2 + x^2}} - \frac{L-x}{\sqrt{h^2 + (L-x)^2}}$$

$$= \frac{x}{d_1} - \frac{L-x}{d_2}$$

$$= \sin(\theta_1) - \sin(\theta_2)$$

$$\sin(\theta_1) = \sin(\theta_2) \rightarrow \theta_1 = \theta_2.$$

(2.2)

We see that for the light to take the path of least time to get from point A to B by reflecting off the mirror surface requires that the angle of incidence, θ_1, is equal to the angle of reflection, θ_2. This simple formula allows us to calculate the path light will take when reflected from a mirror. As we shall see in Chapter 9, Section 3, it is interesting to consider mirrors that are not flat. In this case, the angle of incidence is equal to the angle of reflection, where the angles of incidence and reflection are defined by the local normal to the curved surface.

2.3 Refraction

We can also calculate the path a ray of light will take when it passes through an interface that separates media with two different indices of refraction. The speed of light inside the media with an index of refraction n is c/n, where c is the speed of light in vacuum. Since $n > 1$, the speed of light is always slower in media other than vacuum. Given this speed of light within the media, we can then calculate the time that light takes to travel from a point A in the first medium, with index n_1, to a point B in the second medium, with index n_2, as shown in Figure 2.2.

The time t required for the light to travel from point A to B along this path is given by the following equation:

$$t(x) = \frac{n_1}{c} d_1(x) + \frac{n_2}{c} d_2(L - x) = \frac{1}{c}(n_1 d_1(x) + n_2 d_2(L - x))$$

$$= \frac{1}{c}\left(n_1 \sqrt{h^2 + x^2} + n_2 \sqrt{h^2 + (L - x)^2}\right)$$

To find the minimum time, we take the derivative of t with respect to x and set it equal to zero, which gives:

$$\frac{dt(x)}{dx} = \frac{1}{c}\left(n_1 \frac{1}{2}\left(h^2 + x^2\right)^{-\frac{1}{2}}(2x) + n_2 \frac{1}{2}\left(\left(h^2 + (L - x)^2\right)^{-\frac{1}{2}}(-2(L - x))\right)\right) = 0$$

$$= n_1 \frac{x}{\sqrt{h^2 + x^2}} - n_2 \frac{L - x}{\sqrt{h^2 + (L - x)^2}}$$

$$= n_1 \frac{x}{d_1} - n_2 \frac{L - x}{d_2}$$ (2.3)

$$= n_1 \sin(\theta_1) - n_2 \sin(\theta_2)$$

$$n_1 \sin(\theta_1) = n_2 \sin(\theta_2)$$

We recognize this as Snell's law from Chapter 1.

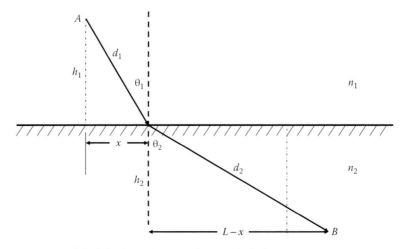

FIGURE 2.2 Refraction of the light through an interface. The light from point A is incident at an angle θ_1 on an interface between two media. The upper media has an index of refraction n_1 and the lower media has an index of refraction n_2. The change in the index bends the path of the light at an angle θ_2, causing it to pass through point B.

2.4 Paraxial Lens Equation

It is also useful to consider refraction from a curved surface, such as the circular arc shown in Figure 2.3. When the arc separates two regions of different indices of refraction, n_1 and n_2, a ray of light will be refracted at the interface according to Snell's law. If we choose the correct shape for the interface, we can cause light that originates from a point O along the axis of the arc to pass through a point I that is also along the axis. This is the geometry of a lens and it enables an object at one point in an optical system to be imaged at a different point in the system.

To find the equation that describes the action of the lens, we consider the fact that a light ray traveling from position O to position I must take the same amount of time regardless of the path. This means that the time taken for light to travel from point O to P and then from point P to I must equal the amount of time taken for light to travel directly from point O to I. Of course, the spatial distance $OP + PI$ is greater than the distance OI, but if n_1 is smaller than n_2, then light will travel slower inside the media with index n_2 and thus will take the same amount of time along the direct path OI as it would along the path $OP + PI$. This is because more of the path of light is inside the media with higher index.

To estimate these travel times, it is useful to make an approximation for light rays that are close to the optical axis (Feynman 1966). These are called paraxial rays. Consider the right triangle with sides of length $s > d > h$ shown in Figure 2.4.

From the Pythagorean theorem, we know that

$$s^2 = h^2 + d^2 \rightarrow h^2 = s^2 - d^2 = (s-d)(s+d)$$

To simplify this, we can approximate that $s \approx d$, so that $s + d \approx 2s$ and substitute $\Delta = (s - d)$ to obtain

$$h^2 = (s-d)(s+d) \approx 2\Delta s \rightarrow \Delta \approx \frac{h^2}{2s} \tag{2.4}$$

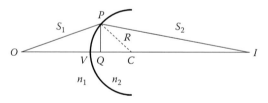

FIGURE 2.3 Refraction at a curved interface. The light traveling from point O (object) to point P is refracted at the interface between two media with indices of refraction n_1 and n_2. The refracted light intersects the optical axes at point I (image). The curved surface has a radius of curvature R.

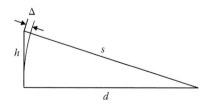

FIGURE 2.4 The paraxial approximation to find the difference in optical path difference for two rays that are close to the optical axis.

We can use this paraxial approximation to find the travel times for light along the different paths shown in Figure 2.3. The travel times along the paths from point O to P and from point P to I are given by

$$t_{OP} = \frac{n_1}{c}\overline{OP} \quad t_{PI} = \frac{n_2}{c}\overline{PI}$$

The travel times along the path from point O to I are given by

$$t_{OV} = \frac{n_1}{c}\overline{OV} \quad t_{VQ} = \frac{n_2}{c}\overline{VQ} \quad t_{QC} = \frac{n_2}{c}\overline{QC} \quad t_{CI} = \frac{n_2}{c}\overline{CI}$$

Then, the total travel times along the paths OPI and OI are as follows:

$$t_{OPI} = \frac{n_1}{c}\overline{OP} + \frac{n_2}{c}\overline{PI}s$$

$$t_{OI} = \frac{n_1}{c}\overline{OV} + \frac{n_2}{c}\left(\overline{VQ} + \overline{QC} + \overline{CI}\right)$$

We can then use the paraxial equation to simplify this equation:

$$\overline{OP} = \overline{OQ} + \Delta_1 = \overline{OV} + \overline{VQ} + \Delta_1$$

$$= \overline{OV} + \overline{VQ} + \frac{h^2}{2s_1}$$

$$\overline{PI} = \overline{QI} + \Delta_2 = \overline{QC} + \overline{CI} + \Delta_2$$

$$= \overline{QC} + \overline{CI} + \frac{h^2}{2s_2}$$

The difference in the travel times along the two routes would be

$$t_{OPI} - t_{OI} = \frac{n_1}{c}\left(\overline{OP} - \overline{OV}\right) + \frac{n_2}{c}\left(\overline{PI} - \overline{VQ} - \overline{QC} - \overline{CI}\right)$$

$$= \frac{n_1}{c}\left(\overline{VQ} + \frac{h^2}{2s_1}\right) + \frac{n_2}{c}\left(\frac{h^2}{2s_2} - \overline{VQ}\right)$$

For the travel times along the two routes to be equal, we need

$$\frac{n_1}{c}\left(\overline{VQ} + \frac{h^2}{2s_1}\right) = \frac{n_2}{c}\left(\overline{VQ} - \frac{h^2}{2s_2}\right)$$

$$\frac{n_1}{c}\frac{h^2}{2s_1} + \frac{n_2}{c}\frac{h^2}{2s_2} = \overline{VQ}\left(\frac{n_2}{c} - \frac{n_1}{c}\right)$$

$$\frac{n_1}{s_1} + \frac{n_2}{s_2} = \frac{2\overline{VQ}}{h^2}(n_2 - n_1)$$

We can apply the paraxial approximation (Equation 2.4) to find the length VQ:

$$R = \overline{QC} + \Delta_3$$

$$= \overline{QC} + \frac{h^2}{2R}$$

$$R - \overline{QC} = \overline{VQ} = \frac{h^2}{2R}$$

Substituting for VQ then gives

$$\frac{n_1}{s_1} + \frac{n_2}{s_2} = \frac{(n_2 - n_1)}{R}$$

Consider what happens when the distance s_1 becomes very large, that is, for an object at a very long distance. In the limit that $s_1 \to \infty$, we have

$$\frac{n_2}{s_2} = \frac{(n_2 - n_1)}{R} \to s_2 = \frac{n_2 R}{(n_2 - n_1)} = f'$$

Here, the light comes to a focus at a set distance f' into the media with index n_2. This is defined as the focal point within medium 2. Since the light is coming from infinity, the light rays would be parallel and the wavefronts would be plane waves. If $s_2 \to \infty$, we have

$$\frac{n_1}{s_1} = \frac{(n_2 - n_1)}{R} \to s_1 = \frac{n_1 R}{(n_2 - n_1)} = f$$

Here, the light comes to a focus at a set distance f into the media with index n_1. This is the focal point within medium 1.

2.5 Thin Lens Equation

In most optical systems, we would like to have the light pass from one point s_1 to another point s_2, both of which are in air rather than inside some material such as glass. This is accomplished using a thin lens that has two curved surfaces, as shown in Figure 2.5. Since a lens is usually used in air, we have set $n_1 = 1$.

When describing the object and image distances, and the radii of curvature of the two surfaces of the lens, the following conventions for the signs of the distances and radii of curvature are used (Feynman 1966):

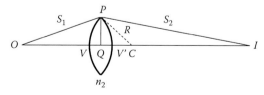

FIGURE 2.5   Refraction at two curved interfaces to form a biconvex lens.

1. The object distance s_1 is positive if the point O is to the left of the lens and is negative if the point is to the right of lens.
2. The image distance s_2 is positive if the point I is to the right of the lens and is negative if the point is to the left of the lens.
3. The radius of curvature of the lens is positive if the center of the radius of curvature is to the right of the lens and is negative if the center is to the left of the lens.

For the example shown in Figure 2.5, both s_1 and s_2 are positive. The radius of curvature on the left-hand side of the lens is positive and that on the right-hand side of the lens is negative. Since this lens has two convex surfaces, it is called a biconvex lens.

To solve the lens equation, we again calculate the travel times for two different paths—direct from point O to I and from point O to P and from P to I—and equate them. The travel time along the straight path between O and I is given by the sum of the travel times along segments OV, VQ, QV', and $V'I$:

$$t_{OV} = \frac{1}{c}\overline{OV} \quad t_{VQ} = \frac{n}{c}\overline{VQ} \quad t_{QV'} = \frac{n}{c}\overline{QV'} \quad t_{V'I} = \frac{1}{c}\overline{V'I}$$

Then the total travel times along paths OPI and OI are

$$t_{OPI} = \frac{1}{c}\overline{OP} + \frac{1}{c}\overline{PI}$$

$$t_{OI} = \frac{1}{c}\left(\overline{OV} + \overline{V'I}\right) + \frac{n}{c}\left(\overline{VQ} + \overline{QV'}\right)$$

We can then use the paraxial approximation (Equation 2.4) to simplify this equation:

$$\overline{OP} = \overline{OQ} + \Delta_1 = \overline{OV} + \overline{VQ} + \Delta_1$$

$$= \overline{OV} + \overline{VQ} + \frac{h^2}{2s_1}$$

$$\overline{PI} = \overline{QI} + \Delta_2 = \overline{QV'} + \overline{V'I} + \Delta_2$$

$$= \overline{QV'} + \overline{V'I} + \frac{h^2}{2s_2}$$

The difference in the travel times along the two different routes would be

$$t_{OPI} - t_{OI} = \frac{1}{c}\left(\overline{OP} + \overline{PI} - \overline{OV} - \overline{V'I}\right) - \frac{n}{c}\left(\overline{VQ} + \overline{QV'}\right)$$

$$= \frac{1}{c}\left(\frac{h^2}{2s_1} + \frac{h^2}{2s_2}\right) - \frac{n-1}{c}\left(\overline{VQ} + \overline{QV'}\right)$$

If the magnitude of the radius of curvature on both sides of the lens is the same, then we have

$$\overline{VQ} = \overline{QV'} = \frac{h^2}{2R}$$

The difference in the travel time can then be written as

$$t_{OPI} - t_{OI} = \frac{1}{c}\left(\frac{h^2}{2s_1} + \frac{h^2}{2s_2}\right) - \frac{n-1}{c}\left(\frac{h^2}{R}\right)$$

For the travel times along the two routes to be equal, we need

$$\frac{1}{c}\left(\frac{h^2}{2s_1} + \frac{h^2}{2s_2}\right) = \frac{n-1}{c}\left(\frac{h^2}{R}\right)$$

$$\left(\frac{1}{2s_1} + \frac{1}{2s_2}\right) = (n-1)\left(\frac{1}{R}\right) \tag{2.5}$$

$$\frac{1}{s_1} + \frac{1}{s_2} = (n-1)\left(\frac{2}{R}\right)$$

In the limit that $s_1 \rightarrow \infty$, we have

$$\frac{1}{s_2} = (n-1)\left(\frac{2}{R}\right)$$

$$s_2 = \frac{1}{n-1}\left(\frac{R}{2}\right) = f$$

In the limit that $s_2 \rightarrow \infty$, we have

$$s_1 = \frac{1}{n-1}\left(\frac{R}{2}\right) = f$$

Therefore, the focal lengths would be the same on either side of the lens, and the light that comes in from infinity is brought to a focus at a distance f from the lens. If the radii of curvature R_1 and R_2 on either side of the lens are not equal, and using the convention that the radius of curvature is positive if the center of the radius of curvature is to the right of the lens (R_1) and is negative if the center is to the left of the lens (R_2), then we would have

$$\overline{VQ} = \frac{h^2}{2R_1} \quad \overline{QV'} = -\frac{h^2}{2R_2}$$

and

$$\frac{1}{s_1} + \frac{1}{s_2} = (n-1)\left(\frac{1}{R_1} - \frac{1}{R_2}\right)$$

In the limit that $s_1 \rightarrow \infty$, we have

$$\frac{1}{s_2} = (n-1)\left(\frac{1}{R_1} - \frac{1}{R_2}\right)$$

$$s_2 = \frac{1}{n-1}\left(\frac{R_1 R_2}{R_2 - R_1}\right) = f.$$

Substituting for the focal length f in Equation 2.5, we have

$$\frac{1}{s_1} + \frac{1}{s_2} = \frac{1}{f} \qquad (2.6)$$

This is called the lensmaker's equation, which provides a relationship between the focal length f and the distances to the object and the image. If both the object and the image distances s_1 and s_2, respectively, are equal, then for $s_1 = s_2 = s$:

$$\frac{1}{s} + \frac{1}{s} = \frac{2}{s} = \frac{1}{f} \rightarrow s = 2f$$

In general, the object and the image may not be points, as shown in Figure 2.5, but rather extended objects, as shown in Figure 2.6. Here the object is shown as an arrow that extends above the optical axis. We can find out where the image is positioned by considering two principal light rays, one from the tip of the object that is parallel to the optical axis and one from the tip of the object that passes through the center of the lens. This is called ray tracing. The light ray from the tip of the object that is parallel to the optical axis is equivalent to a light ray from infinity, and therefore, after passing through the lens, it will intersect the optical axis on the right-hand side of the lens at the focal point f. The light ray that passes through the center of the lens has a symmetric optical path through the thin lens, and therefore it can be drawn as a straight line. This ray will be refracted by a certain amount in traveling from air, with $n = n_1 = 1$, into the glass with $n = n_2$, with the deflection being given by Snell's law, as described in Equation 2.3. This light ray will be refracted a second time when it exits the lens, and therefore it will continue along the same direction after passing through the lens. For a thin lens, we can draw that ray as a straight line. Where these two rays intersect on the right-hand side of the lens, an image will be formed. Since this image is on the opposite side of the lens from the object, it is called a real image. If a screen were to be placed at this location, an image of the object would be seen, although the image is inverted (i.e., upside down), and the arrow in the image may be different in length from the arrow shown as the object. We will discuss this in Section 2.6, when we consider the magnification of the lens.

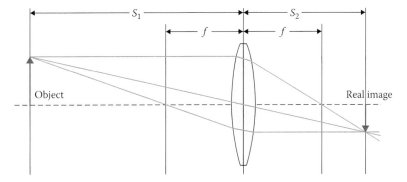

FIGURE 2.6 Ray tracing to find the object and the image for a thin biconvex lens. The object is located at a distance S_1 to the left of the lens, and the image is located at a distance S_2 to the right of the lens. The position of the image can be found by finding the intersection of two light rays emanating from the tip of the object. Three principal light rays are shown. One passes through the focal point on the left-hand side of the lens, while another passes through the focal point on the right-hand side of the lens. These light rays appear to come in from infinity. A third light ray passes through the center of the lens. Since the lens is considered to be a thin lens, this light ray is drawn as a straight line from the tip of the object to the tip of the image. (Credit: Wiki.)

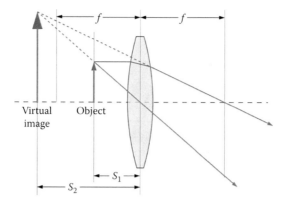

FIGURE 2.7   A virtual image formed from an object at S_1 that is closer than one focal length f to the surface of the lens. In this case, a real image is not formed to the right of the lens, but rather a virtual image is formed to the left of the lens. The light rays on the right-hand side of the lens appear to be emanating from this virtual image. (Credit: Wiki.)

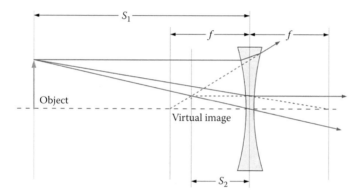

FIGURE 2.8   Biconcave lens composed of two concave lens surfaces. (Credit: Wiki.)

We note there is a third ray that can be traced, as shown in Figure 2.6. This ray passes through the focal point on the left-hand side of the lens. From Equation 2.6, we know that a light ray coming from infinity toward the lens from the right-hand side would pass through the focal point on the left-hand side of the lens, and therefore we can draw this light ray parallel to the optical axis on the image side of the lens. We see that this light ray also intersects the real image where the first two light rays intersect. Since we only need to find the position of the real image at the intersection of any of these lines, any two light rays are sufficient for finding the object location and size.

Consider what happens when the object distance S_1 is closer to the lens than the focal length f. This situation is shown in Figure 2.7. We can draw the light ray from the tip of the object that is parallel to the optical axis. It intersects the optical axis at a distance f to the right of the lens. We can also draw the light ray that goes through the center of the lens, but we see that these two light rays do not intersect at any point on the right-hand side of the lens. If we extend these two lines to the left of the object, as shown by the dashed lines in Figure 2.7, we see that they intersect at a distance S_2 to the left of the lens. From convention 2 above, S_2 has a negative value since it is to the left of the lens. The light rays from the object appear as though they were coming from a virtual image at a distance S_2 to the left of the lens. This image is called a virtual image since it would not be visible if a screen were to be located at that point.

In addition to having convex surfaces, a lens can also be ground to have concave surfaces, as shown in Figure 2.8. A lens with two convex surfaces is called a "biconcave" lens. Using ray tracing from an object located at a distance S_1 to the left of the lens, which is further than one focal length f to the left of the lens, will form a virtual image on the same side of the lens at S_2, which is closer than the focal length. It is also possible to have a lens with one convex surface and one concave surface. This form of a lens is called a meniscus lens. It is also possible to have a lens with one side planar and the other side either convex or concave. These are called plano-convex and plano-concave lenses, respectively.

2.6 Magnification

Again from Figure 2.6, we can determine the ratio of heights of the object and the image. We have redrawn this figure with some similar triangles, where the height of the right triangle that includes the object is y, and the height of the right triangle that includes the image is y' (Figure 2.9).

From the similar right triangles on the left-hand side of the lens, we can find the magnification M:

$$\frac{y}{S_1 - f} = \frac{y'}{f} \rightarrow M \equiv \frac{y'}{y} = \frac{f}{S_1 - f}$$

We can also use similar triangles on the right-hand side of the lens to find that

$$\frac{y}{f} = \frac{y'}{S_2 - f} \rightarrow M \equiv \frac{y'}{y} = \frac{S_2 - f}{f}$$

Since y' is below the optical axis, it is a negative quantity by convention, we say that the magnification is negative, and it results in an inverted image. The virtual image in Figure 2.8 is positive, and therefore, in this case, the magnification is positive and the virtual image is not inverted.

Combining these two equations, we find

$$M \equiv \frac{y'}{y} = \frac{f}{S_1 - f} = \frac{S_2 - f}{f} \rightarrow (S_1 - f)(S_2 - f) = f^2$$

$$S_1 S_2 - S_1 f - S_2 f + f^2 = f^2 \rightarrow S_1 S_2 = S_1 f + S_2 f = f(S_1 + S_2)$$

$$\frac{S_1 S_2}{S_1 + S} = f \rightarrow \frac{1}{S_1} + \frac{1}{S_2} = \frac{1}{f}$$

We recover Equation 2.6, the lensmaker's formula.

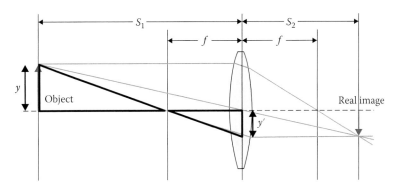

FIGURE 2.9 Magnification of a lens. The height of the object is y and that of the image is y'. (Credit: Wiki.)

2.7 Aberrations

The restriction of the thin lens analysis to monochromatic paraxial rays leads to optical aberrations for real lenses. For an extended lens that admits light rays away from the axis, the paraxial approximation used in Equation 2.4 is no longer valid. In general, the optical surface obtained from the principle of least time is no longer a spherical surface, but rather a higher-order surface than a sphere. Nonetheless, a spherical surface is much easier to fabricate by grinding and polishing than an aspherical surface, and therefore, lenses with spherical surfaces are often used. As the light rays become further removed from the optical axis, they no longer come to a focus at one point, but rather focus at different points depending on their distance from the axis. The resulting aberration is known as spherical aberration, since it results from the spherical surface of the lens. An example is shown in Figure 2.10. The light rays at the edge of the lens come to a focus closer to the lens than those from the paraxial region near the optical axis.

In addition to the spherical shape of the lens, spherical aberration can arise from index mismatches between the lens and the sample. When the light travels between regions of different indices of refraction, the light rays can be bent at the interface according to Snell's law, as described in Equation 2.3. This aberration can be the dominant aberration when imaging deeply into a specimen since it increases with depth into the sample.

Another aberration is caused by the wavelength dependence of the index of refraction n. Chromatic aberration arises when multiwavelength light is refracted at an interface between two regions with different indices of refraction. The different colors of light can be refracted by differing amounts, as shown in Figure 2.11. A familiar example is the decomposition of white light by a prism or water droplet, which forms the familiar colored rainbow when sunlight is decomposed into its spectral components. The cause for the spectral decomposition is the wavelength dependence of the index of refraction. The

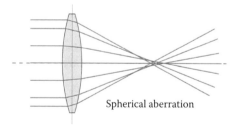

Spherical aberration

FIGURE 2.10   Spherical aberration. The light rays that are farther from the optical axis come to a focus at different distances from the lens. (Credit: Wiki.)

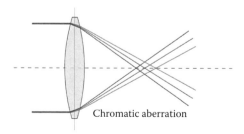

Chromatic aberration

FIGURE 2.11   Chromatic aberration. Different colors of light are refracted by different amounts when passing through regions with different indices of refraction. This is because the index of refraction depends on the wavelength of light. (Credit: Wiki.)

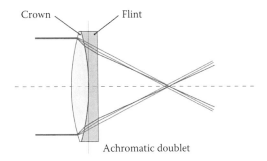

FIGURE 2.12 Achromatic doublet lens to compensate for chromatic aberration due to the wavelength dependence of the indices of refraction. Here one lens is biconvex and is made of crown glass with one index of refraction, and the other lens is plano convex and is made of flint glass with a different index of refraction. The lens pair tends to compensate for the dispersion of multiwavelength light.

wavelength dependence of the index of refraction is called dispersion, since it causes light to be dispersed into its spectral components.

A solution to overcome chromatic aberration is to use two different lenses with two different shapes (e.g., convex and concave radii of curvature) made with optical materials with different dispersion characteristics (e.g., crown and flint glasses). An example is shown in Figure 2.12.

In addition, lenses can have other optical aberrations such as coma, astigmatism, field curvature, and distortions (barrel and pincushion). Misalignment between optical elements can cause further aberration as discussed in Chapters 4 and 9. While the primary goal of adaptive optics in biological imaging is to overcome sample-induced aberrations, as discussed in Chapter 3, the adaptive optical system can also overcome optical system–induced aberrations due to aberrations in the optical components and misalignments between components. In some situations, system aberrations can dominate specimen-induced aberrations. Here, adaptive optics can be used to relax system tolerances to bring down the cost of high-performance optical systems or to improve the performance of lower-performance optical systems.

References

Feynman, R. P., R. B. Leighton, and M. Sands, *The Feynman Lectures on Physics*, Addison-Wesley, Reading, MA, 1966.

Born M. and E. Wolf, *Principles of Optics*, 7th ed., Cambidge: Cambridge University Press, 2006.

3

Theory of Image Formation

Michael Schwertner
confovis GmbH

3.1 Introduction: Imaging Properties and Point Spread Function (PSF)

This chapter briefly mentions some aspects of the scalar theory[*] of the diffraction of light and image formation. One of the main results is the point spread function (PSF) of a lens which in turn is important in the context of lateral and axial resolution and the effects of aberrations in confocal microscopy.

If we want to understand and model an imaging system, we have to ask what optical field (or intensity distribution/image) it will produce for a given input (object). In confocal microscopy we are interested in the quality of the focal spot. It turns out that the PSF, which is the image of a single object point (see Figure 3.1), characterizes the system completely. This is because the output of the optical system can be understood as the superposition of the optical fields caused by the object points. In the following sections we will give expressions for the PSF of a lens. These may be derived from first principles.

3.2 Diffraction of Light

The theory of diffraction for coherent light is based on the Huygens-Fresnel principle as discussed in Chapter 1, stating that every unobstructed point of a wavefront serves as a source of spherical secondary wavelets of the same frequency. Now the amplitude of the optical field at any point beyond is the superposition of all these wavelets taking their phase and amplitude into account (Hecht 1987). This is mathematically equivalent to Goodman (1996):

$$U_2(x_2, y_2) \approx \int_{-\infty}^{\infty} \int_{-\infty}^{\infty} U_1(x_1, y_1) \frac{\exp(-jkR)}{\lambda jR} dx_1 dy_1 \qquad (3.1)$$

---

[*] A transversal electromagnetic wave, such as light, is described by its amplitude, phase, and plane of polarization. Scalar theory means that only the phase and amplitude are considered, and polarization effects are neglected.

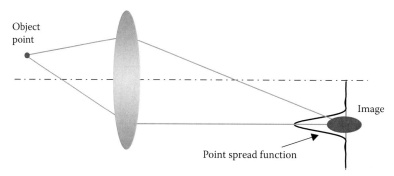

FIGURE 3.1 The point spread function is the image of an ideal point and describes the imaging properties of the optical system.

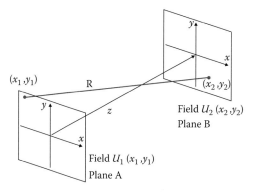

FIGURE 3.2 The geometry considered in Equation 3.1. The electromagnetic field in plane *B* may be calculated if information on the field in plane *A* is available.

which describes the electromagnetic (complex) field $U_2(x_2,y_2)$ in the plane *B* as a result of the propagation from the known field distribution $U_1(x_1,y_1)$ in the plane *A* (see Figure 3.2). Here *R* denotes the geometric distance between two points (x_1,y_1) and (x_2,y_2) and $k = 2\pi/\lambda$ is the wavenumber. Every element of the wavefront, U_1 is seen to cause a spherical wave of a magnitude proportional to U_1, which is the implementation of Huygen's principle. Making further approximations taking into account that $z \gg Max(x_1,y_1)$, one may derive the Fresnel diffraction integral (Wilson and Sheppard 1984):

$$U_2(x_2,y_2) = \frac{\exp(-jkz)}{\lambda jz} \int\limits_{-\infty}^{\infty} \int\limits_{-\infty}^{\infty} U_1(x_1,y_1)\exp\left[\frac{-jk}{2z}((x_1-x_2)^2 + (y_1-y_2)^2)\right]dx_1 dy_1 \qquad (3.2)$$

In case the maximum distance from the *z* axis within the plane *A* also obeys the stricter condition

$$z \gg \frac{1}{2}k(x_1^2 + x_1^2)_{max} \qquad (3.3)$$

we can make further simplifications to yield the Fraunhofer approximation by neglecting terms containing x_1^2 and x_2^2:

$$U_2(x_2, y_2) = \frac{\exp(-jkz)}{\lambda jz} \exp\left[-\frac{jk}{2z}(x_2^2 + y_2^2)\right]$$

$$\times \int_{-\infty}^{\infty} \int_{-\infty}^{\infty} U_1(x_1, y_1) \exp\left[\frac{+jk}{z}(x_1x_2 + y_1y_2)\right] dx_1 dy_1$$

(3.4)

Now we see that, apart from multiplying phase factors, the field distribution $U_2(x_2,y_2)$ is a Fourier transform of the input distribution $U_1(x_1,y_1)$ in plane A evaluated at spatial frequencies $f_x = x_2/\lambda z$ and $f_y = y_2/\lambda z$. The diffraction integrals are useful in the following discussion of the intensity distribution in the focal region of a lens.

3.3 Amplitude PSF of a Thin Lens

We are now concerned with the properties of the focal spot produced by a thin lens as sketched in Figure 3.3. The effect of an ideal lens is to convert a plane wavefront into a spherical one, which converges into a single point. The complex amplitude directly behind the thin[†] lens is given by $U_1(x,y)$. If one illuminates the lens with a uniform plane wave, the effect of the lens may be described by the introduction of a phase factor (Wilson and Sheppard 1984):

$$U_1(x_1, y_1) = P(x_1, y_1) \exp\left[\frac{jk}{2f}(x_1^2 + y_1^2)\right]$$

(3.5)

where the exponential factor gives rise to a spherical wave that converges into a point at a distance f behind the lens.

It is important to be aware that the complex pupil function $P(x_1,y_1)$ of the lens takes the finite size of the lens as well as lens imperfections and aberrations into account. In the next chapters aberrations will be modelled by such pupil functions.

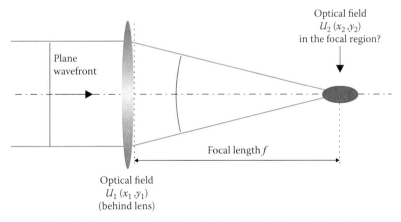

FIGURE 3.3 We seek to calculate the optical field $U_2(x_2,y_2)$ in the focal region from the field $U_1(x_1,y_1)$ directly behind the lens.

[†] We neglect the displacement of the rays while traversing the lens.

We may calculate the optical field $U_2(x,y)$ in the focal plane at the distance f from the lens using the Fresnel diffraction integral (3.2) and Equation 3.5. If we neglect pre-multiplying factors, this leads directly to the amplitude PSF given by

$$h(x_2,y_2) = \int\limits_{-\infty}^{\infty} \int\limits_{-\infty}^{\infty} P(x_1,y_1)\exp\left[\frac{jk}{f}(x_1x_2 + y_1y_2)\right]dx_1dy_1 \tag{3.6}$$

We may incorporate an additional phase factor $(1/2)u\rho^2$ to model defocus and obtain the full three-dimensional amplitude PSF of the lens (Wilson and Sheppard 1984; Booth 2001):

$$h(u,v,\phi) = \int\limits_{0}^{2\pi} \int\limits_{0}^{1} P(\rho,\phi)\exp\left[\frac{1}{2}ju\rho^2 + jv\rho\cos(\phi-\theta)\right]\rho\,d\rho\,d\theta \tag{3.7}$$

where we have introduced the normalized optical coordinates

$$v = kr_2\sin\alpha \approx \frac{2\pi r_2 a}{\lambda f}; \quad r_2^2 = x_2^2 + y_2^2; \quad \text{N.A.} = n\sin\alpha \tag{3.8}$$

and

$$u \approx \frac{k\delta z a^2}{f^2} \approx 4k\delta z \sin^2\left(\frac{\alpha}{2}\right) \tag{3.9}$$

where δz denotes the distance from the focal plane, and α is the semi angle of the lens aperture as defined for the numerical aperture N.A. $= n\sin\alpha$. The coordinate $\rho = r_2/a$ (a is the physical radius of the lens) has been introduced to normalize the radial range of integration.

The intensity in the focal region is described by the intensity PSF, which is the square of the amplitude PSF. Examples for ideal and aberrated intensity PSFs were calculated using Equation 3.7 and the corresponding pupil functions and are shown in Figure 4.3.

3.4 Effective PSF and Image Formation in a Confocal Microscope

In the preceding section, we developed an expression for the amplitude PSF of a single lens. But how does this relate to the confocal image? In a confocal system, we will have an illumination lens with the amplitude PSF $h_1(u,v,\varphi)$ and the detection system, is described by the amplitude PSF $h_2(u,v,\varphi)$. If the object function (which may represent the transmittance, reflectance, or dye concentration) is denoted by t, the coherent image formation process, using an infinitely small pinhole, yields the image for a confocal setup (Wilson and Sheppard 1984):

$$I = |(h_1h_2)\otimes t|^2 \tag{3.10}$$

Here $\otimes$ is the convolution operator and we see that the image is the absolute square of the convolution of the object function t with an effective PSF of $h_{eff} = h_1h_2$ of the system. In reflection mode, the illumination and the collecting lens will be the same, hence we get

$$I_{reflection} = |h_1^2 \otimes t|^2 \tag{3.11}$$

The imaging process in fluorescence mode is completely different, because the coherence is destroyed during the process of excitation and emission of the dye molecule. Let f denote the distribution of the fluorescent dye. Then the excitation creates an incoherent fluorescent field of an intensity $|h_1|^2 f$. Following Wilson and Sheppard (1984) and Wilson (1989), the successive incoherent imaging of this field yields

$$I_{fluorescence} = |h_1(u,v,\phi)h_2(u/\beta,v/\beta,\phi)|^2 \otimes f \tag{3.12}$$

where $\beta = \lambda_2/\lambda_1$ is the ratio between the emitted fluorescence wavelength λ_2 and the excitation wavelength λ_1. Again, the imaging process is modeled by a convolution of some effective PSF and the object function. This insight is the starting point for so-called deconvolution techniques that aim to calculate a closer approximation of the object by taking into account the properties of the imaging device, that is, the PSF.

3.5 PSF for an Ideal Lens and Diffraction Limited Imaging

Let us consider a symmetric pupil function $P(r_1)$ for an ideal lens that has uniform phase and uniform amplitude and is unity for a radius of $r_1 \leq a$. Using Equation 3.7, we can state the radial variation of the intensity distribution in the focal plane of the lens as

$$I(r_2) = \frac{|h(r_2)|^2}{\lambda^2 f^2} \propto \left[\frac{2J_1(v)}{v}\right]^2 \tag{3.13}$$

Here $J_1(.)$ denotes the first-order Bessel function of the first kind (Abramowitz and Stegun 1972). This radially symmetric intensity distribution is also known as the Airy pattern and is plotted in the left panel of Figure 3.4.

The axial intensity distribution for an ideal lens may also be calculated from Equation 3.7, and has the form

$$I(0,u) = |h(u,0,0)|^2 \propto \left[\frac{\sin(u/4)}{u/4}\right]^2 \tag{3.14}$$

and is shown in the right panel of Figure 3.4. Again the definition (3.9) for the axial normalized optical coordinate was used.

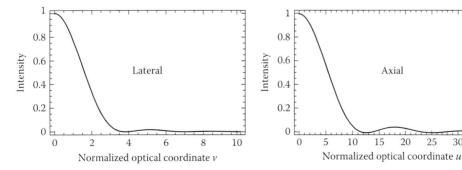

FIGURE 3.4 Intensity distribution in the focal region of an ideal lens. Left: lateral profile through focus. This function is also known as Airy pattern. Right: axial profile through focus. For the definition of the normalized optical coordinates, see Equations 3.8 and 3.9.

These intensity distributions have been calculated for the ideal lens when there are no aberrations present. Then the size and shape of the focal spot take optimum values and we talk of diffraction limited imaging. In a real world, system aberrations are encountered that may originate from different sources. It is the goal of adaptive optics to correct for aberrations and hence restore optimum, diffraction limited imaging.

Bibliography

Abramowitz, M. and I. Stegun (1972). *AAAA Handbook of Mathematical Functions with Formulas, Graphs, and Mathematical Tables* (9th ed.). Dover Publishing, New York.

Booth, M. J. (2001). *Adaptive Optics for Confocal Microscopy*. PhD thesis, University of Oxford.

Goodman, J. W. (1996). *Introduction to Fourier Optics* (2nd ed.). McGraw Hill, New York.

Hecht, E. (1987). *Optics* (2nd ed.). Addison Wesley, New York.

Wilson, T. (1989). Optical sectioning in confocal fluorescent microscopes. *Journal of Microscopy 154*(2), 143–156.

Wilson, T. and C. Sheppard (1984). *Theory and Practice of Scanning Optical Microscopy*. Academic Press, London.

II

Methods

4

Aberrations and the Benefit of Their Correction in Confocal Microscopy

Michael Schwertner

confovis GmbH

4.1 Introduction: Why Measure Specimen-Induced Aberration?

The ultimate goal is to design a microscope that fully corrects aberrations and thus can deliver optimum performance and diffraction-limited imaging in confocal microscopy. This requires knowledge about the nature of aberrations encountered under conditions found in biomedical imaging.

Before we start the design of an adaptive optics system, we need to address the following questions:

- What type and amount of the specimen-induced aberration do we have to expect from typical biological specimen?
- How many Zernike modes do we need to correct to reduce aberrations to an acceptable level?
- Can we optimize the correction process by limiting the sensing and correction to a few Zernike modes while neglecting modes that are either not caused by the specimen or not transmitted by

the optical system? Can this be done without restricting the adaptive optics system to a particular class of specimen?

- What active wavefront correction element can supply the required modes and is most suitable?

The answers to all these questions are crucial for the design of an adaptive optics system and can be inferred from quantitative aberration measurements. Therefore, a system for the measurement of the specimen-induced wavefront aberration was built and is described in this chapter.

4.2 Definition of Aberrations

A confocal microscope (CFM) in fluorescence mode probes the three-dimensional distribution of fluorescent molecules within the sample. In the ideal case, the excitation light is focused to a diffraction-limited focal spot; the size of this spot defines the resolution of the image. Apart from the variation in fluorophore concentration imaged, a typical biological specimen also shows a variation in refractive index, $n(\tilde{x},\tilde{y},\tilde{z})$, as well. The absolute phase ϕ of a point at the location $L(x,y,z)$ within the beam path of the optical system with respect to the focal point is given by

$$\phi(L(x,y,z)) = \int_{F}^{L(x,y,z)} \frac{2\pi}{\lambda} n(\tilde{x},\tilde{y},\tilde{z}) \mathrm{d}w \tag{4.1}$$

where λ is the free space wavelength, w is the geometric distance traveled, F denotes the position of the focal point of the lens, and the path of integration is governed by Snell's law of refraction. The ideal wavefront within the pupil plane of an imaging lens represents a surface of constant phase. If an object is introduced into the beam path, the function $n(\tilde{x},\tilde{y},\tilde{z})$ changes and the wavefront is altered. The deformation of the wavefront is shown in Figure 4.1, and this deviation of the wavefront from the ideal configuration is called aberration.

The effect of an ideal lens is to convert a plane wavefront into a spherical one, which converges into a single point. Then the size of the focal spot is determined only by the diffraction limit (left side of Figure 4.1). This condition represents the optimum case of "diffraction-limited imaging. The aberration of the wavefront can now be defined as the difference in phase between a wavefront of the diffraction-limited system and that of the aberrated system. It is common to specify the aberrations in the pupil plane of the lens.

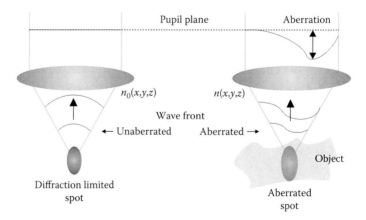

FIGURE 4.1 Aberration of the wavefront. The distortion of the fluorescence emission wavefront (wavy) as it propagates from the focal spot through the specimen. The propagation direction is indicated by the black arrows. During the excitation of the fluorophore (egg-shaped object), the same aberrations cause a blurred focal spot. The perfect, unaberrated case delivers a spot that has a diffraction-limited size (left). The variation in refractive index of the sample may cause aberrations (right). The result is a bigger, aberrated spot.

If we denote the axial position of the pupil plane of the lens with z_0, the refractive index distribution of the unaberrated case with $n_0(\tilde{x},\tilde{y},\tilde{z})$, and the aberrated case with $n(\tilde{x},\tilde{y},\tilde{z})$, we may define the aberration function $\psi(x,y)$ in the pupil plane of the lens by

$$\psi(x,y)= \int_F^{P(x,y,z_0)} \frac{2\pi}{\lambda}\big(n(\tilde{x},\tilde{y},\tilde{z})-n_0(\tilde{x},\tilde{y},\tilde{z})\big)\mathrm{d}w \tag{4.2}$$

It should be noted that the refractive index distributions describe the whole optical system including the lenses. Therefore, the aberration function $\psi(x,y)$, which is measured in the following experiments, takes into account all types of aberrations relative to an ideal optical system. For a perfect system, $\psi(x,y)$ would be constant within the pupil plane of the lens.

4.3 Zernike Mode Analysis of the Aberrated Wavefront

To describe the aberration present in the wavefront, it is convenient to use a Zernike mode representation, which is a decomposition of the aberration function $\psi(x,y)$ into Zernike polynomials that may be expressed as follows:

$$\psi(r,\theta)= \sum_{i=1}^{\infty} M_i Z_i (r,\theta) \tag{4.3}$$

where the modal coefficients M_i describe the strength of each Zernike polynomial within the wavefront. Since the Zernike polynomials $Z_i(r,\theta)$ (Born and Wolf 1983; Noll 1976)* are a set of orthogonal functions defined over the unit circle, we may write

$$M_i = \frac{1}{\pi}\int_0^1\int_0^{2\pi} \psi(r,\theta)Z_i(r,\theta)r\mathrm{d}\theta\mathrm{d}r \tag{4.4}$$

The set of Zernike coefficients M_i fully describes the aberration function $\psi(r,\theta)$. Some Zernike modes correspond to the classical terms of abberations; for instance, astigmatism (modes 5 and 6) or coma (modes 7 and 8). In our data analysis, we characterized the wavefront by the Zernike modes 2 through 22.

Several definitions with different normalization factors and mapping schemes for the indices exist. In this chapter, we use the following definition (Neil et al. 2000):

$$Z_n^m(r,\theta)= \begin{cases} m<0 : \sqrt{2}R_n^{-m}(r)\sin(-m\theta) \\ m=0 : 0 \\ m>0 : \sqrt{2}R_n^m(r)\cos(m\theta) \end{cases} \tag{4.5}$$

$$R_n^m(r)= \sqrt{n+1} \sum_{s=0}^{(n-m)/2} \frac{(-1)^s(n-s)!}{s!((n+m)/2-s)!((n-m)/2-s)!}r^{n-2s}$$

where the indices m and n, restricted to the conditions $n \geq |m|$ and $n - |m|$, are even. The rules to map the double indices (n,m) to the single index i are i starts at 1 for $n = 0$ and rises first with n, then all allowed values of m are ordered with rising magnitude where the positive values come first. The first 22 polynomials $Z_n^m(r,\theta)$ are listed in Table 4.1, and a few modes are plotted in Figure 4.2. Note that some

---

* Note that the normalization and indexing are different between the definitions in Born and Wolf (1983) and Noll (1976).

TABLE 4.1   Definition of the First 22 Zernike Polynomials

i	n	m		Aberration Term
1	0	0	1	Piston
2	1	1	$2r\cos(\theta)$	Tip
3	1	−1	$2r\sin(\theta)$	Tilt
4	2	0	$\sqrt{3}(2r^2-1)$	Defocus
5	2	2	$\sqrt{6}r^2\cos(2\theta)$	Astigmatism
6	2	−2	$\sqrt{6}r^2\sin(2\theta)$	Astigmatism
7	3	1	$2\sqrt{2}(3r^3-2r)\cos(\theta)$	Coma
8	3	−1	$2\sqrt{2}(3r^3-2r)\sin(\theta)$	Coma
9	3	3	$2\sqrt{2}r^3\cos(3\theta)$	
10	3	−3	$2\sqrt{2}r^3\sin(3\theta)$	
11	4	0	$\sqrt{5}(6r^4-6r^2+1)$	Spherical (1st)
12	4	2	$\sqrt{10}(4r^4-3r^2)\cos(2\theta)$	
13	4	−2	$\sqrt{10}(4r^4-3r^2)\sin(2\theta)$	
14	4	4	$\sqrt{10}r^4\cos(4\theta)$	
15	4	−4	$\sqrt{10}r^4\sin(4\theta)$	
16	5	1	$2\sqrt{3}(10r^5-12r^3+3r)\cos(\theta)$	
17	5	−1	$2\sqrt{3}(10r^5-12r^3+3r)\sin(\theta)$	
18	5	3	$2\sqrt{3}(5r^5-4r^3)\cos(3\theta)$	
19	5	−3	$2\sqrt{3}(5r^5-4r^3)\sin(3\theta)$	
20	5	5	$2\sqrt{3}r^5\cos(5\theta)$	
21	5	−5	$2\sqrt{3}r^5\sin(5\theta)$	
22	6	0	$\sqrt{7}(20r^6-30r^4+12r^2-1)$	Spherical (2nd)

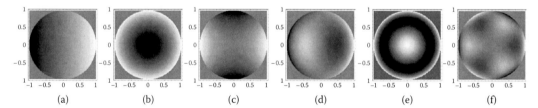

(a)   (b)   (c)   (d)   (e)   (f)

FIGURE 4.2   Polar plots of a few Zernike modes: (a) $Z_2(r,\theta)$, tip; (b) $Z_4(r,\theta)$, defocus; (c) $Z_5(r,\theta)$, astigmatism; (d) $Z_7(r,\theta)$, coma; (e) $Z_{11}(r,\theta)$, first spherical; and (f) $Z_{18}(r,\theta)$, trefoil.

of the polynomials are identical except for a rotation about the origin. For instance, modes 5 and 6 (astigmatism) differ just by a rotation of 45°.

This is, apart from the indexing, equivalent to the definition of Noll (1976) and has the advantage of a normalization such that the mean is zero over the unit circle and the variance is unity (except for $n = 0$, where it is zero). All data in this chapter are given in the Zernike coefficient units according to the definition in Equation 4.5. By calculating the wavefront using Equations 4.3 and 4.5, one directly gets the aberration of the wavefront over the normalized pupil of the system in radians.

Please note that other authors such as Born and Wolf (1983) and Noll (1976) use two indices that relate to the azimuthal and radial orders. In the definition used in Equation 4.5, the orthogonality relation does not depend on *n*, unlike in the definition given in Born and Wolf (1983).

Modes 2 through 4 have the common property that they create geometrical distortions of the image but do not compromise resolution or signal intensity. Mode 2 (tip) represents a linear variation of the wavefront in the *x* direction whereas mode 3 (tilt) is a linear wavefront slope in the *y* direction. These correspond to lateral displacements of the focal spot in the *x* and *y* directions. Similarly, mode 4 (defocus) alters the axial position of the focal spot and has no effect on the shape of the PSF. Therefore, within an adaptive optics system that aims to restore diffraction-limited imaging, it is not necessary to correct for these modes.

The Zernike series expansion of the two-dimensional aberration function is similar to the well-known Fourier analysis, where the base functions are harmonics as opposed to the Zernike polynomials used for wavefront analysis. However, from Fourier analysis we know that most output signals may be represented by a few Fourier coefficients only. This is due to the band-pass characteristics of information processing systems, such as electrical circuits or an optical setup and also because of the limited spectrum of the input signal to the system (in our case, the specific structural properties of the biological specimen). Therefore, we anticipate that only a few selected Zernike modes are relevant for aberration correction. This in turn could greatly simplify the design of the system.

4.4 Sources of Aberrations

Aberrations encountered in a real-world microscope system can be attributed to one of the following categories:

- imperfect design or manufacturing of the optics;
- imperfect alignment of the optical elements; and
- specimen-induced aberrations, where one can distinguish between
  - refractive index mismatch between specimen and immersion medium, and
  - field-dependent variation of the refractive index of the object.

There is nothing much the user can do about the first two items: in an advanced modern microscope, the theoretical design of the optics is close to perfection and one can access only a few controls that alter the alignment of optical elements. In most cases, mainly the spherical contributions to the specimen-induced aberration can be controlled by matching the refractive index of the immersion fluid. Adaptive optics may provide aberration correction.

One can distinguish between two types of specimen-induced aberration: a *field-dependent fraction*, which can be caused by the variations of the refractive index of the specimen, and a *static component*, which is due to a refractive index mismatch between the sample and the embedding medium. Both components may be measured independently, as detailed in Section 4.7.

4.5 Effect of Aberrations on the Imaging Quality of the Confocal Microscope

If the aberration function $\psi(x,y)$ or the equivalent $\psi(r,\theta)$ in polar coordinates is known, the calculation of the amplitude PSF $h(u,v,\phi)$ of the objective lens is straightforward (see Section 3.3). We assume that the lens is illuminated with uniform unit intensity, and in Equation 3.7 the pupil function becomes $P(r,\theta) = \exp(j\psi(r,\theta))$ and we obtain for the amplitude PSF:

$$h(u,v,\phi) = \int_{0}^{2\pi}\int_{0}^{1} \exp(j\psi(r,\theta))\exp\left[j\frac{1}{2}ur^2 + jvr\cos(\theta-\phi)\right] r\,\mathrm{d}r\,\mathrm{d}\theta \qquad (4.6)$$

Again the normalized optical coordinates v and u as defined in Equations 3.8 and 3.9 are used.

From Equation 4.6, we see that the aberrations have a direct effect on the amplitude PSF of the imaging lens. In Section 3.4, it is explained how the PSF relates to the final image of the CFM.

To illustrate the effect of aberrations on the imaging quality of the CFM, we calculated the image of an infinitely small fluorescent object using Equations 4.6 and 3.12 under the assumption that the fluorescence wavelength is equal to the excitation wavelength (i.e., $\beta = 1$). Figure 4.3 shows the result for the unaberrated wavefront (left column) and the aberrated wavefront (right column) that contains 1 Zernike unit of astigmatism (i.e., $M_5 = 1$, other coefficients are zero in Equation 4.3; see Section 4.3). As we see from the experimental data from typical biological specimen, 1 Zernike unit is a rather large amount for an individual aberration mode. For M_5, values as high as 0.5 were observed. However, the combined value across different Zernike modes can easily exceed 1 Zernike unit. In the example of Figure 4.3, the shape of the PSF has completely changed; it has very large side lobes due to the aberration. Please note that the images of the intensity PSF (upper two images in Fig. 4.3) are both normalized, while the plots on the bottom of Figure 4.3 show a cross section through the PSF in the axial direction at the same scale. A drop in intensity of about 80% is visible. Each type of aberration (i.e., Zernike mode) has a distinct effect on the PSF and hence on the imaging performance.

It can be concluded that aberrations degrade both resolution and intensity of the confocal image. The degraded intensity may cause problems in practical setups since the excitation intensity should be kept as low as possible to avoid photobleaching of the dye or sample degradation. Especially the axial extension of the PSF is sensitive to aberrations. Hence, aberrations mainly reduce the axial resolution, which is, even in a perfect system, worse than the lateral resolution.

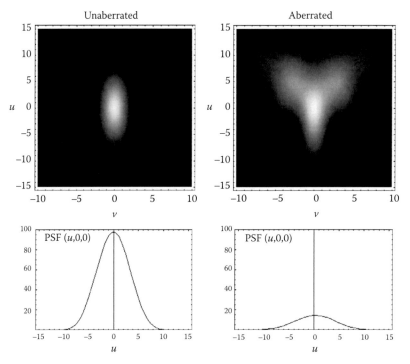

FIGURE 4.3 Effect of aberrations on the imaging performance. The unaberrated intensity PSF is shown on the left and the corresponding aberrated PSF on the right. Shape and intensity of the PSF changed due to the aberrations. Note the different normalization of the two-dimensional plots.

4.6 Experimental System for Aberration Measurement

For the measurement of the specimen-induced aberration, a Mach–Zehnder interferometer type setup was chosen. It essentially compares two wavefronts: a flat reference wavefront and an aberrated wavefront that results from the propagation through the specimen. The interference of the two coherent beams from the reference and the object path delivers an intensity pattern. A phase-shifting technique is used to extract the phase information of the wavefront through the measurement of three interferograms. A schematic drawing of the setup is shown in Figure 4.4. The expanded He-Ne (632 nm) laser beam illuminating the setup was split into a reference path and an object path. A rotation of the λ/2-plate in front of the polarizing beam splitter allows us to adjust the relative intensities of the two paths. The Zeiss LD-Achroplan, 40×, 0.6 numerical aperture (NA) lens, which was used as a condenser, has a correction ring to adjust for the thickness of the microscope slide minimizing spherical aberration.

We used a Zeiss 20×, Plan-Neofluar, 0.5 NA (dry lens without cover glass correction) and an Olympus 20×, UPlanApo, 0.8 NA, oil immersion as objective lenses. When the object beam has traversed the specimen, it is made to interfere with the reference beam and the resulting interference pattern is recorded on a CCD camera. The image plane of the CCD is conjugate to the pupil plane of the objective lens. The phase-stepping unit placed in the reference-beam path takes the form of a flat mirror shifted by a calibrated piezo drive. This allows us to change the relative phase between reference and object beams in well-defined steps. This phase stepping is synchronized with the CCD camera and permits the recording of digital interferogram images at the video rate of 25 Hz using a frame grabber fitted to a Linux PC. The specimen is attached to a piezo-driven stage that can be positioned in three dimensions and feedback control prevents long-term drift of the sample (P-611 NanoCube; Physik Instrumente, Germany). Furthermore, the object path can be switched to operate as a conventional transmission microscope using the additional camera and illumination elements as shown in Figure 4.5.

This bright field illumination mode is used to find regions of interest within the sample prior to the data recording in the interferometer mode and the alignment is such that the focal plane of the transmitted light image coincides with the focal spot of the interferometer.

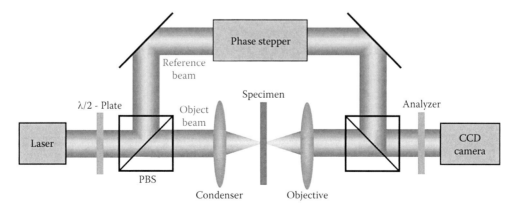

FIGURE 4.4 Configuration of the setup in interferometer mode for wavefront measurements. The pattern resulting from the interference of the two beam paths is recorded on a CCD camera. The phase-stepping unit allows shifting of the relative phase between the paths.

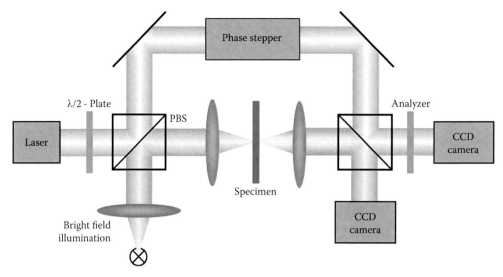

FIGURE 4.5 Configuration of the setup in bright field mode for locating and positioning of the sample prior to the measurement.

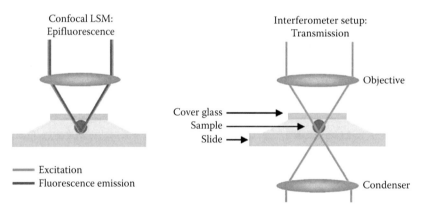

FIGURE 4.6 Comparison between the confocal laser scanning microscopy and the interferometer setup. The two beam paths are equivalent when focused to the bottom of the specimen.

Because we are concerned with aberrations in the CFMs, we have to verify whether this experiment actually models a CFM setup. The phase-stepping interferometer works in transmission mode* while a typical CFM uses epi-detection where the emitted fluorescence light is collected by the same lens. Both setups are compared in Figure 4.6. If the focus is set to the bottom of the specimen, the aberrations introduced within the two beam paths are equivalent because the wavefront leaves the sample below the focus experiencing no further aberration. Similarly, a focal position at the top of the specimen below the coverslip is equivalent to an epi-configuration focused to the top of the specimen from below. Only these two configurations deliver results identical to the epi-setup of a confocal or multiphoton

---

* It does not seem to be practical to build an interferometer that operates in epi-mode. For the interferometry, one would need a reflecting probe at the focal spot to maintain the coherence properties of the wavefront. The reflection of the wavefront would make it travel back to the lens, but it would also cause an inversion of the wavefront symmetry. Therefore, the overall propagation path of the wavefront would not be identical to the one corresponding to the confocal epi-geometry. Solutions to this problem might include a curved mirror or a phase conjugate mirror below the sample but that is likely to lead to serious alignment problems or technical difficulties.

microscope. This is not a severe limitation. We are interested in the maximum aberration that we would have to correct for in an adaptive optics system. It naturally occurs when focusing on the bottom of the sample.

Note that in the case of the CFM, the aberrations are introduced in both the excitation and the emission paths. With the interferometer, we measure the aberrations of exactly this path, but because of the single-pass nature the result of the measurement is lesser by a factor of two.

4.7 Data Acquisition

For the data shown in this chapter, we recorded 256 wavefronts corresponding to the 256 positions within a 16 × 16 grid of focal positions within each specimen. A range of typical biological specimens were chosen. The measurement spot of the interferometer travels across the field of view, and for each point a wavefront is recorded and the Zernike mode analysis is performed. To illustrate the scanning process, a transmitted light image of a mouse blastocyst sample is shown in Figure 4.7. The green dots correspond to positions for which a wavefront was recorded.

One scan takes less than five minutes. The acquisition of the wavefront at each point involves the recording of three interferogram images at relative phase steps of 0°, 120°, and 240°, which are then processed to reconstruct the wavefront as discussed in Section 4.8.

The wavefront recorded for one particular position of the sample contains three contributions to the total aberration, namely the static aberrations of the optical system, the static specimen-induced

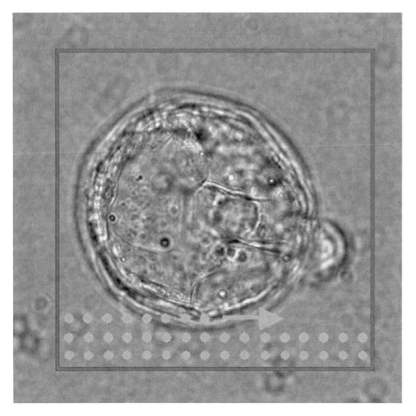

FIGURE 4.7 (See color insert.) Transmitted light image of the mouse blastocyst sample. The recording of the wavefront at different raster positions is shown by the superimposed green dots. Two hundred and fifty-six wavefronts were recorded on a 16 × 16 grid. The size of the scan area indicated by the red frame is 130 × 130 μm.

aberrations, and the field-dependent specimen-induced aberrations. Static aberrations of the optical system may originate from misalignment or slight imperfection of the optics. The specimen-induced static aberrations can be caused by focusing into media with mismatched refractive indices while the field-dependent specimen-induced aberrations are due to the variation in refractive index of the sample.

We are interested in measuring both types of specimen-induced aberrations. It is feasible to separate these contributions to the aberration by a two-step calibration. First, a reference wavefront*, 1, is recorded from a microscopic slide with a coverslip on top without any sample ensuring no additional static specimen-induced aberrations are present. For this measurement, the focus is set directly below the coverslip and the correction ring of the condenser lens is adjusted to minimize remaining aberrations. In a second step, the sample with the actual biological specimen within a water-based solution is inserted without alteration of the optical setting of the lenses. Then a reference wavefront, 2, is recorded at a position where the light traverses the homogenous part of the sample next to the specimen. Then the stage is scanned in the x/y plane and wavefronts are recorded in a raster manner. Now the difference between the reference wavefronts 1 and 2 delivers the static specimen-induced aberration whereas the difference between the wavefronts recorded during the scan and the reference wavefront 2 gives the field-dependent aberration component. Concerning the static fraction of the aberration, it should be mentioned that it depends directly on the difference between the two absolute measurements of the specimen slide and a reference slide. This assumes that both the coverslip and the microscopic slide of the measured specimen have identical thickness and refractive index. Despite nominal standardization, these glass slides may slightly differ.

4.8 Phase Extraction

The complex wavefront, $P(x,y)$, in the pupil plane of the objective can be expressed in terms of its amplitude $A(x,y)$ and phase $\psi(x,y)$ as follows:

$$P(x,y) = A(x,y)\exp(j\psi(x,y)) \tag{4.7}$$

One wavefront measurement consists of a set of three interferograms. Assuming unity amplitude for the reference beam, each interferogram has the form

$$I(x,y,\Delta\psi) = 1 + A^2(x,y)\left[1 + 2\cos(\psi(x,y) + \Delta\psi)\right] \tag{4.8}$$

where $\Delta\psi$ is the phase shift introduced in the reference arm. Since three interferograms $I_1 = I(x,y,0)$, $I_2 = I(x,y,2\pi/3)$, and $I_3 = I(x,y,4\pi/3)$ for known values of $\Delta\psi$ are recorded, one can solve for the phase of the wavefront:

$$\phi(x,y) = \arctan\left[\frac{\sqrt{3}(I_3 - I_2)}{2I_1 - I_2 - I_3}\right] \tag{4.9}$$

Figure 4.8a through c show the three raw images recorded during the phase stepping procedure and Figure 4.8d is a representation of the combined phase and amplitude calculated from the raw images.

The phase is calculated for every pixel position (x,y) and yields the wrapped phase $\phi(x,y)$, which is the absolute phase of the wavefront $\psi(x,y)$ modulo 2π. Now a fast Fourier transform (FFT)–based phase unwrapping technique (Ghiglia and Pritt 1998) is applied to recover the unwrapped phase $\psi(x,y)$ or its equivalent $\psi(r,\theta)$ in polar coordinates. Then the Zernike modal content is extracted from this function through the Zernike mode analysis described in Section 4.3.

---

* The measurement of one wavefront requires the recording of three interferograms as explained in Section 4.8.

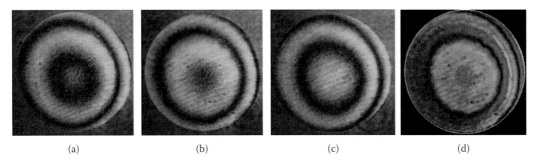

(a)   (b)   (c)   (d)

FIGURE 4.8 **(See color insert.)** Phase stepping. (a through c) Raw interferograms where the relative phase is shifted by 0°, 120°, and 240°, respectively. (d) The wrapped phase (color coded) and amplitude (intensity) that was calculated from the raw images.

4.9 Results for Aberration Measurements at Low Numerical Aperture

The numerical aperture (NA) of a lens specifies the angle of the cone of rays accepted by the lens. In general, optical systems with larger NA feature higher resolution and better light-collection efficiency. Hence they are desirable in confocal microscopy. However, systems of high NA are more susceptible to aberrations, as we will see. Results for the case of low NA are discussed in this section whereas those of high NA systems are discussed in Section 4.11. A comparison between the two cases is given in Section 4.15.

We chose water-based preparations of mouse oocyte cells and blastocysts and *Caenorhabditis elegans* as biological test specimens because of their shape and internal structure. The oocyte cells have a spherical symmetry with only slight variation in refractive index. The shape of the *C. elegans* is cylindrical, and the mouse blastocyst is interesting because it has large variations of refractive index due to many compartments filled with liquids of different refractive indices.

Examples for measured wavefronts are shown in Figure 4.9. The color of the images represents the phase whereas the brightness corresponds to the amplitude. In Figure 4.9a, a wavefront corresponding to a focal position away from the cell is shown. Here the lens positions and the cover-glass correction have been adjusted to minimize phase variation. Thus, the wavefront contains only a small amount of aberration. Figure 4.9b and c show examples of aberrated wavefronts. For Figure 4.9a and b, an objective lens with an NA of 0.5 was used. Since this NA is lower than that of the condenser lens (0.6), the pupil plane image has sharp borders. The white circles mark the pupil diameter of the system over which we performed the Zernike mode fitting. Figure 4.9c was recorded using a lens with an NA of 0.8, which was higher than that of the condenser. Small amounts of light are refracted into angles greater than the aperture angle of the condenser but are still accepted by the larger aperture of the objective. This is indicated by scattered intensities outside the circle, which are not taken into account for the wavefront analysis. Strong scattering of the sample, also inside the pupil, requires robust phase unwrapping techniques.

Each wavefront recorded during the scan across the sample is decomposed into its Zernike modal content. If we make a two-dimensional plot of the modal content for each mode, we can produce a Zernike pseudo-image of the sample for that particular mode. Such plots are shown for the mouse oocyte sample in Figure 4.10.

Here, the sum of the static and field-dependent fractions of the specimen-induced aberration is plotted whereas the static aberration introduced by the optical system was removed using the two-step calibration method described in Section 4.7. The first image within the first line of Figure 4.10 shows the sum of the absolute values of the coefficients 4 through 22 and corresponds to the total aberration apart from tip and tilt. The Zernike coefficient M_2 (tip) corresponds to a linear variation of the wavefront in the horizontal direction whereas M_3 (tilt) represents a linear slope in the vertical direction. All values are given in units that are defined such that one unit is equivalent to a wavefront standard deviation of

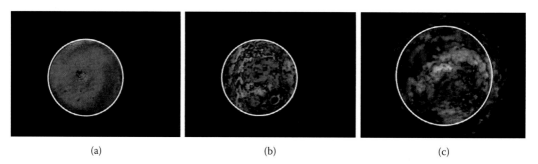

(a)   (b)   (c)

FIGURE 4.9   (See color insert.) Examples for measured complex pupil plane wavefronts. The color represents the (wrapped) phase, while the brightness corresponds to the recorded intensity. (a) Wavefront recorded beside the cell (objective NA = 0.5, dry, condensor NA = 0.6). (b) Aberrating region of the sample (mouse blastocyst, objective NA = 0.5, dry, condensor NA = 0.6). (c) Aberrating region of the sample (mouse oocyte, objective NA = 0.8, oil, condensor NA = 0.6). The circles indicate the limiting aperture of the system.

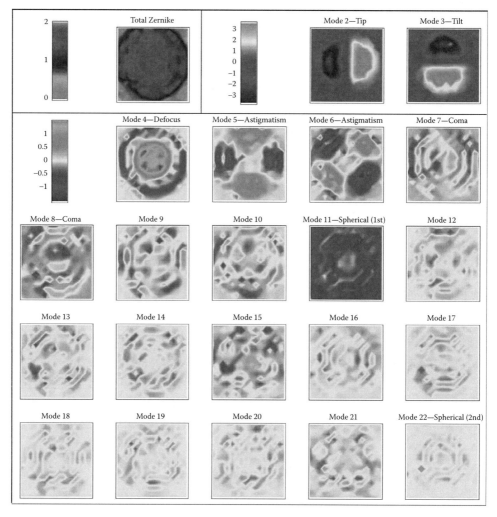

FIGURE 4.10   (See color insert.) Zernike mode plots of the mouse oocyte sample, coefficients 2–22. For these images, the Zernike modal content was extracted from 256 wavefronts. Objective lens: Zeiss Plan-Neofluar 20×, 0.5 NA, dry lens. Condenser lens: Zeiss LD-Achroplan, 40×, 0.6 NA, correction ring. The scanned area was 100 × 100 μm.

one radian (see Section 4.3). The Zernike charts of the mouse oocyte showed rather simple symmetry because of the spherical shape of the cell. Plots of tip and tilt are similar because of the symmetry of the sample and the fact that the corresponding Zernike polynomials are identical apart from a rotation of 90° about the origin. Note also the similarity between the two astigmatism modes 5 and 6. The Zernike polynomials for these modes are identical except for a rotation of 45°.

The next sample, a mouse blastocyst specimen, was less symmetric and showed more-complex patterns within the extracted the Zernike modes. Several cavities filled with liquids of a different refractive index are typical for this kind of specimen. Figure 4.7 shows a bright field image of this sample where the focus is adjusted to the middle of the specimen. The scanning process is indicated as well. Pseudo-images for the sum of the static and field-dependent components of the Zernike modes 2 through 22 of the blastocyst sample are shown in Figure 4.11. Mean and standard deviation are shown in the middle

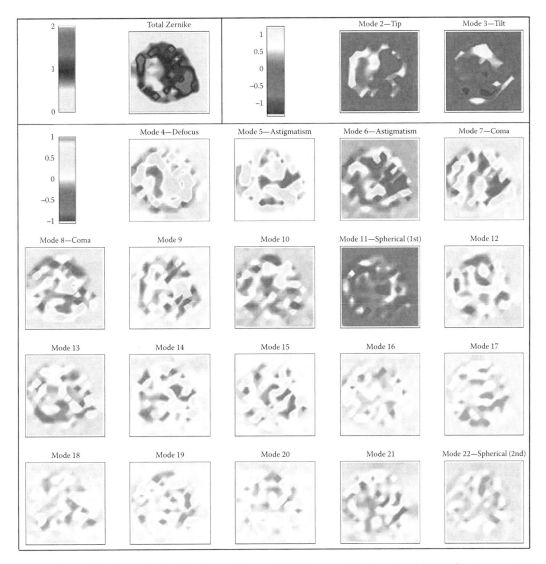

FIGURE 4.11   Zernike mode plots for the coefficients 2–22 of the mouse blastocyst sample. For these images, 256 wavefronts were recorded and the Zernike modal content was extracted as described in the text. Objective lens: Zeiss Plan-Neofluar 20×, 0.5 NA, dry lens. Condenser lens: Zeiss LD-Achroplan, 40×, 0.6 NA, correction ring. The scanned area was 130 × 130 μm (see Figure 4.7).

inset of Figure 4.13. Note that again the variation of the coefficients declines with increasing order of the successive Zernike modes. As with the previous sample, coefficient 11 (spherical) shows a significant offset, as expected. For both the mouse specimens, the objective lens Zeiss Plan-Neofluar 20×, 0.5 NA was used. This dry lens does not have a coverslip correction, but the correction ring of the condenser lens was adjusted during the calibration. Therefore, the additional spherical aberration introduced by the coverslip of the sample was compensated. Only the spherical aberration originating from the liquid layer of the specimen was added to the total aberration measured.

For the last sample, a *C. elegans* specimen of cylindrical shape, the experimental conditions were slightly different: an Olympus 20×, UPlanApo, 0.8 NA, oil immersion was used as the objective lens, and a one-step calibration in respect to the reference wavefront 2 within the specimen slide was carried out. This means that the aberration data shown in Figure 4.12 and the bottom inset of Figure 4.13 contain the field-dependent fraction of the specimen-induced aberration only; the static part caused by focusing

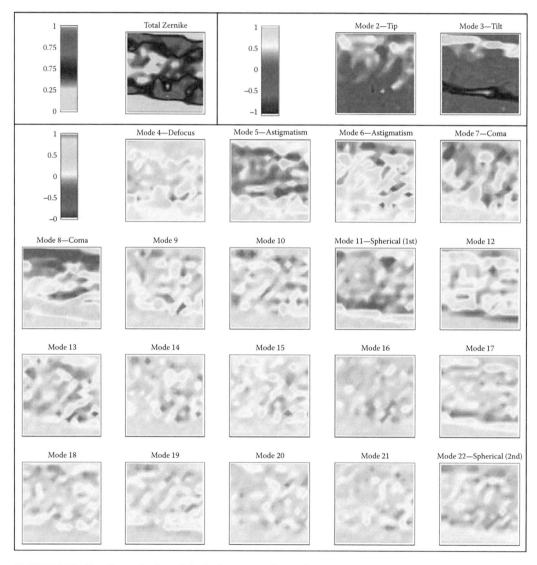

FIGURE 4.12 Zernike mode plots of the *C. elegans* sample, coefficients 2–22. For these images, the Zernike modal content was extracted from 256 wavefronts. Objective lens: Olympus 20×, UPlanApo, 0.8 NA, oil immersion. Condenser lens: Zeiss LD-Achroplan, dry lens, 40×, 0.6 NA, correction ring. The scanned area was 100 × 100 μm.

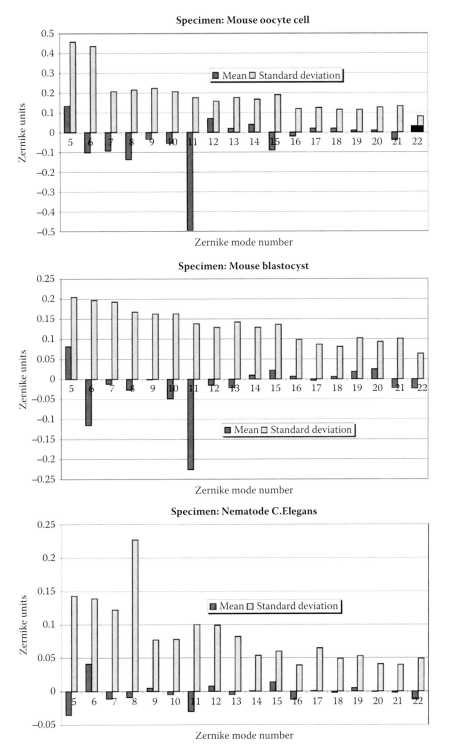

FIGURE 4.13 Mean and standard deviation across the field of view for the extracted Zernike coefficients for 256 measured wavefronts. The standard deviation of the Zernike modes is declining with rising order. Note that the graphs for the mouse oocyte sample (top) and the mouse blastocyst (middle) contain the static- and the field-dependent contributions to the aberration, while the chart for the *C. elegans* (bottom) includes the field-dependent fraction only.

into the sample is not included. Therefore, the magnitude of the mean for the spherical aberration (mode 11) was much lower compared to that of the other samples. Again, we see that the magnitude of the variations is declining as the index of the modes increases.

The mean and the standard deviation of the Zernike data for all samples are shown in Figure 4.13. The large value found for the mean of the spherical aberration (mode 11) agrees with theoretical expectations from focusing into a sample of mismatched refractive index (Booth and Wilson 2000; Booth et al. 1998; Hell et al. 1993). A decrease in the standard deviation of measured Zernike modes occurs with rising order. In addition to this decrease, superimposed steplike changes of the standard deviation can be seen. This interesting effect can be understood by considering the definition of the Zernike polynomials (see Table 4.1). The single indexing scheme maps to the two independent azimuthal and radial indices of the Zernike modes. For example, the wavefronts measured close to the center of the mouse oocyte specimen are expected to contain mainly radial spatial frequencies. Steps are visible between coefficients (6; 7), (10; 11), (15; 16), and (21; 22). These are exactly those pairs of indices i for which the radial frequency of the polynomials changes (see Section 4.3). The rather constant sections between the steplike features of the standard deviations of the Zernike modal content of the mouse oocyte specimen are due to relatively constant azimuthal spatial frequencies, which are underrepresented in an object of this symmetry.

4.10 Setup for Large Numerical Aperture, Specimens Investigated, and Data Acquisition

To measure the aberrations caused by biological specimens under high NA conditions, a few modifications had to be made to the setup depicted in Figure 4.4. Because of the short working distance of high NA lenses, the specimens were mounted between two coverslips and placed between the two opposing objective lenses. This required a new design of the specimen holder as well as the stage. Apart from that, the data acquisition procedure was identical to that for lower NA and is described in Section 4.7. An *XYZ* piezo element scanned the sample in three dimensions. An example for the data acquisition process is shown in Figure 4.14 (Schwertner 2004).

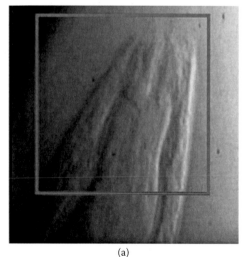

(a)   (b)

FIGURE 4.14 **(See color insert.)** (a) Transmitted light image of the specimen number 5 (*C. elegans*). The red box indicates the scanned region of 50 × 50 μm. (b) Video of the disturbance of the wavefront in the pupil plane of the lens as the focal spot scans across the specimen. Here the complex wavefront consisting of the amplitude $A(r,\theta)$ and the wrapped phase function $\phi(r,\theta)$ is displayed. The color encodes the phase, whereas the brightness corresponds to the amplitude of the wavefront. The green dot within the red frame in the lower left corner of the video indicates the relative position within the scanned area. (AVI video file online at http://www.opticsinfobase.org/oe/viewmedia.cfm?uri=oe-12-26-6540&seq=1.)

The goal of this study was to the quantify the specimen-induced aberrations for a representative variety of biological specimens. Therefore, we had to select a set of common biomedical specimens. Some of them are listed in Table 4.2 and the specimen numbers listed there will be referenced in the following discussion.

4.11 Example Results for High Numerical Aperture

After the described Zernike mode extraction, each coefficient M_i may be represented as a Zernike mode pseudo-image that shows the variation of M_i as one scans across the specimen. Figure 4.15 displays an example for the *C. elegans* specimen. Here modes 5 through 12 are depicted. The total Zernike figure is calculated using $\left(\sum_{i=5}^{22} M_i^2\right)^{1/2}$.

For the purpose of designing and implementing adaptive optics, it is important to consider the variations of the aberrations across the field of view. The mean and the standard deviation for each mode were calculated from maps such as shown in Figure 4.15; a result for the *C. elegans* is shown in

TABLE 4.2 Approximate Thickness, Lateral Scan Range, and Embedding Medium of the Specimens We Investigated

Specimen Number	Description	Embedding	Thickness (µm)	Lateral Scan (µm)
1	Brain tissue, rat	PBS	30	120 × 120
2	Mouse oocyte	PBS	80	125 × 125
3	Liver tissue, mouse	PBS	20	20 × 20
4	Striated muscle, rat	PBS	40	50 × 50
5	*C. elegans*	Agar gel	40	50 × 50
6	Vas deferens, rat	PBS	30	50 × 50

PBS, Phosphate-buffered saline.

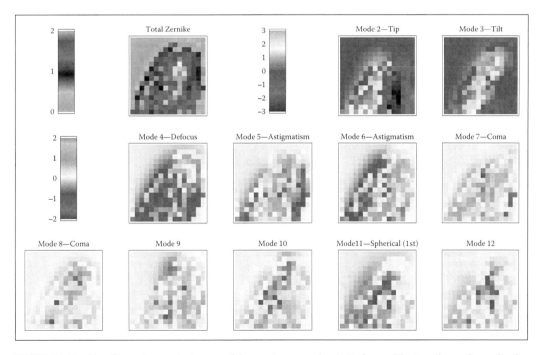

FIGURE 4.15 Zernike mode pseudo-images of the specimen number 5, *C. elegans*. The Zernike mode amplitudes M_i of modes 2 through 12 (for the Zernike mode units, see definition in Equation 4.4) are depicted.

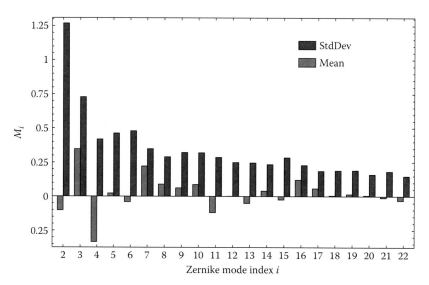

FIGURE 4.16   Mean and standard deviations of the Zernike mode amplitudes, in the Zernike mode units, for the *C. elegans*, specimen 5. Modes 2 through 22 are shown.

Figure 4.16. The Zernike mode standard deviation declines with rising order. This general behavior was found for all the specimens. The aberration effect that is contributed from the individual coefficients is proportional to the square of the Zernike mode amplitude, as can be inferred from Equation 4.12. Thus, the effect of the higher-order modes could be considered to decline even faster than the amplitudes shown in Figure 4.16. The magnitude of the Zernike mode amplitudes of all specimens was within the range of up to 1.5 for modes 5 through 11. Maximum magnitudes smaller than 1 Zernike unit were observed for modes from 12 through 22, and for modes above 22, all magnitudes were smaller than 0.5.

4.12 Simulation of the Zernike Modal Correction

First we assume that the Zernike mode composition was extracted from each wavefront using the procedure detailed in Section 4.3. Then it is a simple matter to mathematically subtract this combination of the Zernike modes from the original wavefront up to a certain Zernike mode order O:

$$\psi_{\text{corr}}(r,\theta,O) = \psi(r,\theta) - \sum_{i=1}^{O} M_i Z_i(r,\theta) \tag{4.10}$$

This simulates the process of the Zernike modal wavefront correction.

Figure 4.17 shows examples of measured wavefronts (top) and their simulated correction (bottom) up to the Zernike mode order $O = 22$. Image (1) shows an interferogram from rat brain tissue, 30 µm, covering half of the pupil. Two effects are visible: a rather strong spherical aberration component (note the colored rings at the edge of the top half of the pupil) is introduced and a modulation of the phase function with high spatial frequencies is visible. The Zernike mode–based approach can correct for the lower-order terms but cannot correct the high spatial frequency components; a partial correction is achieved and can be considered optimal for the modes included in the correction. This method worked well for 30-µm-thick brain slices. However, for thicker brain specimens of about 90 µm

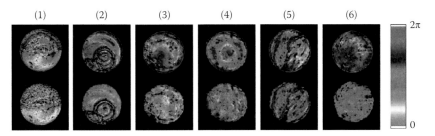

FIGURE 4.17 **(See color insert.)** Specific interferogram examples from a particular position within the 16 × 16 grid that was recorded for each specimen. The color encodes the phase, and the brightness corresponds to the amplitude. The numbers (1)–(6) are the specimen numbers listed in Table 4.2. The upper part shows the measured initial wavefront; the lower part, a simulated correction of the Zernike modes up to $i = 22$.

(data not shown), the aberration appeared to be dominated by high spatial frequency variations that could not be removed by a small number of modes. In image (2), some of the structure of the mouse oocyte is visible in the pupil. There is a sudden change of phase accompanied by amplitude effects at the cell boundaries. Again partial correction was achieved, which can be considered optimal within the modes included. The mouse liver wavefront shown in image (3) exhibits strong but rather smooth disturbance and good correction was achieved. In image (4), the sample of striated muscle is shown. A rather smooth disturbance, which can be well corrected with a low-order Zernike modal approach, is present. In this case, a large amount of spherical aberration, probably arising from the refractive index mismatch, was removed in the calibration step. Image (5) shows a heavily aberrated wavefront from the *C. elegans* specimen. The smooth muscle tissue wavefronts are depicted in image (6); the phase disturbance contained mainly low-order Zernike terms and the almost uniform color (lower part) indicates very good phase correction.

4.13 Wavefront Quality Characterization

The quality of an optical system is often characterized by the Strehl ratio (Born and Wolf 1983), as discussed in Chapter 1, which is defined as the ratio of the maxima of the focal intensity distributions for the aberrated and the unaberrated wavefronts. We define the Strehl ratio as follows:

$$S = \frac{|\int\limits_0^1\int\limits_0^{2\pi} A(r,\theta)\exp(j\psi(r,\theta))r\,dr\,d\theta|^2}{\left(\int\limits_0^1\int\limits_0^{2\pi} A(r,\theta)r\,dr\,d\theta\right)^2} \tag{4.11}$$

This definition is slightly different from that by Born and Wolf (1983) in that it takes the amplitude variations into account. For moderate aberrations, and especially if the wavefront contains only small amounts of the Zernike modes tip, tilt, or defocus, Equation 4.11 is equivalent to the ratio of the maximum focal intensity to the diffraction-limited maximum intensity.

Also, for small aberrations and no amplitude variations, the Strehl ratio may be estimated from the variance of the wavefront (Born and Wolf 1983) and finally the Zernike mode coefficients M_i:

$$S \approx 1 - \mathrm{Var}(\psi(r,\theta)) = 1 - \sum_{i=5}^{\infty} M_i^2 \tag{4.12}$$

The maximum intensity in the focal spot is proportional to the Strehl ratio. In a CFM, the coherence between excitation and emission light is destroyed in the fluorescence process. If one makes the approximation that the fluorescence excitation and emission have the same wavelength, the double-pass system introduces the same amount of aberration on the way to and from the focal spot (Booth et al. 2002). Hence, we define the signal improvement factor, F_{sig}, for a confocal fluorescence microscope system to be

$$F_{\text{sig}} = \left(\frac{S_{\text{corr}}}{S_{\text{ini}}} \right)^2 \qquad (4.13)$$

This is equivalent to the improvement in signal when adaptive aberration correction is applied to the excitation and the emission paths for the case that the CFM uses an infinitely small pinhole. This simple model of the imaging process allows us to estimate the potential benefit of adaptive optics in a straightforward manner. For small but finite-sized pinholes, F_{sig} is expected to be similar.

We note that the Zernike coefficients are a function of wavelength. If no dispersion is present, the measured Zernike coefficients scale inversely with the wavelength. In this case, it would be a simple matter to recalculate the factor of improvement for other wavelengths. Furthermore, the expression is also valid for pinhole-less two-photon microscopes where the M_i should be scaled using the excitation wavelength (Booth et al. 2002).

4.14 Which Level of Correction Is Sensible?

The technical effort in an active correction system increases with the quality of the correction. Thus, we have to find the best compromise between correction effort and the anticipated effect. There are also other things to consider that may limit the performance: an adaptive optics system requires measurement of the wavefront aberrations. In a CFM system, this measurement is based on fluorescence intensity and can therefore harm (e.g., photobleach) the specimen under investigation. Therefore, the quality of the measurement is also limited by the fluorescence light budget.

For the wavefront measurement procedure described in this chapter, the number of photons available for the measurement does not limit the wavefront detection accuracy because a transmission geometry is used. This gives access to the full wavefront data. We can predict the wavefront quality for different correction levels by simulating the wavefront correction for a certain mode order O according to Equation 4.10 and evaluating the quality of the obtained wavefront $\psi_{\text{corr}}(r,\theta,O)$ using the Strehl ratio and signal factor calculation described in Section 4.13. Details of this procedure and the results are given in the following paragraphs.

We quantified the Zernike mode–based correction by calculating the Strehl ratios of the wavefronts before (S_{ini}) and after the correction (S_{corr}) for the correction up to the mode orders $O = 12, 18, 22$, and 37. The Zernike modes 2 through 4 do not affect the signal in an epi-confocal system as they correspond to a shift of the focal spot only but do not affect its shape. Therefore, the Strehl ratio S_{ini} was calculated for a wavefront corrected up to mode order $O = 4$.

The results for the mean initial Strehl ratio, the mean-corrected Strehl ratio, and the mean and median of the confocal F_{sig} are summarized in Table 4.3 whereas the specimen numbers used are defined in Table 4.2. To investigate the interplay between the correction quality and the number of corrected Zernike modes, we calculated simulations up to modes 12, 18, 22, and 37 for all the specimens. In general, the correction of up to 37 modes was found to give only moderate improvement in the Strehl ratio compared to that of 22 modes. It appears that for these specimens, the correction of 22 or even 18 Zernike modes is a good compromise between the effort required for the correction and the improvement in Strehl ratio. Maps of the Strehl parameters are shown in Figure 4.18; the specimens are again

TABLE 4.3   Correction Benefit for Different Degrees of Correction

Specimen Number	Correction Up to mode	Mean S_{ini}	Mean S_{corr}	Mean F_{sig}	Median F_{sig}
1	12	0.40	0.49	4.47	1.80
	18	0.40	0.54	5.50	2.24
	22	0.40	0.65	7.57	3.48
	37	0.40	0.66	7.88	3.81
2	12	0.47	0.51	4.00	1.10
	18	0.47	0.58	6.13	1.38
	22	0.47	0.62	6.27	1.63
	37	0.47	0.65	7.02	1.78
3	12	0.32	0.52	5.30	2.90
	18	0.32	0.57	6.35	3.36
	22	0.32	0.65	7.84	4.39
	37	0.32	0.72	8.98	5.79
4	12	0.48	0.63	2.15	1.86
	18	0.48	0.78	3.13	2.58
	22	0.48	0.81	3.35	2.76
	37	0.48	0.84	3.66	2.97
5	12	0.46	0.59	7.31	1.29
	18	0.46	0.60	7.13	1.39
	22	0.46	0.71	9.65	1.98
	37	0.46	0.78	12.66	2.28
6	12	0.47	0.58	1.83	1.47
	18	0.47	0.62	2.11	1.69
	22	0.47	0.72	2.80	2.27
	37	0.47	0.77	3.21	2.64

denoted by the numbers listed in Table 4.2. Significant spatial variations of the initial Strehl ratio are evident in all specimens. It is important to note this phenomenon. The measured intensity displayed in a standard CFM image is normally regarded as a quantity that represents fluorophore concentration. The results of this study indicate that aberrations, and not just the fluorophore concentration, affect the measured intensity. This has important implications for quantitative fluorescence microscopy.

We note that in some cases the specimen does not fill the whole field of view and F_{sig} is smaller in the areas adjacent to the specimen. The calculated mean F_{sig} was typically in the range between 2 and 10 for a correction of the Zernike modes 5 through 22. The median value is listed as this tends to be less sensitive to extreme values.

4.15 Effect of the Numerical Aperture on Aberrations

Another aspect investigated was the effect of the NA on the size of the aberrations and the benefit of modal adaptive correction. Wavefront data for different NAs can be extracted from the interferograms by analyzing an aperture subregion of smaller radius. Our data sets were recorded with a physical NA of 1.2 and data analysis was performed for the NAs 1.2, 0.9, and 0.6 for a correction up to Zernike mode 37; the results are shown in Table 4.4 (see Table 4.2 for specimen numbers). The magnitude of aberrations, especially higher orders, increases with NA. As an example, the phase function corresponding to spherical aberration (Booth and Wilson 2000; Török et al. 1995a,b) rises sharply toward the edge of a high NA

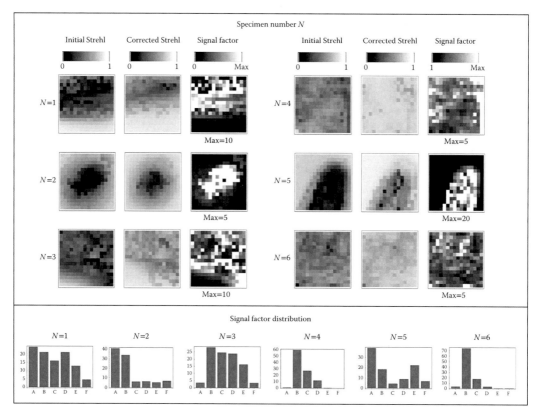

FIGURE 4.18 Maps of the initial Strehl ratio S_{ini}, the Strehl ratio S_{corr} after correction up to Zernike mode 22, and the derived signal correction factor F_{sig}. The distribution of F_{sig} is shown in a histogram for each of the specimens. The nonuniform histogram intervals are A:[0, 1.5); B:[1.5, 3); C:[3, 5); D:[5, 10); E:[10, 40); and F: [40, ∞]. The vertical axis shows percentage of pixels within the range. The maximum of the range for each F_{sig} plot is shown below the plot and values larger than this maximum are shown in white.

pupil. Since a lower NA objective accepts only the central portion, the effects of spherical aberration are correspondingly reduced. The dependence of other, higher-order aberration modes is conceptually similar. Therefore, in a low NA system, the aberrations tend to be smaller in amplitude and the initial Strehl ratio is correspondingly higher. This is supported by the experimental results. As a consequence, the benefit of correction for lower NA was found to be smaller. For the particular case of the mouse oocyte (specimen 2), the mean of F_{sig} for NA = 0.6 is largest but this is because of a few locations with very large values of F_{sig}. However, the median value of the distribution for NA = 0.6 is lower than that for the higher NAs.

4.16 Discussion and Conclusion

We can conclude that the specimen-induced aberrations lead to reduced signal levels and deterioration in image quality in optical microscopy, especially in confocal microscopy and two-photon microscopy (TPM). The specimen-induced aberrations that occur with various biological specimens have been classified and quantified for the most relevant condition of high NA. The above approach provides detailed information about the variation of each Zernike coefficient across the scan. Our calculation of the correction benefit is based on the assumption that a correction would be applied at every position within the scanned area. The feasibility of this assumption depends on the bandwidth of the wavefront sensing and correction devices and the scan speed. In some cases, it may be required to either reduce the scan speed or update the aberration correction every few pixels only.

TABLE 4.4 Correction Benefit at Different Numerical Apertures

Specimen Number	NA	Mean S_{ini}	Mean S_{corr}	Mean F_{sig}	Median F_{sig}
1	1.2	0.40	0.66	7.88	3.81
	0.9	0.62	0.76	2.24	1.37
	0.6	0.70	0.82	1.81	1.18
2	1.2	0.47	0.65	7.02	1.78
	0.9	0.55	0.73	6.18	1.58
	0.6	0.60	0.79	9.02	1.49
3	1.2	0.32	0.72	8.98	5.79
	0.9	0.63	0.85	2.16	1.59
	0.6	0.78	0.90	1.42	1.33
4	1.2	0.48	0.84	3.66	2.97
	0.9	0.60	0.85	2.19	1.98
	0.6	0.75	0.92	1.57	1.49
5	1.2	0.46	0.78	12.66	2.28
	0.9	0.64	0.87	4.15	1.68
	0.6	0.75	0.90	2.32	1.07
6	1.2	0.47	0.77	3.21	2.64
	0.9	0.66	0.84	1.71	1.58
	0.6	0.76	0.89	1.40	1.29

The measurements indicate significant variations of the uncorrected Strehl ratio throughout the specimen, which could influence quantitative fluorescence measurements in an uncorrected system. We showed that low-order aberration correction based on the Zernike modes provide significant recovery of signal levels in confocal microscopy and TPM, even if the diffraction limit is not restored. For the six specimens examined, the mean F_{sig} was in the range between 2 and 10 for a correction of the Zernike modes 5 through 22 at an NA of 1.2. Note that the quoted values refer to frame averages and the factors in specific areas might be even higher. It should be pointed out that the set of biological specimens investigated is necessarily incomplete and differences between biological specimens can be large. However, the benefits for these specimens would be highly significant in confocal microscopy and TPM since light budgets are typically tight and efficient use of available photons is crucial to minimize photobleaching and phototoxic effects (Manders and Cook 1999).

As expected from theory, lower NA systems are less susceptible to aberrations than high NA systems under otherwise similar conditions. Low-order correction would still provide benefits, even though the initial aberrations are smaller. The results presented here quantify the benefit of adaptive optics for biological microscopy and provide the bounds within which these systems must operate.

References

Booth, M., M. Neil, and T. Wilson (2002). New modal wave-front sensor: application to adaptive confocal fluorescence microscopy and two-photon excitation fluorescence microscopy. *J. Opt. Soc. Am. A 19* (10), 2112–2120.

———(1998). Aberration correction for confocal imaging in refractive-index-mismatched media. *J. Microsc. 192* (2), 90–98.

Booth, M. and T. Wilson (2000). Strategies for the compensation of specimen induced aberration in confocal microscopy of skin. *J. Microsc. 200* (1), 68–74.

Born, M. and E. Wolf (1983). *Principles of Optics* (6th ed.). Oxford: Pergamon Press.

Ghiglia, D. and M. Pritt (1998). *Two-Dimensional Phase Unwrapping: Theory, Algorithms and Software.* New York: Wiley.

Hell, S., G. Reiner, C. Cremer, and E. Stelzer (1993). Aberrations in confocal fluorescence microscopy induced by mismatches in refractive index. *J. Microsc. 169* (3), 391–405.

Manders, E. M. M., H. Kimura and P. Cook (1999). Direct imaging of DNA in living cells reveals the dynamics of chromosome formation. *J. Cell Biol. 144*, 813–821.

Neil, M. A. A., M. J. Booth, and T. Wilson (2000). New modal wavefront sensor: a theoretical analysis. *J. Opt. Soc. Am. A 17* (6), 1098–1107.

Noll, R. (1976). Zernike polynomials and atmospheric turbulence. *J. Opt. Soc. Am. 66*, 207–277.

Schwertner, M., M. Booth, and T. Wilson (2004). Characterizing specimen induced aberrations for high NA adaptive optical microscopy. *Optics Express* 12 (26), 6540–6552.

Török, P., P. Varga, Z. Laczik, and G. R. Booker (1995b). Electromagnetic diffraction of light focused through a planar interface between materials of mismatched refractive indices: an integral representation. *J. Opt. Soc. Am. A 12*, 325.

5

Specimen-Induced Geometrical Distortions

Michael Schwertner
confovis GmbH

5.1 Introduction and Overview

Specimen-induced aberrations affect the imaging properties in optical three-dimensional (3D) microscopy, especially when high-numerical-aperture (NA) lenses are used. In Chapter 4, we have measured aberrations to understand their effect toward image quality parameters, such as intensity and resolution. Apart from these effects, aberrations can also introduce geometric image distortions, a fact that is often overlooked. This is because it cannot be observed directly in the reciprocal epi-configuration of a common fluorescence microscope. Still the effect of specimen-induced distortions can lead to substantial errors for spatial measurements.

In this chapter, transmission-mode interferometric data are used to quantify the geometric distortions associated with the specimen-induced aberrations. This information is otherwise inaccessible but allows the estimation of the accuracy of spatial measurements within acquired 3D data sets. An assessment for a range of biological specimens shows that spatial measurements can be significantly compromised by the specimen-induced aberrations.

5.2 Problem Description

In an ideal microscope, spherical wavefronts converge toward an optimum, diffraction-limited focal spot at the nominal focal position (NFP). Unfortunately, the specimen under examination can introduce deviations from this ideal wavefront, which lead to an aberrated focal spot. The effect of these aberrations can be divided into two categories. The first category produces a blurred and enlarged focal spot, which leads to a loss in resolution and image intensity. This effect is caused by the Zernike aberration modes higher than 4 (see Table 4.1 of Chapter 4), and this is investigated in detail in the same chapter.

The second category is investigated in this chapter. The associated effects are related to the three lower-order Zernike modes: 2 (tip), 3 (tilt), and 4 (defocus). They are not aberrations in the classical sense and have no influence on the intensity distribution of the focal spot but create a displacement between the actual focal position (AFP) and the NFP as shown in Figure 5.1. A standard 3D microscope system maps

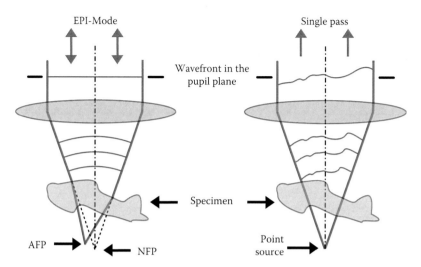

FIGURE 5.1 *Left*: Specimen-induced aberrations and their effect in epi-configuration. Because of the variation in refractive index within the specimen, the actual focal position (AFP) of the focal spot can be different from the nominal focal position (NFP). *Right*: Wavefront aberrations can be measured in the pupil plane of the lens in a single pass configuration when a point source is placed in the nominal focal spot position. The measured wavefront information allows one to model the position and shape of the focal spot in the equivalent epi-configuration.

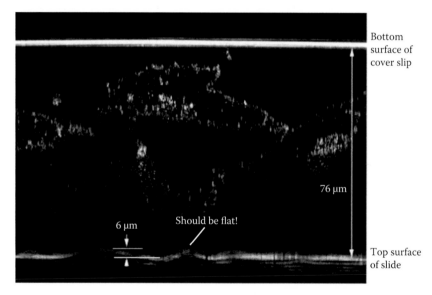

FIGURE 5.2 Illustration of the specimen-induced distortions: confocal axial X/Z scan of cheek cells above the interface between embedding medium and the microscope slice. The reflection from the interface is marked. In the scan, because of the discussed effect, the interface appears to be bent even though the object is known to be flat. From Pawley, J. (2002). Limitations on optical sectioning in live-cell confocal microscopy. *Scanning 24*, 241–246, © FAMS Inc.

the intensity sampled at the AFP into a 3D data set using the information that was actually collected at the NFP. This leads to local spatial distortions in the data set and can cause measurement inaccuracies. A striking example was presented by Pawley (2002). An image from this article is shown in Figure 5.2, where the reflection image from the flat interface between the embedding medium of the specimen and the microscope slide is distorted because of the refractive index structure of the cell above.

5.3 Experimental Setup

In the standard epi-configuration that is common in fluorescence microscopy, the lateral displacement cannot be inferred from the wavefront originating from the focal spot because of the reciprocal geometry. Therefore, to measure the distortion introduced by a number of specimens, the single pass transmission geometry shown on the right of Figure 5.1 as a part of the interferometer described in section Section 4.6 was used to measure the wavefront in the pupil plane of the lens. Subsequently, a Zernike mode analysis as discussed in Section 4.3 was performed on the measured wavefronts to obtain the coefficients of the Zernike mode orders 2, 3, and 4. We explain further how these Zernike coefficients relate to the geometric distortions we intend to quantify.

5.4 Relation between Zernike Aberration Modes and Geometric Distortion

It is assumed that the phase $\Psi(\xi,\eta)$ of the aberration function in the pupil plane of the lens was measured during the previously discussed interferometric experiment. The Cartesian coordinates (ξ,η) refer to the pupil plane of the lens, and the radius of the circular pupil is normalized to unity. The wavefront aberrations can then be described by the pupil function

$$P(\xi,\eta) = \exp[\, j\psi(\xi,\eta)] \tag{5.1}$$

where $j = \sqrt{-1}$. Here the amplitude term has been dropped because amplitude is assumed to be constant across the pupil. The phase function can also be represented by a series of Zernike polynomials:

$$\Psi(\xi,\eta) = \sum_{i=1}^{N} M_i\, Z_i(\xi,\eta) \tag{5.2}$$

where Z_i is the Zernike polynomial with index i, and M_i, the corresponding mode amplitude. Generally, this method allows one to extract the mode coefficients, including the higher-order terms. However, the information related to the geometric distortion is contained within the three lower-order Zernike terms tip, tilt, and defocus corresponding to the Zernike coefficients M_2, M_3, and M_4. To relate the measured Zernike coefficients to displacements in the focal region, we look at the intensity distribution at the focus (Wilson and Sheppard 1984):

$$I(t,w,u) = \left| \int_0^{2\pi} \int_0^1 P(\xi,\eta)\exp\left[j\frac{u}{2}(\xi^2 + \eta^2) - j(\xi t + \eta w) \right] \mathrm{d}\xi\mathrm{d}\eta \right|^2 \tag{5.3}$$

Here we have used the normalized optical coordinates

$$u = \frac{8\pi}{\lambda}nz\sin^2\left(\frac{\alpha}{2}\right) \tag{5.4}$$

in the axial direction and

$$t = \frac{2\pi}{\lambda}xn\sin\alpha; \quad w = \frac{2\pi}{\lambda}yn\sin\alpha \tag{5.5}$$

in the lateral direction. We note that the polar coordinates (ρ,θ) are related to ξ, η by $\rho^2 = \xi^2 + \eta^2$, $\xi = \rho\cos\theta$, and $\eta = \rho\sin\theta$. The term $n\sin\alpha$ refers to the NA, and λ is the wavelength. The variables x, y, and z denote the actual radial and axial coordinates in the focal region. Now we consider a specific pupil function that has uniform amplitude $A(\xi,\eta)=1$ and contains the Zernike terms corresponding to tip, tilt, and defocus only:

$$P(\xi,\eta) = \exp\left[j\left(M_2 Z_2(\xi,\eta) + M_3 Z_3(\xi,\eta) + M_4 Z_4(\xi,\eta) \right) \right] \tag{5.6}$$

Here the corresponding Zernike polynomials in Cartesian coordinates are defined as

$$Z_2 = 2\xi, \quad Z_3 = 2\eta, \quad \text{and} \quad Z_4 = \sqrt{3}(2(\xi^2 + \eta^2)-1) \tag{5.7}$$

When we introduce Equation 5.6 into Equation 5.3 and compare the coefficients in the arguments of the exponentials, we find that the aberration modes (Equation 5.7) effectively introduce a shift, Δ, in the x, y, and z directions, given by

$$\Delta x = -\frac{M_2 \lambda}{\pi n \sin\alpha}, \quad \Delta y = -\frac{M_3 \lambda}{\pi n \sin\alpha}, \quad \Delta z = \frac{M_4 \sqrt{3}\lambda}{2\pi n \sin^2(\alpha/2)} \tag{5.8}$$

The above expressions allow one to calculate the displacement of the focal spot directly from the Zernike mode coefficients for tip, tilt, and defocus extracted from $\Psi(\xi,\eta)$.

The calculation of the spot shape and position from the wavefront as expressed in Equation 5.3 is an approximation that is valid only for relatively small variations in refractive index. We also note that, for high-NA lenses, defocus is represented by a pupil function consisting of even-order polynomials including the terms in ρ^4 and higher orders. In Equation 5.3, we have approximated defocus by the dominant quadratic term.

5.5 Results and Discussion

To quantify the specimen-induced distortions, we processed wavefront data from a variety of biological specimens; the specimen characteristics are listed in Table 4.2 of Chapter 4. We chose examples that represent common specimen categories. For confocal or multiphoton techniques, high-NA lenses are usually preferred because of their superior light collection efficiency and resolution. However, some applications may require lower-NA lenses, and therefore we investigated the effect of specimen-induced distortions at lower NA as well. In both cases, the wavefront information was extracted from the data sets recorded with a 1.2 NA water-immersion lens. For the lower NA, a smaller circular subregion of the interferograms corresponding to an NA of 0.6 was analyzed, and the Zernike modes were fitted to this subregion using the same procedure.

The local displacements of the focal spot calculated using the aforementioned method are displayed in Figure 5.3 for the NA of 1.2 and in Figure 5.4 for the lower NA of 0.6. For each specimen, the deviations in the x, y, and z directions are shown separately. Statistical values, such as the individual standard deviations in the X, Y, and Z directions, are listed in Table 5.1. The maximum value of the total spatial deviation $d = \sqrt{\Delta x^2 + \Delta y^2 + \Delta z^2}$ for each specimen is also given in Table 5.1.

The total 3D displacement of the spot, depending on the specimen structure and location within the specimen, can exceed 1 μm. The patterns observed correlate with the specimen structures as expected. For example, the scan of specimen no. 1 contains the horizontal edge of the tissue, which causes distortion in the y and z directions but leaves the x component almost unaffected. The largest distortion value in the order of 2 μm was found within a particular region of the mouse oocyte cell (specimen no. 2) at an NA of 0.6, where the axial component was dominant.

To understand the influence of the NA on the specimen-induced distortions, we turn to the illustrations of Figure 5.5. Here cross sections $\Psi(r)$ of two-dimensional phase functions $\Psi(r,\theta)$ in the pupil plane of the lens are shown for three different situations. Using a lens of larger NA is equivalent to capturing

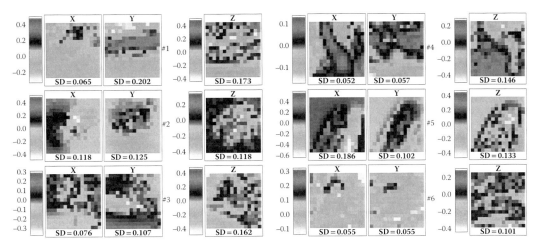

FIGURE 5.3   (See color insert.) Specimen-induced distortion at a numerical aperture (NA) of 1.2. The specimen numbers 1–6 refer to the description in Table 4.2 of Chapter 4. The abbreviation SD denotes the standard deviation across the field of view (units are in micrometers).

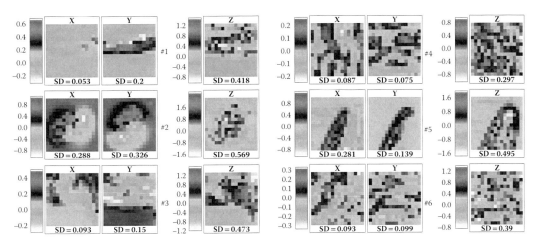

FIGURE 5.4   (See color insert.) Specimen-induced distortion at an NA of 0.6. The specimen numbers 1 through 6 refer to the description in Table 4.2 of Chapter 4. The abbreviation SD denotes the standard deviation across the field of view (units are in micrometers).

a larger range of the same wavefront or phase function. The simplest case is depicted in Figure 5.5a, where the phase function is a constant slope function. Then the average wavefront tilt across the pupil is the same for the lower NA of 0.6 and the higher NA of 1.2, and focal spot displacements are identical for both NAs*. However, typical aberration patterns tend to be more complicated, which can lead to a dependence of the specimen-induced distortion on the NA. Such an example is shown in Figure 5.5b for lateral distortion: the same phase function leads to different average wavefront slopes across the pupil for different NAs. A similar situation is shown in Figure 5.5c for axial specimen-induced distortion. The source of axial distortion is a quadratic term in the phase function (defocus). From Figure 5.5c, it is clear that the displayed phase function has a dominant first-order spherical aberration term at the higher NA

---

* Note that the Zernike mode coefficients are calculated for a pupil normalized to unity. Therefore, the NA of 1.2 would yield twice the tip/tilt Zernike coefficients compared to the NA of 0.6, but this is compensated for by the NA dependence of the conversion between Zernike mode units and spot displacement/distortion (see Equation 5.8).

TABLE 5.1   Statistical Properties of the Measured Specimen-Induced Focal Spot Displacements

Specimen No.	NA	Std Dev (X)	Std Dev (Y)	Std Dev (Z)	Max Dev (total)
1	1.2	0.065	0.202	0.173	0.521
	0.6	0.053	0.200	0.418	1.563
2	1.2	0.118	0.125	0.118	0.672
	0.6	0.288	0.326	0.569	2.080
3	1.2	0.076	0.107	0.162	0.491
	0.6	0.093	0.150	0.473	1.440
4	1.2	0.052	0.057	0.146	0.457
	0.6	0.087	0.075	0.297	1.090
5	1.2	0.186	0.102	0.133	0.713
	0.6	0.281	0.139	0.495	1.911
6	1.2	0.076	0.107	0.162	0.423
	0.6	0.093	0.099	0.390	1.270

Units are in micrometers

– – Average slope across pupil at NA = 1.2   – – Spherical aberr. (1st) dominates at NA = 1.2
····· Average slope across pupil at NA = 0.6   ····· Defocus dominates at NA = 0.6

Pupil size

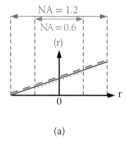
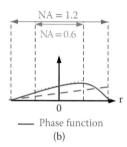
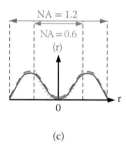

—— Phase function

(a)   (b)   (c)

FIGURE 5.5   Illustration of the dependence of the specimen-induced distortions and other aberrations on the NA. Please see text. (a) If the phase function is a constant slope function the average wavefront tilt across the pupil is the same for the lower NA of 0.6 and the higher NA of 1.2 (b) A more typical aberration pattern can lead to a dependence of the specimen-induced distortion on the NA shown here for lateral distortion. (c) For axial distortion.

of 1.2, while the quadratic term, indicating axial displacement, is only minor. This is exactly opposite at lower NA, where the axial displacement (quadratic component) dominates while the spherical aberration component is very small. This example also coincides with the general observation that larger NAs lead to higher spatial frequencies in the aberration function and, therefore, require higher aberration orders for the appropriate description and/or correction.

Another representation of the results for an NA of 0.6 is also shown in Figure 5.6. Here the imaging of a regular square grid underneath the specimen is simulated and the mesh knots of the regular grid are displaced according to the measured deviations. Note that the effect is exaggerated in the drawing—the mesh knot displacement is drawn five times larger at the scale of the grid.

When a relative measurement within a 3D data set is done, both the absolute errors of the two individual spot position measurements contribute to the inaccuracy of the relative position measurement. In the worst case, the individual distortions will add up. The data imply that the effect of specimen-induced aberrations can easily lead to a distance measurement inaccuracy on the order of 1–2 µm. Considering that typical cell dimensions are in the range from 5 to 20 µm, this can lead to large relative errors. If, for example, a distance of approximately 5 µm has to be measured within a cell and the uncertainty of the distance measurement is 1 µm, the expected relative error is 20% and hence these effects cannot be neglected.

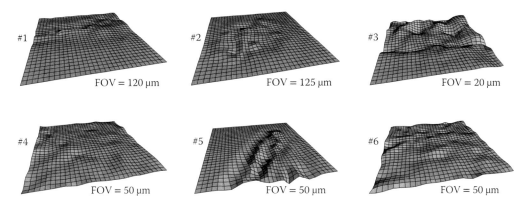

FIGURE 5.6 Visualization of the specimen-induced distortions at an NA of 0.6: simulation of imaging a regular grid placed underneath the specimen. The measured distortions are plotted as a deviation from the regular mesh positions of the grid and are enlarged by a factor of five. The specimen numbers refer to Table 4.2 of Chapter 4, and the field of view (FOV) covered by the scan is also given for each specimen.

5.6 Summary and Conclusion

A method to obtain quantitative data for the spatial measurement inaccuracies due to the specimen-induced aberrations has been presented. The approach uses interferometry to record wavefront aberration data from the pupil plane of the microscope lens to extract information about local distortion. To the best of our knowledge, this is the first attempt to quantify the effects of specimen-induced image distortion in microscopy. This information is usually not accessible and cannot be observed directly in a standard epi-type microscope. However, our results show that the specimen-induced distortions can significantly influence the spatial measurement accuracy that can be achieved with an otherwise perfect and calibrated microscope. Investigators should consider these effects whenever spatial measurements of high accuracy are required.

Acknowledgment

It is noted that these results have also been presented in the publication (Schwertner et al., 2007).

References

Pawley, J. (2002). Limitations on optical sectioning in live-cell confocal microscopy. *Scanning 24*, 241–246.

Schwertner, M., M. J. Booth, and T. Wilson (2007). Specimen-induced distortions in light microscopy. *J. Microsc.* 228(1), 97–102.

Wilson, T. and C. Sheppard (1984). *Theory and Practice of Scanning Optical Microscopy*. Academic Press, London.

6

Simulation of Aberrations

Michael Schwertner
confovis GmbH

6.1 Introduction

It has been suggested to correct for aberrations of the wavefront using adaptive optics, as is discussed in Chapters 4, 7 and 8 on adaptive correction. There the idea is to feed a pre-aberrated wavefront to the optical system that contains aberrations opposite to those generated within the system. Hence, aberrations will cancel out and diffraction-limited imaging is accomplished again. A standard adaptive optics system includes a wavefront-sensing element and operates in a closed loop to correct aberration actively. Another approach is to use predictive aberration correction, sometimes referred to as open loop control. Here, the aberration conditions are inferred from secondary parameters—for example, the position of the focal spot within the specimen—and corrected accordingly. Therefore, the simulation and prediction of aberrations is the key to predictive correction.

The effect of refractive index mismatch in fluorescence microscopy was first theoretically analyzed by Hell et al. (1993). The spherical aberration components caused by focusing through a layer of mismatched refractive index can be calculated analytically if the refractive index and thickness of each layer is known (Török et al. 1995; Booth et al. 1998).

Many biological specimens can be approximated by cylindrical or spherical object shapes (e.g., cells) of uniform refractive index or superpositions of several of these structures. In this chapter, we calculate the specimen-induced deformations of the wavefront for spherical and cylindrical objects. A ray-tracing method is used to obtain the pupil phase function and subsequently the Zernike mode content of the wavefront using numerical integration. A mouse oocyte cell and an optical fiber are modeled and the simulations are compared to experimental results. The interferometer setup used for the direct measurements of the specimen-induced aberrations has been described in Chapter 3.

6.2 Wavefront Aberration Simulation for a Spherical Object

As we have said, the wavefront in the pupil plane of the lens can be described by a complex pupil function (Wilson and Sheppard 1984)

$$P(r,\theta) = A(r,\theta)\exp[i(\psi_0 + \psi(r,\theta))] \tag{6.1}$$

where ψ_0 is an arbitrary offset of the phase and $\psi(r,\theta)$ the change in phase induced by the specimen. In the ideal, unaberrated case, $P(r,\theta)$ would be constant. Our model specimen shows a variation in refractive index, but no absorption. We assume uniform illumination of the pupil, such that the amplitude $A(r,\theta)$ is unity, while the phase $\psi(r,\theta)$ varies. The specimen is approximated by a spherical region, which has an absolute difference of $\Delta n_1 = n_1 - n_0$ in refractive index between the spherical region (n_1) and the homogenous embedding medium (n_0). It is assumed that the small change in refractive index does not cause a deviation of the direction of the traced ray. A schematic diagram of the simulation model is shown in Figure 6.1. Here γ represents the half angle of the cone of marginal rays defined by the numerical aperture (NA), $n \sin(\alpha)$, of the objective lens. If the virtual specimen were immersed in a substance of refractive index $n_0 = n$, for which the lens has been designed, then $\gamma = \alpha$. Otherwise, if the specimen were immersed in a medium of refractive index, $n_0 = n'$, γ would be given by Snell's law: $\gamma = \arcsin(n\sin(\alpha)/n')$. The ray being traced is defined by the unit vector $\mathbf{t}$ and the point P through which it passes; β is the inclination angle of the ray to the optical axis (see Figure 6.1). The functions $P(r,\theta)$ and $\psi(r,\theta)$ are defined over a normalized pupil of radius $r_{max} = 1$. Assuming the objective lens obeys Abbe's sine condition (Born and Wolf 1983), the ray is mapped to the radial coordinate of the pupil by $r = \sin(\beta)/\sin(\gamma)$. If the length of the ray section within the sphere is denoted by $a(r,\theta)$, the phase function is given by

$$\psi(r,\theta) = \frac{2\pi}{\lambda}\Delta n\, a(r,\theta) = \frac{2\pi}{\lambda}\Delta n\, a(\mathbf{t},\mathbf{m}) \tag{6.2}$$

where λ is the wavelength and the coordinates (r,θ) in the pupil plane may be expressed in terms of the vectors $\mathbf{m}$ and $\mathbf{t}$ as defined in Figure 6.2. The vector $\mathbf{m}$ points from the focus P to the center C of the sphere and $\mathbf{t}$ is a unit-length direction vector with the components.

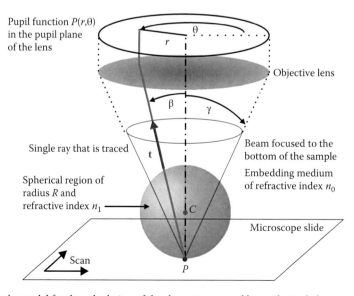

FIGURE 6.1 On the model for the calculation of the aberration caused by a spherical object.

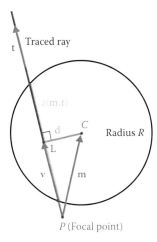

FIGURE 6.2 Illustration of the plane that contains the center *C* of the sphere and the traced ray passing the focal point *P* in a direction defined by the vector **t**.

$$t_x = r\sin(\gamma)\cos(\theta)$$
$$t_y = r\sin(\gamma)\sin(\theta)$$
$$t_z = \sqrt{1 - r^2 \sin^2(\gamma)}$$
(6.3)

The value of *a*, which is the section the ray travels within the sphere, can be calculated from the distance *d* between the ray and the center *C* of the sphere by

$$(\mathbf{m},\mathbf{t}) = \begin{cases} 2\sqrt{R^2 - d^2(\mathbf{m},\mathbf{t})} & \text{for } d < R \\ 0 & \text{otherwise} \end{cases}$$
(6.4)

From Figure 6.2 we see that *d* follows from

$$d = |\overrightarrow{LC}| = |\mathbf{m} - \mathbf{v}| = |\mathbf{m} - \mathbf{t}\cdot(\mathbf{m}\cdot\mathbf{t})|$$
(6.5)

If the vector **m** has the components (m_x, m_y, m_z), we get

$$d^2 = (m_x - t_x(m_x t_x + m_y t_y + m_z t_z))^2$$
$$+ (m_y - t_y(m_x t_x + m_y t_y + m_z t_z))^2$$
$$+ (m_z - t_z(m_x t_x + m_y t_y + m_z t_z))^2$$
(6.6)

Now the phase function $\psi(r,\theta)$ can be calculated using Equations 6.2 through 6.6. The scanning across the virtual sample may be implemented by altering the coordinates of the center *C* of the sphere, and the focusing depth can be changed by modifying the coordinates of the focal point *P* of the model.

Since the raw phase function itself is hard to interpret, it is useful to represent the phase function of the aberrated wavefront by its Zernike mode content. The Zernike polynomials are a set of orthogonal functions over the unit circle; some of the lower-order modes correspond directly to the classic aberration terms, as for instance, astigmatism and coma. It is also convenient to base correction schemes of adaptive optics systems on Zernike modes since, in many cases, it is sufficient to correct for a few lower-order Zernike modes only. Because of the orthogonality of the polynomials, the decomposition of the phase function into Zernike mode amplitudes can be written as

$$M_i = \frac{1}{\pi} \int_0^1 \int_0^{2\pi} \psi(r,\theta) Z_i(r,\theta) r \, d\theta \, dr \tag{6.7}$$

where $Z_i(r,\theta)$ is the Zernike polynomial of order i, and M_i, the corresponding amplitude of that mode. For a listing of lower-order modes, see Table 4.1 of Chapter 4.

Figure 6.3 shows the simulated Zernike mode variations of the aberrated wavefront while "scanning" across the virtual spherical sample. The simulations (Figure 6.3a through 6.3c) differ in focusing depth. The plane probed in Figure 6.3a contains the center of the sphere and shows the lowest magnitude of aberrations, and no tip or tilt is present because of the symmetry conditions for the cone of rays focused to this plane. For an intermediate focal position between center and bottom (Figure 6.3b) and for the bottom focal plane (Figure 6.3c), higher-aberration magnitudes are found; nonzero values for the tip/tilt modes and slightly changed shapes of some of the Zernike aberration patterns occur.

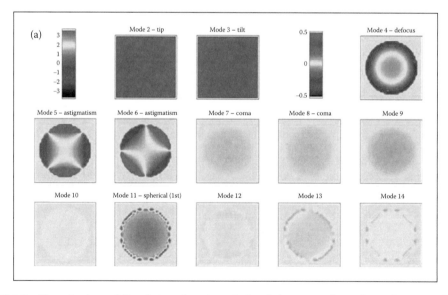

FIGURE 6.3   (**See color insert.**) Simulation of specimen-induced aberrations for a sphere of 80 μm in diameter and uniform refractive index $n_1 = 1.35$, embedded in water ($n_0 = 1.33$) and dry lens with the numerical aperture (NA) = 0.5. The virtual sample was scanned on a 32 × 32 grid covering a field of view of 100 μm × 100 μm. For each of the grid positions, a wavefront is calculated using the described ray-tracing method. Three focal depths are shown: (a) center of the cell—40 μm above bottom, (b) intermediate plane—20 μm above the bottom, and (c) bottom of the cell. Note the different scales among the simulations.

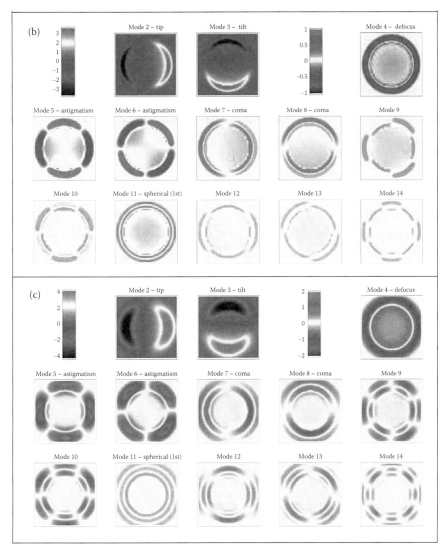

FIGURE 6.3 (*Continued*)

This simulated sphere could be regarded as a very simple model for a spherical cell within a water-based sample. The model sphere had a refractive index of 1.35 and cell cytoplasm has typically a refractive index in the range of 1.35–1.38 (Dunn 1998).

6.3 Modeling of an Oocyte Cell and Comparison to Experimental Results

The aberrations caused by a mouse oocyte cell are modeled to verify the experimental results obtained by interferometric wavefront measurements.

Since this particular type of cell has a basic shape, wavefront simulations can be based on a simple model. The transmitted light image of the cell is shown in Figure 6.4 next to the model used for the simulations. The

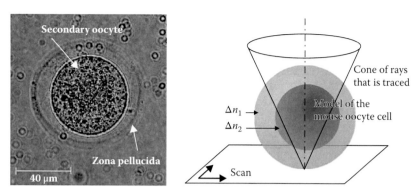

FIGURE 6.4   Transmitted light image of the mouse oocyte cell (left) and the geometric model used for the simulation of wavefront aberrations (right).

real cell had a diameter of approximately 80 μm, and a range of 100 μm was scanned on a 16 × 16 grid. The simulation calculates the field-dependent component of the aberration only. Additional aberration terms due to the refractive index mismatch between immersion medium and embedding medium are not considered here but have been calculated by Török et al. (1995) and Booth et al. (1998). The simulation assumed diameters of the two concentric spheres of 80 and 52 μm as inferred from a transmitted light image of the cell and used the values for the NA (dry lens, NA = 0.5) and the "scanned" field (100 μm × 100 μm) as determined from the experiment. The difference in the values of refractive index between embedding medium and the biological specimen could not be determined by a complementary method, but $\Delta n_1 = 0.016$ and $\Delta n_2 = 0.030$ (inner sphere) were found to indicate a good agreement with the experimental data. In the experiment, the focus was set to the supposed bottom of the cell with an accuracy better than 5 μm. A setting of 10 μm above the lower edge of the outer sphere of the simulation was found to give the best reproduction of the experimental data. This could be explained by the fact that the cell is not perfectly spherical and the cross section in the z-direction is smaller than that in the x-direction.

Figure 6.5 shows the simulation and the interferometric wavefront measurement results next to each other. The Zernike mode charts for both the data sets are displayed at the same scale and show the variation of the different modes across the scanned field of view. The experimental setup uses a phase-stepping interferometer in transmission geometry in which the sample is mounted between two opposing objective lenses contained in the measurement arm of the interferometer. A mirror mounted in the reference path is moved by a piezo actuator and performs phase stepping at videorate. The sample is mounted on an x–y stage and translated perpendicular to the optical axis and wavefronts are recorded on a 16 × 16 grid across the specimen. For experiment and simulation, Zernike modes 2 through 22 were extracted from the wavefronts but the figures show modes 2 through 12 only since the higher-order modes have relatively low amplitudes. It should be mentioned that an adaptive optics aberration correction system would not correct the Zernike modes tip (2), tilt (3), or defocus (4) since these aberration modes correspond to lateral or longitudinal translation of the focal spot but leave signal intensity and resolution unaffected. The presence of these modes in the aberrated wavefront causes geometric distortions in the obtained 3D data set because there is a difference between the actual and the predicted positions of the focal spot. However, correction would require a detailed knowledge of the refractive index distribution of the specimen.

Note that the experimental measurement shown in Figure 6.5 contains the static aberration component caused by the refractive index mismatch in addition while the simulation covers the field-dependent fraction only. The static component mainly affects mode 11 (first spherical) and causes an additional Zernike amplitude offset for this mode. The agreement between experiment and simulation is good, and the main features of the Zernike mode variations across the field are reproduced by the simulations.

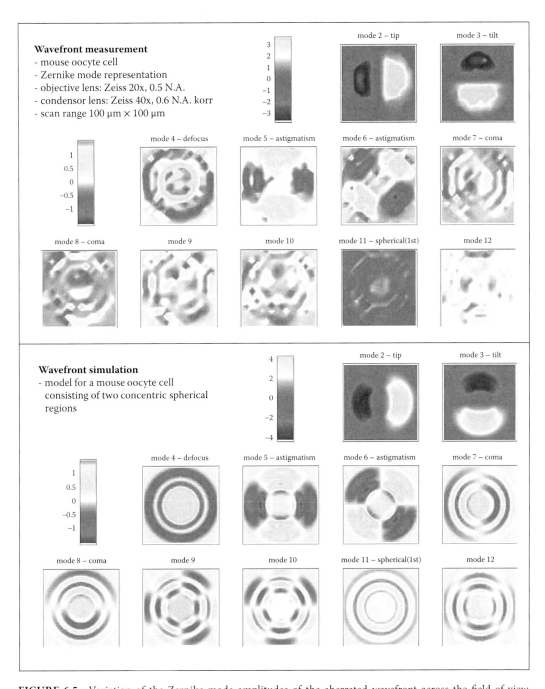

FIGURE 6.5 Variation of the Zernike-mode amplitudes of the aberrated wavefront across the field of view. Experimental results for the wavefront measurements of a mouse oocyte cell (top) and corresponding simulation results (bottom). In the experiment, the specimen was translated and wavefronts were recorded on a grid of 16 × 16 points across the specimen. For the simulation, four times the number of data points are calculated on a 32 × 32 grid.

6.4 Wavefront Aberration Simulation for a Cylindrical Object

The simulation model for wavefront aberrations of cylindrical objects is very similar to that used for spherical objects. Again, we would like to find the pupil function $P(r,\theta)$ of Equation 6.1, depending on the focal position and the geometry of the specimen. A sketch of the simulation model is depicted in Figure 6.6. The unit direction vector w defines the direction of the cylinder axis, C is a point on the cylinder axis, R denotes the radius, and $\Delta n_c = n_c - n_0$ is the absolute difference in refractive index of the cylinder in respect to the immersion medium. The unit-direction vector **t** again describes the direction of the traced ray and is obtained from the polar coordinates of the pupil plane using Equation 6.4. The vector denoted by **m** points from the focus P to C on the cylinder axis. If $a(\mathbf{m},\mathbf{t},\mathbf{w})$ denotes the distance the ray travels within the cylinder, we find

$$a(\mathbf{m},\mathbf{t},\mathbf{w}) = \begin{cases} \dfrac{2\sqrt{R^2 - d^2(\mathbf{m},\mathbf{t},\mathbf{w})}}{|\mathbf{t}\times\mathbf{w}|} & \text{for } d < R \\ 0 & \text{otherwise} \end{cases} \tag{6.8}$$

Here, d refers to the distance between the passing ray and the axis of the cylinder. If, furthermore, the vector **w** is normalized to unity, the distance d is given by

$$d = \frac{|\mathbf{m}\cdot(\mathbf{w}\times\mathbf{t})|}{|\mathbf{w}\times\mathbf{t}|} \tag{6.9}$$

Using the vector components, we get

$$d = \sqrt{\frac{(m_x(w_y t_z - w_z t_y))^2 + (m_y(w_z t_x - w_x t_z))^2 + (m_z(w_x t_y - w_y t_x))^2}{(w_y t_z - w_z t_y)^2 + (w_z t_x - w_x t_z)^2 + (w_x t_y - w_y t_x)^2}} \tag{6.10}$$

Now the set of Equations 6.2, 6.4, 6.8, and 6.10 allows one to calculate the phase function that contains the aberration due to the wavefront distortions imposed by the object.

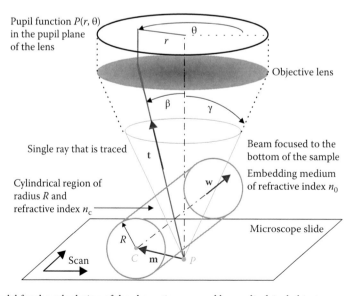

FIGURE 6.6 Model for the calculation of the aberrations caused by a cylindrical object.

6.5 Optical Fiber Simulation and Comparison to Experimental Results

An optical fiber has a defined geometry and a well-known refractive index. Therefore, it is ideal for wavefront aberration experiments and corresponding simulations. For the experiments, we chose a multimode fiber made of fused silica. The polymer fiber jacket was carefully stripped off using a scalpel and the remaining glass fiber was mounted in immersion oil (Zeiss; $n_0 = 1.518$ at 23°C). A transmitted light image of the fiber is shown in Figure 6.7; its outer diameter was measured to be 71 μm. The refractive index of fused silica for the used wavelength of 632.8 nm (HeNe Laser) is 1.49, thus the difference in the values of refractive index is expected to be $\Delta n_c = -0.028$. The wavefront measurement and subsequent Zernike-mode extraction were performed using Equation 6.7. Again, details on data acquisition and the setup can be found in the work by Schwertner et al. (2004).

The experimental results and the corresponding simulations are shown in Figure 6.8. Since the condenser lens was equipped with a correction collar, it was feasible to compensate for the remaining static fraction of the aberration due to the refractive index mismatch. This occured because a dry lens was used to focus into an oil-based sample. As a result of this compensation adjustment, the experimental data contain the field-dependent part of the aberration only.

For the simulations, the refractive index difference of $\Delta n_c = -0.028$, the refractive index $n_0 = 1.518$ of the immersion medium, the measured fiber diameter of 71 μm, the position of the focal spot 15 μm below the bottom of the fiber, the NA 0.5, and a field of view of 130 μm were used corresponding to the parameters of the experiment. The core of the optical fiber is visible in Figure 6.7. It was not included in the simulations since this doped region has only a small difference in refractive index and the core volume can be neglected compared to the volume of the whole fiber.

The experimental results for the fiber simulation are shown in the lower part of Figure 6.8 and agree well with the experimental results shown in the upper part of the figure. The main features of the measured

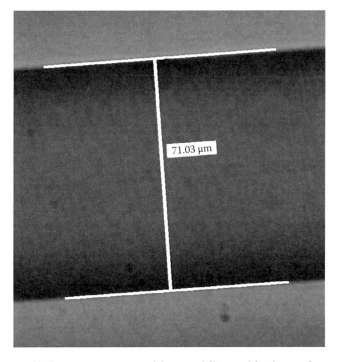

FIGURE 6.7 Transmitted light microscope image of the optical fiber used for the interferometric wavefront measurements. A diameter-measurement bar indicates the scale; this value was used for the simulations.

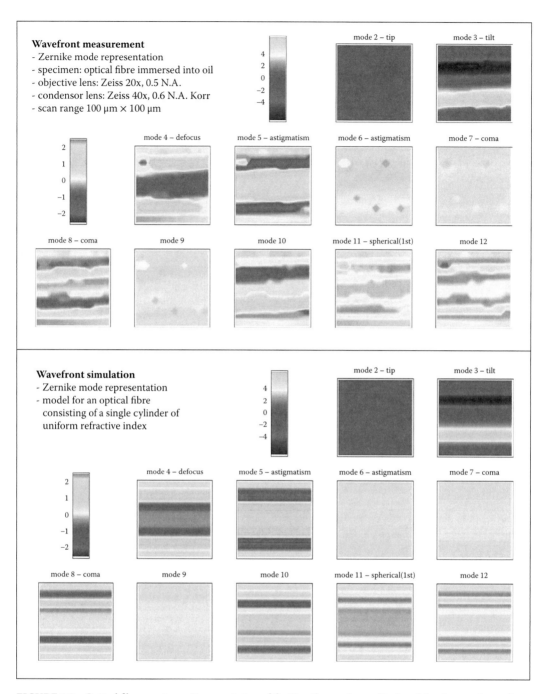

FIGURE 6.8 Optical fiber specimen. Representation of the Zernike-mode amplitudes of the aberrated wavefront when the optical fiber is imaged. Experimental results from the interferometric wavefront measurements (top) and simulation results for a cylinder of uniform refractive index ($n_c = 1.49$) embedded in oil ($n_0 = 1.518$) (bottom).

Zernike mode variations across the field of view are reproduced by the simulation. There are deviations from the simulation for some modes in the top left corner of the fiber (for instance, visible in mode 5). This was due to total internal reflection effects close to the edge of the fiber, which caused deflection of intensity out of the pupil, which in turn triggered problems for the Zernike-mode fitting for these particular wavefronts. Only very few data points of the experimental data were affected since the fiber was not horizontally aligned, only a very narrow range of positions in respect to the fiber axis show the effect, and wavefront measurements were recorded on a rather coarse grid corresponding to steps of about 9 μm.

6.6 Discussion and Conclusion

Ray-tracing simulations of the specimen-induced aberrations caused by cylindrical and spherical objects were performed. Simulation examples for focusing into a spherical object at different depths were given. A sphere of refractive index 1.35 immersed in water can be regarded as a very simple model of a cell within a water-based sample. One has to be aware that the simulations are carried out for a wavefront propagating through the whole sample—as this is the case for our interferometric wavefront measurement setup. A confocal microscope works in epi-mode and the fluorescence excitation and emission beams propagate between the objective lens and the focal spot only. Therefore, the simulation results are strictly equivalent to the epi-mode only in case the focus is set to the bottom or top of the sample.

Simulations for a mouse oocyte cell and an optical fiber were calculated and compared to experimental results obtained by interferometric wavefront measurements. The Zernike-mode data from the ray tracing are in good agreement with the measured experimental data for both the modeled samples; the characteristic features of the Zernike amplitude distribution across the field of view were reproduced by the simulation. Furthermore, an approximation of the difference in refractive index between the biological cell and the embedding medium and the focal position was inferred from the simulation due to the best approximation found for the experimental data.

Again, the simulations showed that the aberrations can be described by a relatively low number of Zernike modes. This means that it is sufficient in adaptive optics systems to correct for a small number of modes only.

A limitation of the current ray-tracing model is the assumption of a straight propagation of the rays. This approximation holds for small and rather smooth variations of the refractive index and is justified for the samples under investigation. The relative differences in refractive index of about 2% were rather small, but values of that magnitude are expected for biological samples embedded in water-based solutions.

Acknowledgement

It is noted that these results have also been presented in the publication Schwertner et al. (2004a).

References

Booth, M., M. Neil, and T. Wilson (1998). Aberration correction for confocal imaging in refractive-index-mismatched media. *J. Microsc. 192* (2), 90–98.

Born, M. and E. Wolf (1983). *Principles of Optics* (6th ed.). Pergamon Press.

Dunn, A. (1998). Light scattering properties of cells. Dissertation, University of Texas, Austin, http://www.nmr.mgh.harvard.edu/~adunn/papers/dissertation/node7.html.

Hell, S., G. Reiner, C. Cremer, and E. Stelzer (1993). Aberrations in confocal fluorescence microscopy induced by mismatches in refractive index. *J. Microsc. 169* (3), 391–405.

Schwertner, M., M. J. Booth, M. Neil, and T. Wilson (2004). Measurement of specimen-induced aberrations of biological samples using phase stepping interferometry. *J. Microsc. 213* (1), 11–19.

Schwertner, M., M. J. Booth, and T. Wilson (2004a). Simulation of specimen-induced aberrations for objects with spherical and cylindrical symmetry. *J. Microsc. 215* (3), 271–280.

Török, P., P. Varga, and G. Németh (1995). Analytic solution of the diffraction integrals and interpretation of wave-front distortion when light is focussed throug a planar interface between materials of mismatched refractive indices. *J. Opt. Soc. Am. 12*, 2660–2671.

Wilson, T. and C. Sheppard (1984). *Theory and practice of Scanning Optical Microscopy*. Academic Press, London.

7

Overview of Adaptive Optics in Biological Imaging

Elijah Y. S. Yew
Singapore-MIT Alliance for Research and Technology (SMART)

Peter T. C. So
Massachusetts Institute of Technology

7.1 Biological Imaging and Adaptive Optics

The problem of biological imaging, like all other imaging problems, is the formation of an image that is as close as possible to the original object. Like all other imaging problems, the image quality depends very much on the quality of the optics, the assembly, the imaging medium, and the object itself. While modern-day optics are close to being perfect, often the sources of aberrations lie in the assembly, such as in custom-built imaging systems; the imaging medium, such as the turbulent atmosphere in astronomy; and the object itself, such as in refractive index heterogeneity in many biological specimens. The way in which imaging quality is restored under such circumstances is through the use of adaptive optics (AO), a field that is relatively established in astronomy. The use of adaptive optics has been applied to biological imaging, in particular the eye, as early as 1989 [1], although it really took off with the implementation of the Shack–Hartmann wavefront sensor (SHWS) to form a feedback loop for active correction [2].

The various parts of an AO system consist primarily of a wavefront sensing device, an active wavefront shaping element, a controller that links the sensing device to the active element, and the feedback to monitor the image quality.

7.1.1 Wavefront Sensing

The detection of aberrations is important since with that information the active element of the AO system can implement the appropriate corrections to the deformation to compensate for the detected aberrations. However, the detection of aberrations need not be limited to AO systems as the image quality of the fundus of an aberrated eye needs to be determined accurately for a corrective surgery to be successful.

To a certain extent, AO systems for biological imaging of the eye can be said to have taken off with the implementation of the Shack–Hartmann wavefront sensor (SHWS) and subsequent demonstration of the improvement to the images of the fundus with AO [2, 3]. Perhaps the most popular among wavefront sensing devices is the SHWS. There are, however, other modes of sensing the wavefront and the following section touches on some of these various methods.

7.1.1.1 Shack–Hartmann Wavefront Sensor

Currently, one of the most popular methods of measuring the aberrations of the eye in the field of AO is with an SHWS. The SHWS is a simple device consisting of a lenslet array that focuses the incident beam onto a charge-coupled device (CCD) placed at the focal plane of the lenslet array. Each lenslet produces a focal spot that is detected over a region on the CCD and is a group of pixels (subdetection area) corresponding to the aperture of the lenslet. If the incident wavefront is planar, the foci are all centered on the appropriate pixels. If the wavefront is aberrated, the foci shift across the subdetection area, according to the type of aberration. The slope of the wavefront is given by the ratio of the shift over the focal length of the lenslet.

The first implementation of an SHWS in ophthalmology was in 1994 by Liang et al. [2]. In this experiment, they used a 1.5 mW laser at 632.8 nm focused onto the retina. The reflected light and wavefront aberrations imparted by the eye were then detected by a home-built SHWS. As the retina was acting as a diffuse reflecting source, it was not possible to retrieve the wavefront errors arising from the first pass through the eye. At the same time, they reported that the double pass by reflection meant that coma was canceled although they found that the third-order coma could not be ignored in their results [2].

7.1.1.2 Curvature Sensor

Curvature sensing has been used in astronomical research but was first introduced to examining the aberrations of the eye in 2006 by Diaz-Douton et al. [4]. Curvature sensing is based on the principle that changes in the local wavefront curvature result in variations in the local intensity as a detector traverses the focus. In the paraxial approximation, the amount of curvature at the input aperture bears a linear relationship to the intensity changes at two different planes of defocus (one before and the other after the focus). Since the defocus between the two planes is small, the detected variations are considered to be well localized over the same cluster of pixels on the detector. The advantages of the curvature sensor are that it often costs less than an SHWS and has a higher dynamic range.

7.1.1.3 Pyramid Sensor

The pyramid sensor is another novel method of measuring the wavefront aberrations of the eye. First developed for the sensing element in an astronomical telescope by Ragazzoni and Farinato [5], it was first applied in measuring the aberrations of the human eye in 2002 by Iglesias et al. [6]. In this version, the pyramid wavefront sensor was modified such that it was able to sense the wavefront variations from an extended source instead of a point source, thereby obviating the need to oscillate the pyramid. Unlike the SHWS, the pyramid sensor can adjust its sampling rate and dynamic range through the oscillations of the pyramid [6], allowing for a higher update frequency to the active element. In 2006, Chamot, Dainty, and Esposito [7] built a pyramid-sensing-based AO system for the eye, operating at a frame rate of 55 Hz over a 6 mm pupil. More recently, Daly and Dainty [8] have achieved a frame rate of 83 Hz in both open- and closed-loop operations, with RMS errors below 0.1 μm for a 6 mm pupil.

In principle, the pyramid wavefront sensor is similar to that of the Foucault knife edge test. A pyramid of transmissive material is placed such that the apex of the pyramid is at the focus of the beam whose wavefront variation over the aperture is what is to be studied. Using geometrical optics, the converging rays are split into four images, which are detected on different parts of a CCD. Each image is therefore an image of the aperture. The variations in intensity over the pixels and for each quadrant will allow the wavefront to be reconstructed. One drawback of the pyramid wavefront sensor is that it is not suitable for large aberrations as the relationship between the local slope and the wavefront becomes nonlinear. However, comparing results obtained from geometrical optics, rigorous diffraction theory, and experiments yields similar conclusions [9]. This means that the geometrical approximation can be used to speed up the entire wavefront reconstruction process.

7.1.1.4 Digital Reconstruction and Lensless Approaches

A newer method recently proposed by Hattori and Komatsu [11] and Sekine et al. [12] involves the dispensation of the lenslet array found in the SHWS. The advantage is that there is more efficiency as compared to using a microlens array in a conventional SHWS [10]. Instead, a two-dimensional grating replaces the lenslet array. Another variation of this principle is the development of a digital SHWS by replacing the lenslet array with a spatial light modulator (SLM) [13]. A new technique based on in-line holography has been proposed by Lai, King, and Neifield [14]. In principle, all the above methods can be used to sense the aberrations of the eye, and thereby provide continuous feedback to the active element for aberration correction.

7.1.2 Wavefront Correction

The active element imparts the correction to the aberrated input wavefront. This is achieved by physically changing the optical path, as with a deformable mirror (DM), or through adjusting the phase, as with a liquid crystal spatial light modulator (LC-SLM). This section briefly characterizes the principle of operation behind the various methods of wavefront correction.

7.1.2.1 Deformable Mirrors

The category of DMs covers a wide range of instruments. For DMs, the incident wavefront is changed through physically altering the optical path length by means of a deformable surface that is reflective and mounted on actuators. There are two main classes of DMs—segmented and continuous mirrors, depending on the construction.

In the segmented DM, the DM consists of small mirrors, each mounted on an actuator. The actuators can generate solely piston (push/pull) motions or have an added tip/tilt capability. Naturally, the DMs with added tip/tilt actuators are better able to correct the aberrations. Segmented DMs have the advantage of being free from the coupling that affects continuous sheet DMs as each actuator is able to move independently from its neighbor. For the same reason, the stroke of a segmented DM is often larger than that of a continuous sheet DM and is often better suited for correcting higher-order modes of high frequencies, although the versions with tip/tilt are well placed to correct both high- and low-order modes. However, segmented DMs suffer from diffraction effects that occur due to the edges between the mirror segments. At the same time, the relatively large size of a DM means that the pupil of the eye needs to be magnified significantly. This results in a larger overall size of the optical setup, which is a disadvantage for a clinical instrument.

Continuous DMs consist of a thin, continuous sheet with actuators on the underside that cause a conformational change. In such devices, each actuator is never truly independent of the others because of an influence function, although this effect can be minimized. This function describes the coupling between actuators and how a conformational change of one actuator affects the entire shape of the continuous sheet. As with segmented mirrors, the conformational change in the mirror is effected through actuators. These need not be piston like actuators and can be of the bimorph type [15], wherein two strips of dissimilar metals that react differently to an applied voltage are bonded together; electrostatic actuators that rely on electrostatic attraction [16]; and also magnetic actuators [17].

Newer technology exists, based on microelectromechanical mirrors (MEMs). MEMs have the advantage that the DM can be built smaller, faster, and cheaper than conventional DMs. The technology is based on established silicon surface micromachining. This confers the advantage of cost, reliability, and a smaller aperture size that is more suited for AO work, especially in ophthalmic applications. Some of the earliest instances of MEMs being used as the active element in an AO are the works of Zhu et al. [18],

Bifano and coworkers [19–21], and Vdovin, Sarro, and Middelhoek [16]. A comparison of the various types of DMs can be found in Devaney et al. [22] and Loktev, Monteiro, and Vdovin [23].

7.1.2.2 Liquid Crystal Spatial Light Modulators

The previous methods relied on changing the surface profile of the mirror to account for the physical optical path length. The optical path length corresponds to an additional phase term, which can be introduced with an LC-SLM operating in the phase-only mode.

The earliest reported works on the application of liquid crystals in wavefront correction were from Russia [24, 25]. In general, the LC-SLM consists of a thin layer of birefringent liquid crystals sandwiched between two glass plates with a particular "brushing" on both glass plates to ensure the alignment of the liquid crystals. The liquid crystals twist to realign their long axis to match the direction of the electric field and, by doing so, cause a change in the refractive index. The entire surface of the LC-SLM is typically divided into pixels, each individually addressed. The methods of addressing each pixel can be optical or electronic through conventional LCD display technology. The optically addressed systems are pixel-less and do not suffer from diffraction effects whereas the electronically addressed versions have a fill-factor, and certain types have fill-factors of 90% or even equivalent to 100%, when we take into account how the electric field varies between pixels.

The fact that the liquid crystals are birefringent imposes the condition that the incident beam has to be of a specific polarization for the LC-SLM to operate efficiently. This does limit the use of LC-SLMs in adaptive optics, especially for ophthalmic applications, since the back-reflection from the retina is unpolarized. Recent work [26] has shown that the incident polarization state does not affect the results, and the work undertaken by Shirai et al. [27] has demonstrated that the depolarized light reflected from the retina can still be phase modulated, albeit with a loss of intensity that they circumvented through the use of highly sensitive detectors.

The advantages of using an LC-SLM is that new technology, such as the liquid-crystal-on-silicon spatial light modulators (LCOS-SLM), utilizes CMOS and current LCD television technology, therefore making them cheaper, with better modulation characteristics and a higher pixel count. The most recent LCOS-SLM as at the time of writing is 1920×1080 pixels, a full-HD SLM.

The disadvantages of SLMs in general are that they require a polarized input for efficient phase modulation, are not broadband and require calibration for different wavelengths, and suffer from diffraction effects due to the pixelated nature of the technology. This, in turn, reduces the amount of power reflected back into the system. There is also a need to carry out phase wrapping to achieve phase modulation above 2π. The refresh rate is also locked at 60 Hz since the SLM is often driven through the video card of a computer. While this may be less of a concern for ophthalmic applications, it is insufficient for scanning microscopy.

7.2 Biological Imaging in the Eye

The eye is an imaging system and consists of a lens (the cornea) and a detector (retina). The lens can accommodate various imaging conditions, being able to adjust the power through the contraction of the ciliary body. This causes the image to form on the retina, where the photoreceptors convert light to electrical impulses, which are then relayed to the brain. Like man-made imaging systems, the eye is not perfect because the cornea may contain imperfections that give rise to higher-order aberrations. At the same time, the ciliary body may be unable to adjust and bring the image to the surface of the retina, as in the common conditions of myopia or hypermetropia. Added to this, the retina is curved and the most sensitive part of the retina is typically at 5° off the central axis. As an optical system, the eye is always far from perfect. While spectacles and Lasik can help correct for myopia or hypermetropia, they often leave the higher-order aberrations (which are always present) uncorrected for. The result is that while spectacles and Lasik may be sufficient for day-to-day use, the imaging of the retina, which is important for the clinical diagnosis of eye diseases, calls for a high level of imaging quality. To address this issue, AO has been applied to the imaging of the eye, with remarkable results. Many of the above active elements have been incorporated into the AO imaging of the eye and research has also gone into the optimization of the requirements needed to optimize the active element for retinal imaging [28, 29].

7.2.1 Adaptive Optics Ophthalmoscope

The ophthalmoscope is a device used to study the fundus of the eye. It is therefore an indispensible tool for tracking and diagnosing the state of health and diseases of the eye. The scanning laser ophthalmoscope [30, 31] and the confocal scanning laser ophthalmoscope (cSLO) [32, 33] represent a quantum leap in the speed and efficiency over which the images of the fundus could be imaged. The use of a laser scanning confocal system allows for sectioning of retina to study the microstructure at different planes because the confocal pinhole provides the optical sectioning required for depth discrimination. It is to be noted that this is effectively performing laser scanning confocal microscopy, and the examining beam is effectively focused ideally on a diffraction-limited spot size at the retina. This requires the scanning to build up an image of the retina, but it is also possible to use detectors of higher sensitivity, such as photomultiplier tubes, to increase the signal detection.

The introduction of wavefront sensing and AO demonstrated that much better images of the fundus could be obtained, allowing for a more precise and detailed study of the eye [3, 34]. The first AO scanning laser ophthalmoscope (AOSLO) was introduced in 2002 by Roorda et al. [35], although a much earlier work by Dreher, Bille, and Weinreb in 1989 [1] implemented a scanning confocal system for imaging the retina using an active optical system for axial scanning and the correction of aberrations in the eye with the name "laser tomographic scanner." In their work, Roorda et al. demonstrated the ability to optically section the retina in vivo while imaging photoreceptors, nerve fibers, and the flow of white blood cells within the capillaries of the retina. Using adaptive optical compensation of the aberrations introduced by the eye, they were able to improve the axial resolution to around 100 μm, compared to 300 μm for cSLO. The transverse resolution was also improved by a factor of two.

It is also known from population studies that the aberrations of the eye consist of lower- and higher-order modes [36, 37] and that there is benefit (in terms of visual quality) in correcting them [38]. Due to the finite number of actuators on the active element as well as the limitations to the stroke of each actuator, a single active element will sometimes be unable to compensate for all the significant orders, especially if there is a mix of low and high orders. To resolve this issue, multiple active elements may be used by splitting the correction for lower and higher order modes over individual active elements. Chen et al. [39] have used this approach and used two DMs—one, a bimorph mirror with a large actuation stroke for correcting large-stroke aberrations (up to 18 μm stroke over a 10 mm pupil) and the other, a MEMs continuous sheet DM (up to 1.5 μm over a 3.3×3.3 μm^2 area). Both DMs were placed at a plane conjugate to the pupil and the wavefront sensor. The entire setup fitted onto a 600 mm × 900 mm optical breadboard, small enough to be mounted on castors and used in an ophthalmology clinic.

There is also another implementation, more akin to conventional ophthalmoscopes—the adaptive optical floodlit ophthalmoscope. In this setup, a larger portion of the retina is illuminated with a source, such as fiber-coupled, strobing superluminescent and laser diodes. The illuminated patch, however, is kept to a certain size because of the need to satisfy the isoplanatic condition (i.e., where the aberrations do vary significantly to cause image degradation) [40–42]. The size of the illuminated patch is still larger than that in the SLO and the detection of a floodlit ophthalmoscope requires a CCD. Currently, such setups can capture images of the retina at 60Hz and these studies have revealed rapid fluctuations of cone reflectivity in dark adapted eyes [43].

Current research in AOSLO is still ongoing, with a thrust toward compactness and wide-field imaging of the retina. Examples of such work can be found in Refs. [44–46].

7.2.2 Optical Coherence Tomography

Optical coherence tomography (OCT) was developed by Huang et al. [47], using an in vitro retinal sample. Since then, OCT has seen widespread use as an in vivo ophthalmic imaging device for tracking macular-related diseases [48–53]. The major advantage of OCT is that the axial and transverse resolutions are decoupled and different strategies can be adopted to maximize both. This is not true for imaging systems that rely on conventional focusing to achieve high spatial resolutions, where the transverse resolution, axial resolution, and depth of focus vary with the numerical aperture as well as wavelength. There are two modes in which OCT works. One is the time-domain OCT and the second is the Fourier domain OCT. Currently, AO has been incorporated into both types of OCT, with the first instance of AO time-domain OCT in 2003 [54].

Since the transverse the axial resolution is inversely proportional to the wavelength spread of the source, whereas the transverse resolution is determined independently, it possible to introduce a broadband source to improve the axial resolution while maintaining the transverse resolution and the depth of focus. This led to the development of ultrahigh-resolution OCT (UHR-OCT), where the first version used a pulsed titanium–sapphire laser with a spectral bandwidth of 350 nm [55].

In 2004, Hermann et al. [56] demonstrated an adaptive optics system based on a commercial UHR-OCT, wherein the transverse resolution of the system reached 5–10 μm and a 3 μm axial resolution with a 130 nm spectral bandwidth. In this implementation, they utilized an AO setup consisting of an SHWS and a MEMs membrane DM, all fitting within a 300 mm × 300 mm footprint. Unlike the commercial UHR-OCT used, they were able to use an input beam diameter of 3.8 mm (from the original 1 mm) while still being able to obtain an improvement in the transverse resolution.

While AO is capable of correcting monochromatic aberrations at wider pupil diameters, it still runs up short against chromatic aberrations, which are aberrations that occur as a function of wavelength. This becomes important as ultrawide broadband light sources have been developed and their application to UHR-OCT becomes obvious [57]. To this end, Fernandez et al. [58] have investigated the correction of polychromatic light (as used in UHR-OCT) using an LC-SLM as the active element. They found that they were able to correct for very high-frequency aberrations due to the density of the pixels on the LC-SLM, but the drawback was that the speed of correction was still below that of conventional MEMs DM.

The Fourier domain OCT is faster than the time-domain OCT in terms of acquisition. It is also more sensitive and was first demonstrated for retinal imaging by Wojtkowski et al. [59]. Because of this improvement in speed and sensitivity, it becomes possible to image live cells in vivo. Such a system has been implemented by Zhang et al. [60]. In their work, they combined an AO floodlit OCT (for locating their imaged areas with higher precision) with an AO Fourier domain OCT. In this system, they were able to image single cells with three-dimensional (3D) resolutions of 3 × 3 × 5.7 μm and speeds of up to 100 A-scans/ms in a single shot. They were thus able to demonstrate, for the first time, simultaneously resolved images in the laterally and axially directions of the interface between the outer and inner segments of individual cones. Some limitations they encountered included the presence of speckle, which has an average size close to that of the theoretical point spread function (PSF) of the imaging system. As a result, objects of interest that are close to the PSF of the system suffer more from speckle. However, speckle contrast is inversely correlated to the diffusion speed of the scatterers. The temporal averaging of images is, therefore, a possible way to overcome this limitation at the expense of the acquisition speed [61], and image-processing techniques have been described in the removal of the motion artifacts incurred by the moving eye [62].

7.2.3 Multiconjugate Adaptive Optics for Wide Field of View

As mentioned previously, the ability to image a large portion of the retina simultaneously faces several challenges even with AO. Good correction and imaging are often limited because the isoplanatic patch is limited in the human eye. This means that if the two objects are spaced so as to be within the isoplanatic patch, the aberrations incurred are similar and a single correction at the active element is sufficient to correct the image. Conversely, if the two objects have a spatial separation greater than the isoplanatic patch, the rays emanating from the objects will see different aberrations as they pass back through the eye to the detector. Since the aberrations differ, it will not be possible to correct for both objects using only one active element. Although research has been carried out so that a single active element adopts a best fit correction, the increased field of view of the retina is only at 3° [63]. It is, of course, possible to image small sections of the retina and, by creating a montage, build up a whole picture of the retina. This is difficult in in vivo studies as not all patients can hold their eyes steady for the extended period of time required to build the montage. There is, therefore, a drive toward extending this field of view, which can be corrected by AO.

The current solution to this problem is multiconjugate adaptive optics (MCAO) and this has its roots in astronomy [64, 65]. The key idea behind MCAO is that the different layers of the aberrating medium will have a corresponding active element at its conjugate plane. In a study using a modeled human eye, Bedggood et al. [40, 66] found that with five calibration points positioned at various parts of the eye, it was possible to identify the aberrations in the various assumed aberrating layers. These planes corresponded to the anterior corneal plane, which has the greatest variation in the index of refraction and thereby contributes most to aberrations, the pupil of the eye, and other arbitrarily chosen positions in anticipation of the ease of conjugating the active elements to the selected planes. It was found through modeling that by placing five active elements conjugate to these positions, it was possible to increase the size of the isoplanatic patch by up to six times. The practical implementation of five correctors in their work would have been difficult, so they gave a general rule of thumb that for a two-corrector system, the best locations would be the anterior corneal position and the posterior surface of the lens.

7.3 Biological Imaging with Microscopes

Yet another application exists in incorporating an AO system into a microscope for biological imaging. The primary aim of an AO microscope is to correct for either the system-induced aberrations, the specimen-induced aberrations, or both. Various approaches have been devised and this section covers some of these approaches.

7.3.1 Wavefront Sensorless

This chapter has touched on many various aspects of AO systems in biological imaging. One common thread that runs through them is the presence of a wavefront sensor to detect and feed the measured wavefront to the active element and together they form a feedback loop. Curiously enough, AO in microscopy tends not to incorporate a wavefront sensor [67]. This is because the detection of the reflected or backscattered light causes problems with the aberration detection, simply because the aberrations of even orders are added, while the odd orders are canceled [67, 68]. The net result is, of course, under- or over correction. At the same time, most of the aberrations are specimen-induced and the correction of lower order modes is sufficient to improve the image quality [69].

To determine the appropriate amount of correction to apply via the active element, most of these wavefront sensorless methods rely on the maximization of the detected fluorescence, often through an optimization algorithm [70, 71]. Such algorithms include genetic algorithms, hill-climbing, random search, and adaptive random search. What the system does is basically impose a sequence of aberrations as initial "guesses" on the DM and detect the subsequent change in fluorescence. Recent research into the optimization of such wavefront sensorless techniques has focused on the decomposition of the correcting aberration imposed onto the DM in terms of modes, such as the Zernike polynomials. Booth [72] has proposed an optimization approach using model-based sphere packing that, under certain conditions, requires only $N + 1$ measurements for N modes. Alternatively, it is possible to expand as Lukosz rather than Zernike modes, and a detailed exposition can be found in the works by Debarre, Booth, and Wilson [73] and Booth [74].

In an implementation of a smart microscope, Albert et al. [71] used a genetic algorithm to determine the optimal shape of the DM only through maximizing the detected two-photon fluorescence. The setup utilized an *f*/1 parabolic mirror to focus an expanded beam from a pulsed titanium sapphire laser. Since the intensity of the two-photon signal scales quadratically with the intensity at the focus, the closer the DM was to correcting for the unknown aberrations, the greater the two-photon signal detected. In this manner, the system was able to learn beforehand the needed corrections for every position of the scanned beam. With certain assumptions and modifications to the algorithm, the smart microscope was able to learn the necessary corrections in one minute, and thereby extend the scanning range of the parabolic mirror ninefold. The optimization of fluorescence signal is well suited to laser scanning microscopy because unlike other modes that form the image directly on the detector, only the signal strength is detected with the image formed electronically.

A recent study [75] compared four methods of optimizing the shape of the deformation through maximizing the signal intensity. In this study, the four methods compared were genetic algorithm (GA), hill-climbing algorithm (HC), random search (RS), and adaptive random search (ARS). These algorithms were tested for the repeatability and reliability of the various algorithms, the time taken to reach a solution corresponding to a set level of signal recovery, and the final axial resolution of the system. For a predetermined aberration, it was found that each of the different algorithms resulted in a different shape of the DM, thereby indicating that there was no one unique solution. However, repeated runs of the algorithm with a fixed amount of aberration to correct indicated that of the four, only the ARS appeared to not reach the predetermined level of improvement required. Of the three algorithms capable of reaching an optimal solution repeatedly, HC was the quickest at a constant 48 seconds, compared to a maximum of 12 min 30 s for GA, with the others falling somewhere in between. Among the methods, the algorithm that gave the best axial resolution was RS. The drawback of using RS was that it took between 2 and 6 minutes to reach a solution. It is obvious that the use of optimization routines limits the applicability, especially if imaging speed is a concern.

Since the imaging of biological samples often requires a 3D stack to be built up, the signal strength also varies throughout the sample, making the use of signal intensity as the optimizing metric ill-suited. Oliver, Debarre, and Beaurepaire [76] recently proposed the use of image sharpness as an alternative metric. Using image sharpness to build an influence matrix, they found that they were able to apply the previously obtained influence matrix to samples different from the initial sample. This implies that the technique was relatively insensitive to intersample variations and could mean a faster AO system since the influence matrix could be predetermined beforehand and still be used for subsequent samples.

One current example of AO using a RS algorithm that may be of particular significance in biological imaging is that of signal-enhanced coherent Stokes Raman spectroscopy (CARS) microscopy. In this work, Wright et al. [77] used adaptive optics with CARS microscopy and demonstrated that they could correct for both the system and the specimen aberrations. With a test sample, they reported that they were able to obtain signals up to 700 μm deep, with a trebling of the signal. When the same system was applied to muscles, they reported a more modest imaging depth of 260 μm, but a sixfold increase in signal.

7.3.2 Wavefront Sensors and Guide Stars

Although many implementations of AO in microscopes adopt a sensorless approach, there are systems that incorporate a wavefront sensor. One of the reasons why wavefront sensors like the SHWS have not been incorporated is because these rely on a guide star to determine the aberrations picked up by the rays as they travel through the intervening medium. This places restrictions on using the backscattered excitation beam as the input beam for the SHWFS for the reason that odd aberrations are canceled on specular reflection and that even aberrations are summed [68]. Another alternative is to use a fluorescent source, which, because it is sensitive only to the incoming excitation beam (first pass), can be used as a measure of the aberrations [67]. Such a system has been described in a wide-field fluorescence microscope utilizing a fluorescent bead as a guide star [78] embedded in a *Drosophila* embryo. The specimen-induced aberrations were measured by detecting the emitted fluorescence with a SHWFS. The field of view was relatively small due to the isoplanatic angle imposed by the objective used.

Another study using an SHWFS, but relying on backscattered light, has recently been reported by Denk and coworkers [10, 79]. This method utilizes the backscattered light to determine the aberrations. Since the primary interest is in biological tissues that are highly scattering, a few issues arise with the use of the backscattered excitation light. The first is that the portion of the backscattered light that carries the information about the aberrations in the focal plane forms only a small fraction of the total backscattered light. The second is that due to the coherent nature of the backscattered light near the focus and the random nature of the scatterers, speckle is formed. The third is that scattering is also polarization-dependent. To address these three issues, Denk and coworkers realized that the most important backscattered photons to be collected were those near the focus, with minimal scattering. To isolate these photons, a coherence-gated scheme was implemented to detect only the ballistic photons. The issue of speckle was addressed by averaging the speckle through slightly moving the sample, which reduces the speckle contrast. It is also known that in a reflecting sample, such as the retina, coherent light picks up aberrations while both passing into and out of the eye [68, 80]. The result is that odd aberrations are lost because they are replicated in the return pass in the wrong position, while even aberrations are doubled, although this is not true with incoherent light. To overcome this issue, Denk and coworkers made the assumption that even though the backscattered light is coherent, the random nature and the density of the distribution of the scatterers and the speckle averaging have the effect of rendering the detected backscattered light effectively incoherent. This meant that the reflected (backscattered) light was effectively incoherent, and thus it carried with it the information about the aberrations only on the return path, thereby making an SHWS approach possible.

More recently, Cha, Ballesta, and So [81] used the back-reflected light from the sample to perform AO multiphoton microscopy. In this study, Cha, Ballesta, and So used a two-photon microscope to image the brain, heart, and tongue of a mouse. A pinhole as a spatial filter was employed before the SHWS, with the pinhole size being just above the size that would transmit the maximum spatial frequency attainable with the DM. This method is an alternative to using an interferometric setup for coherence gating, with the pinhole effectively accepting only the light from near the focal region. In this work, aberration correction was done only for a single position corresponding to the center of the image. In general, adopting a single configuration is practically useful since the aberrations present are often a result of imaging depth (i.e., specimen-induced or refractive index–induced aberrations) [69, 82–84]. In this study, it was found that the imaging depth was typically at 100–200 µm and the signal strength increased on the order of 20–70%. In conclusion, the study found that it was scattering that was important in determining the imaging depth as it degraded the image on an exponential scale, whereas degradations associated with aberrations were approximately linear.

Up till this point, all the methods have been based on expressing the aberrations in terms of Zernike or Lukosz modes. This is often done because of their orthogonality, and that they express most of the aberrations very well and can be easily modeled by current actuator-based active elements. However, newer active elements, such as the SLM, utilize many separate and individual elements, each being independent of the other. This, of course, allows for a departure from Zernike or Lukosz polynomials. In a recent study [85], a pupil-segmentation approach was adopted over that of implementing a wavefront sensor or decomposition into orthogonal modes, such as Zernike or Lukosz modes. Instead, a geometrical optics approach was used, where the pupil of the objective was subdivided into smaller subzones that were each individually corrected using phase ramps such that the rays focused by that subzone were shifted to cross at the focus. In general, it was found that fewer than 100 subzones were required to recover diffraction-limited performance. This was achieved via two methods—the first being a "direct" measurement method in which a reference illumination beam (thin pencil of rays) was brought to focus along with another beam focused along a small subzone of the pupil. The variations in intensity (due to constructive or destructive interference between the reference and the measurement beam) were measured by changing the applied phase to the subzone N times till the maximum was reached, and the applied phase was recorded for that particular subzone. The second method was through a reconstruction of the phase by noting that changing the phase implies a focus shift in the rays focused by the subzone. It is therefore possible to reconstruct the appropriate phase applied such that all the foci belonging to each subzone would intersect at the focus. The former method results in a smaller RMS wavefront error, although at a cost of extra computation required by changing and applying N number of phase changes to determine the maximum signal. This method was shown to be successful at improving the signal level recorded while imaging through a slice of mouse-brain tissue 250–450 μm in thickness. An image correlation method was used to average the aberrations over the various zones, thereby resulting in only one correction being used over the entire scanned field (45×165 μm).

In a final example of the recent developments in AO, a recent development utilizing AO in imaging is not so much for correcting for a good image but rather for a large scanned field of view [86]. In this particular case, the system was designed to have a large scanned field. This of course meant that the image quality was compromised by aberrations near the edges. To correct for this, Potsaid et al. designed the system from scratch using ray-tracing software, allowing for sufficient aberrations that were corrected by incorporating AO rather than going for broke with a diffraction-limited design. The system was called the adaptive scanning optical microscope (ASOM) and was, at one point, commercialized by Thorlabs™.

7.4 Conclusion

Adaptive optics has been used successfully in the area of ophthalmologic examination of the eye. This is evident by the various techniques that exist and use AO. Some examples of these techniques are the AOSLM and the AO-OCT. However, the implementation in microscopy is less straightforward as it is difficult to insert a wavefront sensor within the imaging system. This can be overcome through various methods of defining an image metric or by using the backscattered light and detecting with a wavefront sensor. The high scan speeds of laser scanning microscopy also mean that feedback and optimization of the aberrations based on various imaging metrics are slower by several orders of magnitude. While lookup tables and various algorithms can help speed up the optimization, recent research has also indicated that as the aberrations induced are often dependent on imaging depth rather than the field, it is possible to implement a single correction for a particular scanned area without having to continually change the mirror shape. In this manner, the demand on the refresh rate of the DM can be decreased.

References

1. Dreher AW, Bille JF, Weinreb RN. Active optical depth resolution improvement of the laser tomographic scanner. *Appl Optics*. 28(4), 804–8 (1989).
2. Liang J, Grimm B, Goelz S, Bille JF. Objective measurement of wave aberrations of the human eye with the use of a Hartmann–Shack wave-front sensor. *J Opt Soc Am A Opt Image Sci Vis*. 11(7), 1949–57 (1994).
3. Liang J, Williams DR, Miller DT. Supernormal vision and high-resolution retinal imaging through adaptive optics. *J Opt Soc Am A Opt Image Sci Vis*. 14(11), 2884–92 (1997).
4. Diaz-Douton F, Pujol J, Arjona M, Luque SO. Curvature sensor for ocular wavefront measurement. *Opt Lett*. 31(15), 2245–7 (2006).
5. Ragazzoni R, Farinato J. Sensitivity of a pyramidic wave front sensor in closed loop adaptive optics. *Astron Astrophys*. 350(2), L23–L6 (1999).
6. Iglesias I, Ragazzoni R, Julien Y, Artal P. Extended source pyramid wave-front sensor for the human eye. *Opt Express*. 10(9), 419–28 (2002).
7. Chamot SR, Dainty C, Esposito S. Adaptive optics for ophthalmic applications using a pyramid wavefront sensor. *Opt Express*. 14(2), 518–26 (2006).

8. Daly EM, Dainty C. Ophthalmic wavefront measurements using a versatile pyramid sensor. *Appl Optics*. 49(31), G67–G77 (2010).

9. Burvall A, Daly E, Chamot SR, Dainty C. Linearity of the pyramid wavefront sensor. *Opt Express*. 14(25), 11925–34 (2006).

10. Rueckel M, Mack-Bucher JA, Denk W. Adaptive wavefront correction in two-photon microscopy using coherence-gated wavefront sensing. *Proc Natl Acad Sci U S A*. 103(46), 17137–42 (2006).

11. Hattori M, Komatsu S. An exact formulation of a filter for rotations in phase gradients and its applications to wavefront reconstruction problems. *J Mod Optic*. 50(11), 19 (2003).

12. Sekine R, Shinuya T, Ukai K, Komatsu S, Hattori M, Mihashi T, Nakazawa N, Hirohara Y. Measurement of wavefront aberration of human eye using Talbot image of two-dimensional grating. *Opt Rev*. 13(4), 5 (2006).

13. Zhao L, Bai N, Li X, Ong LS, Fang ZP, Asundi AK. Efficient implementation of a spatial light modulator as a diffractive optical microlens array in a digital Shack–Hartmann wavefront sensor. *Appl Opt*. 45(1), 90–4 (2006).

14. Lai SC, King B, Neifeld MA. Wave front reconstruction by means of phase-shifting digital in-line holography. *Opt Commun*. 173(1–6), 155–60 (2000).

15. Dainty JC, Koryabin AV, Kudryashov AV. Low-order adaptive deformable mirror. *Appl Opt*. 37(21), 4663–8 (1998).

16. Vdovin G, Sarro PM, Middelhoek S. Technology and applications of micromachined adaptive mirrors. *J Micromech Microeng*. 9(2), R8–R19 (1999).

17. Fernandez EJ, Vabre L, Hermann B, Unterhuber A, Povazay B, Drexler W. Adaptive optics with a magnetic deformable mirror: Applications in the human eye. *Opt Express*. 14(20), 8900–17 (2006).

18. Zhu LJ, Sun PC, Bartsch DU, Freeman WR, Fainman Y. Adaptive control of a micromachined continuous-membrane deformable mirror for aberration compensation. *Appl Opt*. 38(1), 168–76 (1999).

19. Bifano TG, Perreault J, Mali RK, Horenstein MN. Microelectromechanical deformable mirrors. *IEEE J Sel Top Quant*. 5(1), 83–9 (1999).

20. Perreault JA, Bifano TG, Levine BM, Horenstein MN. Adaptive optic correction using microelectromechanical deformable mirrors. *Opt Eng*. 41(3), 561–6 (2002).

21. Gonglewski JD, Vorontsov MA, Gruneisen MT (Editors). High-Resolution Wavefront Control: Methods, Devices, and Applications III, Proceedings of SPIE Vol. 4493 (2002).

22. Devaney N, Coburn D, Coleman C, Dainty CJ, Dalimier E, Farrell T, et al. In: Olivier SS, Bifano TG, Kubby JA, editors. Characterisation of MEMs mirrors for use in atmospheric and ocular wavefront correction. MEMS adaptive optics II; Tuesday 22 January 2008; San Jose, CA, USA. SPIE; 2008. p. 688802.

23. Loktev M, Monteiro DWD, Vdovin G. Comparison study of the performance of piston, thin plate and membrane mirrors for correction of turbulence-induced phase distortions. *Opt Commun*. 192(1–2), 91–9 (2001).

24. Vasilev AA, Naumov AF, Schmalgauzen VI. Wavefront correction by liquid-crystal devices. *Sov J Quantum Electron*. 16(4), 4 (1986).

25. Vasilev AA, Vorontsov MA, Koryabin AV, Naumov AF, Schmalgauzen VI. Computer controlled wavefront corrector. *Sov J Quantum Electron*. 19(3), 4 (1989).

26. Marcos S, Diaz-Santana L, Llorente L, Dainty C. Ocular aberrations with ray tracing and Shack–Hartmann wave-front sensors: Does polarization play a role? *J Opt Soc Am A*. 19(6), 1063–72 (2002).

27. Shirai T, Takeno K, Arimoto H, Furukawa H. Adaptive optics with a liquid- crystal-on-silicon spatial light modulator and its behavior in retinal imaging. *Jpn J Appl Phys*. 48(7), 070213–15, (2009).

28. Miller DT, Thibos LN, Hong X. Requirements for segmented correctors for diffraction-limited performance in the human eye. *Opt Express*. 13(1), 275–89 (2005).

29. Doble N, Miller DT, Yoon G, Williams DR. Requirements for discrete actuator and segmented wavefront correctors for aberration compensation in two large populations of human eyes. *Appl Opt*. 46(20), 4501–14 (2007).

30. Webb RH, Hughes GW. Scanning laser ophthalmoscope. *IEEE Trans Biomed Eng.* 28(7), 488–92 (1981).

31. Webb RH, Hughes GW, Pomerantzeff O. Flying spot TV ophthalmoscope. *Appl Opt.* 19(17), 2991–7 (1980).

32. Webb RH, Hughes GW, Delori FC. Confocal scanning laser ophthalmoscope. *Appl Opt.* 26(8), 1492–9 (1987).

33. Wornson DP, Hughes GW, Webb RH. Fundus tracking with the scanning laser ophthalmoscope. *Appl Opt.* 26(8), 1500–4 (1987).

34. Roorda A, Williams DR. The arrangement of the three cone classes in the living human eye. *Nature.* 397(6719), 520–2 (1999).

35. Roorda A, Romero-Borja F, Donnelly Iii W, Queener H, Hebert T, Campbell M. Adaptive optics scanning laser ophthalmoscopy. *Opt Express.* 10(9), 405–12 (2002).

36. Thibos LN, Hong X, Bradley A, Cheng X. Statistical variation of aberration structure and image quality in a normal population of healthy eyes. *J Opt Soc Am A Opt Image Sci Vis.* 19(12), 2329–48 (2002).

37. Porter J, Guirao A, Cox IG, Williams DR. Monochromatic aberrations of the human eye in a large population. *J Opt Soc Am A Opt Image Sci Vis.* 18(8), 1793–803 (2001).

38. Williams D, Yoon GY, Porter J, Guirao A, Hofer H, Cox I. Visual benefit of correcting higher order aberrations of the eye. *J Refract Surg.* 16(5), S554–9 (2000).

39. Chen DC, Jones SM, Silva DA, Olivier SS. High-resolution adaptive optics scanning laser ophthalmoscope with dual deformable mirrors. *J Opt Soc Am A Opt Image Sci Vis.* 24(5), 1305–12 (2007).

40. Bedggood P, Daaboul M, Ashman R, Smith G, Metha A. Characteristics of the human isoplanatic patch and implications for adaptive optics retinal imaging. *J Biomed Opt.* 13(2), 024008 (2008).

41. Fried DL. Anisoplanatism in adaptive optics. *J Opt Soc Am.* 72(1), 52–61 (1982).

42. Jankevics AJ, Wirth A. Wide-field-of-view adaptive optics. Proc. SPIE 1543, Active and Adaptive Optical Components, 438 (January 13, 1992). doi:10.1117/12.51199.

43. Rha J, Jonnal RS, Thorn KE, Qu JL, Zhang Y, Miller DT. Adaptive optics flood-illumination camera for high speed retinal imaging. *Opt Express.* 14(10), 4552–69 (2006).

44. Mujat M, Ferguson RD, Iftimia N, Hammer DX. Compact adaptive optics line scanning ophthalmoscope. *Opt Express.* 17(12), 10242–58 (2009).

45. Ferguson RD, Zhong ZY, Hammer DX, Mujat M, Patel AH, Deng C, Zou WY, Burns SA. Adaptive optics scanning laser ophthalmoscope with integrated wide-field retinal imaging and tracking. *J Opt Soc Am A.* 27(11), A265–A77 (2010).

46. Burns SA, Tumbar R, Elsner AE, Ferguson D, Hammer DX. Large-field-of-view, modular, stabilized, adaptive-optics-based scanning laser ophthalmoscope. *J Opt Soc Am A.* 24(5), 1313–26 (2007).

47. Huang D, Swanson EA, Lin CP, Schuman JS, Stinson WG, Chang W, Hee MR, Flotte T, Gregory K, Puliafito CA, et al. Optical coherence tomography. *Science.* 254(5035), 1178–81 (1991).

48. Hee MR, Puliafito CA, Wong C, Duker JS, Reichel E, Rutledge B, Schuman JS, Swanson EA, Fujimoto JG. Quantitative assessment of macular edema with optical coherence tomography. *Arch Ophthalmol.* 113(8), 1019–29 (1995).

49. Hee MR, Puliafito CA, Wong C, Duker JS, Reichel E, Schuman JS, Swanson EA, Fujimoto JG. Optical coherence tomography of macular holes. *Ophthalmol.* 102(5), 748–56 (1995).

50. Hee MR, Puliafito CA, Wong C, Reichel E, Duker JS, Schuman JS, Swanson EA, Fujimoto JG. Optical coherence tomography of central serous chorioretinopathy. *Am J Ophthalmol.* 120(1), 65–74 (1995).

51. Puliafito CA, Hee MR, Lin CP, Reichel E, Schuman JS, Duker JS, Izatt JA, Swanson EA, Fujimoto JG. Imaging of macular diseases with optical coherence tomography. *Ophthalmol.* 102(2), 217–29 (1995).

52. Schuman JS, Hee MR, Arya AV, Pedut-Kloizman T, Puliafito CA, Fujimoto JG, Swanson EA. Optical coherence tomography: A new tool for glaucoma diagnosis. *Curr Opin Ophthalmol.* 6(2), 89–95 (1995).

53. Schuman JS, Hee MR, Puliafito CA, Wong C, Pedut-Kloizman T, Lin CP, Hertzmark E, Izatt JA, Swanson EA, Fujimoto JG. Quantification of nerve fiber layer thickness in normal and glaucomatous eyes using optical coherence tomography. *Arch Ophthalmol.* 113(5), 586–96 (1995).

54. Miller DT, Qu JL, Jonnal RS, Thorn K. Coherence gating and adaptive optics in the eye. *P Soc Photo-Opt Inst.* 4956, 65–72, 378 (2003).

55. Drexler W, Morgner U, Kartner FX, Pitris C, Boppart SA, Li XD, Ippen EP, Fujimoto JG. In vivo ultrahigh-resolution optical coherence tomography. *Opt Lett.* 24(17), 1221–3 (1999).

56. Hermann B, Fernandez EJ, Unterhuber A, Sattmann H, Fercher AF, Drexler W, Prieto PM, Artal P. Adaptive-optics ultrahigh-resolution optical coherence tomography. *Opt Lett.* 29(18), 2142–4 (2004).

57. Drexler W. Ultrahigh-resolution optical coherence tomography. *J Biomed Opt.* 9(1), 47–74 (2004).

58. Fernandez EJ, Povazay B, Hermann B, Unterhuber A, Sattmann H, Prieto PM, Leitgeb R, Ahnelt P, Artal P, Drexler W. Three-dimensional adaptive optics ultrahigh-resolution optical coherence tomography using a liquid crystal spatial light modulator. *Vision Res.* 45(28), 3432–44 (2005).

59. Wojtkowski M, Leitgeb R, Kowalczyk A, Bajraszewski T. In vivo human retinal imaging by Fourier domain optical coherence tomography. *J Biomed Opt.* 7(3), 7 (2002).

60. Zhang Y, Rha J, Jonnal R, Miller D. Adaptive optics parallel spectral domain optical coherence tomography for imaging the living retina. *Opt Express.* 13(12), 4792–811 (2005).

61. Zawadzki RJ, Jones SM, Olivier SS, Zhao M, Bower BA, Izatt JA, Choi S, Laut S, Werner JS. Adaptive-optics optical coherence tomography for high-resolution and high-speed 3D retinal in vivo imaging. *Opt Express.* 13(21), 8532–46 (2005).

62. Zawadzki RJ, Choi SS, Jones SM, Oliver SS, Werner JS. Adaptive optics-optical coherence tomography: optimizing visualization of microscopic retinal structures in three dimensions. *J Opt Soc Am A.* 24(5), 1373–83 (2007).

63. Glanc M, Gendron E, Lacombe F, Lafaille D, le Gargasson J-F, Lena P. Towards wide-field retinal imaging with adaptive optics. *Opt Commun.* 230, 14 (2004).

64. Beckers JM, editor. Increasing the size of the isoplanatic patch with multiconjugate optics. Proceedings of a ESO Conference on Very Large Telescopes and Their Instrumentation; 1988. Garching: European Southern Observatory.

65. Ragazzoni R, Marchetti E, Valente G. Adaptive-optics corrections available for the whole sky. *Nature.* 403(6765), 54–6 (2000).

66. Bedggood PA, Ashman R, Smith G, Metha AB. Multiconjugate adaptive optics applied to an anatomically accurate human eye model. *Opt Express.* 14(18), 8019–30 (2006).

67. Booth MJ. Adaptive optics in microscopy. *Philos Transact A Math Phys Eng Sci.* 365(1861), 16 (2007).

68. Artal P, Marcos S, Navarro R, Williams DR. Odd aberrations and double-pass measurements of retinal image quality. *J Opt Soc Am A Opt Image Sci Vis.* 12(2), 195–201 (1995).

69. Schwertner M, Booth M, Wilson T. Characterizing specimen induced aberrations for high NA adaptive optical microscopy. *Opt Express.* 12(26), 6540–52 (2004).

70. Marsh PN, Burns D, Girkin JM. Practical implementation of adaptive optics in multiphoton microscopy. *Opt Express.* 11, 8 (2003).

71. Albert O, Sherman L, Mourou G, Norris TB, Vdovin G. Smart microscope: An adaptive optics learning system for aberration correction in multiphoton confocal microscopy. *Opt Lett.* 25(1), 52–4 (2000).

72. Booth MJ. Wave front sensor-less adaptive optics: A model-based approach using sphere packings. *Opt Express.* 14(4), 14 (2006).

73. Debarre D, Booth MJ, Wilson T. Image based adaptive optics through optimisation of low spatial frequencies. *Opt Express.* 15(13), 15 (2007).

74. Booth MJ. Wavefront sensorless adaptive optics for large aberrations. *Opt Lett.* 32(1), 3 (2007).

75. Wright AJ, Burns D, Patterson BA, Poland SP, Valentine GJ, Girkin JM. Exploration of the optimisation algorithms used in the implementation of adaptive optics in confocal and multiphoton microscopy. *Microsc Res Tech.* 67(1), 36–44 (2005).

76. Oliver N, Debarre D, Beaurepaire E. Dynamic aberration correction for multiharmonic microscopy. *Opt Lett.* 34(20), 3 (2009).

77. Wright AJ, Poland SP, Girkin JM, Freudiger CW, Evans CL, Xie XS. Adaptive optics for enhanced signal in CARS microscopy. *Opt Express.* 15(26), 18209–19 (2007).

78. Azucena O, Kubby J, Crest J, Cao J, Sullivan W, Kner P, Gavel D, Dillon D, Olivier S. Implementation of a Shack–Hartmann wavefront sensor for the measurement of embryo-induced aberrations using fluorescent microscopy. *Proc SPIE.* 7209, 9 (2009).

79. Feierabend M, Ruckel M, Denk W. Coherence-gated wave-front sensing in strongly scattering samples. *Opt Lett.* 29(19), 2255–7 (2004).

80. Artal P, Iglesias I, Lopez-Gil N, Green DG. Double-pass measurements of the retinal-image quality with unequal entrance and exit pupil sizes and the reversibility of the eye's optical system. *J Opt Soc Am A Opt Image Sci Vis.* 12(10), 2358–66 (1995).

81. Cha JW, Ballesta J, So PTC. Shack–Hartmann wavefront-sensor-based adaptive optics system for multiphoton microscopy. *J Biomed Opt.* 15(4), (2010).

82. Kam Z, Kner P, Agard D, Sedat JW. Modelling the application of adaptive optics to wide-field microscope live imaging. *J Microsc.* 226(Pt 1), 33–42 (2007).

83. Kner P, Sedat JW, Agard DA, Kam Z. High-resolution wide-field microscopy with adaptive optics for spherical aberration correction and motionless focusing. *J Microsc.* 237(2), 136–47 (2010).

84. Booth MJ, Neil MAA, Wilson T. Aberration correction for confocal imaging in refractive-index-mismatched media. *J Microsc.* 192(2), 90–8 (1998).

85. Ji N, Milkie DE, Betzig E. Adaptive optics via pupil segmentation for high-resolution imaging in biological tissues. *Nat Methods.* 7(2), 141–7 (2010).

86. Potsaid B, Bellouard Y, Wen J. Adaptive scanning optical microscope (ASOM): A multidisciplinary optical microscope design for large field of view and high resolution imaging. *Opt Express.* 13(17), 6504–18 (2005).

8

Wavefront Correctors

Joel A. Kubby
*University of California
at Santa Cruz*

8.1 Introduction

As described in Chapter 4, an initially planar wavefront will become aberrated after passing through a media with an inhomogeneous index of refraction. By correcting the wavefront using adaptive optics (AO), the spatial resolution and contrast of an image can be increased. Wavefront correction is accomplished by delaying the leading parts of the wavefront so that the trailing parts have a chance to catch up. As shown in Figure 8.1, if the wavefront is reflected from a deformable mirror (DM), the optical path length can be varied across the mirror surface by deforming it into a shape that is conjugate to the wavefront aberration, so that the wavefront is corrected after reflection.

Another way to correct the wavefront is to use a liquid crystal spatial light modulator to slow down the leading part of the wavefront by changing its velocity relative to the lagging part of the wavefront. The lead edge is ahead by a distance d, so that it precedes the trailing edge in time by d/c, where c is the speed of light. The velocity of the wavefront can be changed by changing the index of refraction of the liquid crystal media through which the wavefront passes (transmission) or in reflection. As shown in Figure 8.2, the index of refraction for the liquid crystal in the central region where the wavefront is leading is n_1 so that the wavefront velocity is c/n_1, and the index in the outer regions where the wavefront

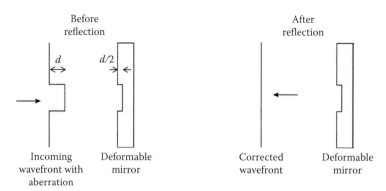

FIGURE 8.1   Wavefront correction using a deformable mirror. Before reflection, the leading edge of the wavefront is at a distance *d* ahead of the trailing edge. If the deformable mirror has an indentation that is *d*/2 deep, the leading edge will have to travel a distance *d* further than the trailing edge after reflection, allowing the trailing edge to catch up to the leading edge. (Courtesy of Lawrence Livermore National Laboratory and NSF Center for Adaptive Optics.)

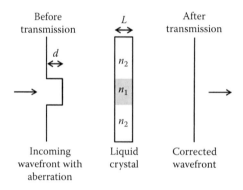

FIGURE 8.2   Wavefront correction using a liquid crystal spatial light modulator. The index of refraction in the center of the spatial light modulator is n_1 and n_2 at the edges. If $n_1 > n_2$, the leading edge of the wavefront will be slowed down more than the trailing edge, enabling the trailing edge to catch up after passing through the liquid crystal spatial light modulator.

is lagging is n_2, so that the wavefront velocity is c/n_2. If the spatial light modulator has a length L, then the trailing edge will be able to catch up with the leading edge when the index difference Δn has been adjusted to be $\Delta n = n_2 - n_1 = d/L$.

8.2 Liquid Crystal Spatial Light Modulators

Liquid crystal spatial light modulators (LC-SLM) operate by changing the orientation of liquid crystals using electrostatic forces. By changing the orientation of the liquid crystal, the index of refraction can be varied spatially across the modulator. Either the phase or amplitude of the wavefront, or both, can be varied in transmission through, or reflection from, the SLM. The liquid-crystal pattern can be controlled directly from a computer-graphics card. Since the index is wavelength and polarization dependent, the use of an SLM for wavefront correction is limited to monochromatic polarized light.

The phase can be varied only from 0 to 2π without phase wrapping, so it is limited in the amount of wavefront correction that it is capable of correcting, usually being used as a "tweeter" to make low-amplitude, high-order wavefront corrections. The change in orientation of the liquid crystal is limited to approximately 100 Hz, so it is limited in bandwidth for rapidly varying wavefront corrections. Nonetheless, the pixel size on an SLM can be very small, allowing for very high-order corrections to be made. In addition, since each pixel can be controlled independently, there is no influence of one actuator on another.

The use of SLMs in optical microscopy has recently been reviewed by Maurer et al. (2011). They have been used for wavefront shaping for both correction of refractive image aberrations and scattering. They have also been used as dynamic spatial Fourier filters in the imaging path.

For correction of refractive image aberrations, LC-SLMs have been used to precorrect the illumination wavefront to overcome shifts in the wavefront as it propagates through the sample (Ji 2009; Milkie 2011). This approach is described in detail in Chapter 13 and is shown schematically in Figure 8.3. For a sample with a homogeneous index of refraction, as shown in Figure 8.3a, a plane wave that fills the back aperture of an objective lens is brought to a focal spot within the sample. The objective lens creates a spherically converging wavefront with each normal to the wavefront interfering constructively at the focus. For a sample with a spatially varying index of refraction, the light rays are refracted and the wavefront phase is shifted by the sample inhomogeneities as shown in Figure 8.3b, so that they no longer reach a common focus in phase for constructive interference. To overcome this, the pupil can be segmented into N subregions by an LC-SLM, and the tilt and phase of each subregion varied to optimize constructive interference at the focus, as shown in Figure 8.3c. When the wavefront tilts and phases have been optimized, the optical rays intersect at the same point with the appropriate phase for constructive interference. To perform the optimization, a reference image using all of the beamlets is first acquired. Then all but one of the beamlets are removed with a binary phase pattern that deflects them toward a field stop. An image is then acquired with the remaining beamlet. Any inhomogeneities along the path of the beamlet that deflect it from the ideal focal point are evidenced as a shift in this image relative to the reference image. Once the image shift has been determined, the deflection angle can be calculated and an equal but opposite angle can be imparted to the beamlet by application of an appropriate phase ramp at the corresponding subregion of the SLM. This is then repeated for each beamlet, one by one, until all of the beamlets intersect at a common focal point. To bring all of the beamlets into phase for constructive interference at the

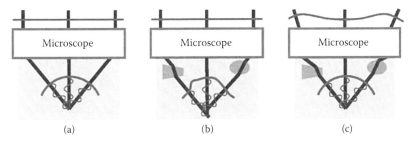

(a)   (b)   (c)

FIGURE 8.3 A simple model for the formation of an optical focus. (a) An ideal microscope converts a planar wavefront to a converging spherical one in a sample. Propagation vectors or "rays", defined by the direction normal to the wavefront, converge at a common point and, being in phase, constructively interfere there to create an optimal focus. Sinusoidal curves denote the phase variation along each ray. (b) Inhomogeneities in the refractive index of the sample change the directions and phases of the rays, leading to a distorted wavefront and an enlarged focal volume with lower peak intensity. (c) Controlling the input wavefront using an active optical element (not shown) can cancel these aberrations, recovering a diffraction-limited focus. (From Ji, N., D. E. Milkie, and E. Betzig, *Nat Methods*, 7, 141–147, 2009.)

focal point, an optimization algorithm is enabled by varying the relative phase between two segments until the signal at the focal point is a maximum when the two beams interfere constructively. This can then be repeated pairwise, until all of the remaining beamlets interfere constructively at the focus. In a related approach, the full pupil is illuminated, rather than individual segments (Milkie 2011), to avoid imaging through the small numerical aperture of a single segment. In this approach, the wavefront is kept fixed in all but one pupil segment, which has a phase ramp applied to it. This procedure is applied sequentially to each pupil segment. Once the tilt has been removed from each of the pupil segments, the phase at each segment is adjusted to obtain constructive interference at the focus.

A related approach, again using pupil segmentation by an LC-SLM, has been used to overcome scattering in the sample rather than refractive image aberrations, as discussed earlier. A coherent source illuminates an LC-SLM that segments the beam into individual beamlets. A region of interest or pattern defined by a CCD camera is used to detect constructive interference of the beamlets. This is shown schematically in Figure 8.4. The phase of each beamlet is varied one by one, or in blocks of $N \times N$, and the intensity in the region of interest is maximized. Once each segment has been optimized, the beamlets interfere constructively at the region of interest on the CCD camera. In a related approach, as shown in Figure 8.5, a fluorescent bead "guide star" is implanted in the sample and the fluorescence from the guide star is measured and optimized as the phase of each segment, or $N \times N$ block of segments, is varied. When all of the segments have been optimized, the beamlets from all of the segments interfere constructively at the guide star.

In both these techniques—pupil segmentation to overcome refractive image aberrations and interferometric focusing to overcome scattering—the high order of the spatial light modulator is an advantage to make high-order corrections; however, the need for optimizing the phase of each segment makes this approach slow. Obtaining an interferometric focus using an LC-SLM can take on the order of 10 minutes. Thus, this approach to AO will be too slow for live imaging, which is required for biologists to dynamically image rapidly changing samples such as the mitosis process in *Drosophila* embryos. Most likely, LC-SLM approaches will be limited in application to imaging either fixed (frozen) tissues or slowly changing live samples, for example, the slow variations that occur in neural cell development.

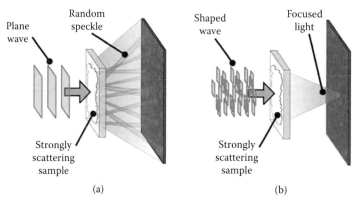

FIGURE 8.4 (a) A plane wave is focused on a disordered medium and a speckle pattern is transmitted. (b) The wavefront of the incident light is shaped so that scattering makes the light focus at a predefined region of interest on a CCD camera. (From Vellekoop, I. M. and A. P. Mosk, *Opt Lett.*, 32, 2309–2311, 2007.)

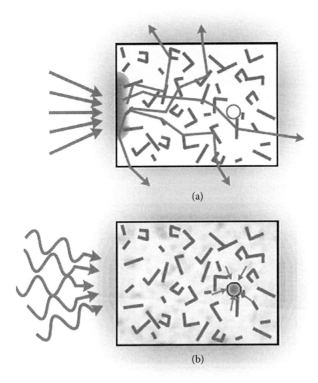

FIGURE 8.5   (a) Conventional way of illuminating a scattering sample with a lens (not shown). The lens causes the light rays to converge outside the sample, but owing to scattering in the sample, the light does not come to a geometric focus. (b) An alternative to geometric focusing, using a lens, is interferometric focusing, using a spatial light modulator. When the phase delays for each of the incident beamlets are adjusted correctly, the beamlets interfere constructively in the sample, coming to an interferometric focus at the fluorescent bead. (From Vellekoop, I. M., E. G. van Putten, A. Lagendijk, and A. P. Mosk, *Opt Express.* 16, 67–80, 2008.)

8.3 History of Deformable Mirrors in Adaptive Optics

It is helpful to discuss the history of AO in astronomy and vision science, since the use of DMs for AO in both these fields has a longer history and has helped to guide the application of DMs for AO in biological imaging systems.

Astronomer Horace W. Babcock at the Mount Wilson and Palomar Observatories first proposed a concept for AO in 1953 to improve astronomical image resolution (Babcock 1953). Light waves from a distant star passing through the earth's atmosphere become aberrated by the dynamic variations in the refractive index of the air. This causes the star image to blur and jitter with time, since the light rays bend to travel along the minimum optical path distance routes through the atmosphere, according to Fermat's principle, as discussed in Chapters 1 and 2. Light rays are refracted when passing through areas of different indices of refraction. A familiar example of bending of light by refraction, and one that is important for biological imaging, is that seen at an interface between air and water. These transmissive media have different indices of refraction (1.0 and 1.5, respectively) and the effect on the light path is clear, as shown in Figure 8.6.

In astronomy, the changes in the index of refraction are due to temperature variations that lead to density variations in the air. These variations are random and fluctuate rapidly according to the laws of

FIGURE 8.6   Refraction of light at an air–water interface. Bending of light rays due to refraction makes the straw appear to be displaced at the air–water interface.

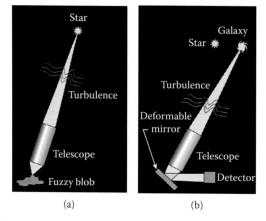

FIGURE 8.7   (a) Optical aberrations in astronomy induced by light from a star passing through turbulence in the earth's atmosphere. The refraction of the light through the time varying index of refraction causes the focus of the light from the telescope to move over time, causing the image, which is collected over time in a CCD camera, to appear as a fuzzy blob. (b) The solution proposed by Horace Babcock was to reflect the starlight off of an Eidophore, an early form of a deformable mirror, and to use the light from a single star as a reference beacon or "guide star." The mirror is deformed to bring the light from the star back into focus as a single point of light. A nearby galaxy, which passes through the same portion of the atmosphere and thus has the same aberration, is reflected off of the deformable mirror, and the image is corrected. (Credit: Lawrence Livermore National Laboratory and NSF Center for Adaptive Optics.)

turbulent gases. Winds aloft move these density patterns at high speeds, depending on altitude, which cause additional rapid variations in light paths. In addition to shifting the image of the star, the density variation changes cause the intensity of the starlight to vary as rays cross and interfere constructively and destructively, making the star appear to twinkle or scintillate. The net effect of the refractive ray bending and the scintillation is to change the sharp point of light from a star into a fuzzy blob, as shown in Figure 8.7a.

Babcock proposed to compensate for the turbulence of the atmosphere by measuring the wavefront aberrations from a guide star and correcting them with an early form of a DM that was then used in an electronic projector called an Eidophor (Hornbeck 1998). The Eidophor used a spatial light modulator principle. It consisted of a thin layer of oil on a conducting mirror. With a raster scanned beam of electrons from a cathode ray tube, a charge can be deposited on the surface of the oil layer in a pattern. The attractive force between the charge and the conducting substrate deforms the surface of the oil as shown in Figure 8.8. The deformation of the oil causes a variation of the thickness above the mirror so that light traversing the oil and reflecting off the mirror experiences a varying optical path length. In the Eidophore projection application, the charging over a portion of the surface of the oil formed a schlieren grating. This phase grating was used to diffract light reflecting off of the mirror below. A dark pixel was created in the projection system when there was no charge deposited and, therefore, no grating and no diffraction and a bright pixel formed when there was diffraction. More recently a liquid deformable mirror has been demonstrated that is based on based on electrocapillary actuation (Vuelban 2006).

The "seeing compensator" that Babcock proposed to use the Eidophor for wavefront correction is shown schematically in Figure 8.9. Light from a star is brought to a focus at F. A field lens images the input pupil of the telescope onto the Eidophor, which is formed on an off-axis parabolic mirror. The reflected light is brought to a focus on a rotating knife edge at the focal point of a second off-axis

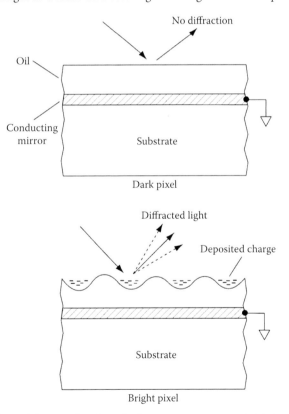

FIGURE 8.8 Spatial light modulator used in the Eidophor projection system. A thin layer of oil is deposited over a grounded conducting mirror that has been deposited on a substrate. *Top*: No charge has been deposited on the oil layer, so the optical path length for all light rays is the same and light reflects off of the mirror, creating a dark pixel in the Eidophor projection system. *Bottom*: When a sinusoidal charge pattern has been deposited with a cathode ray tube, the charge is attracted to the conducting mirror, deforming the surface of the oil layer into a sinusoidal pattern. Light traversing the deformed oil layer and reflecting off of the mirror surface experiences a spatially varying optical path length, causing the light to diffract. The diffracted light forms a bright pixel in the projection system. (Courtesy of Texas Instruments.)

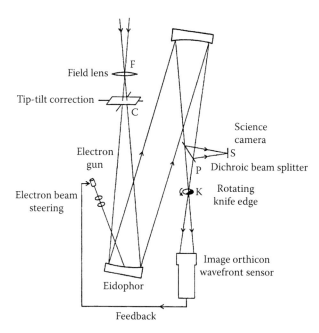

FIGURE 8.9 Schematic diagram of "seeing compensator" using an Eidophor spatial light modulator. The light from the telescope's objective is brought to a focus at F. A field lens images the light onto the Eidophor, the surface of which is controlled by a patterned layer of charge from an electron gun that is raster-scanned by an electron beam steering system under feedback control from a wavefront sensor. The light is then split by a dichroic beam splitter P, with some of the light going to a science camera S and some of the light going to the wavefront sensor formed by a rotating knife edge K at the focal point of the light and the image orthicon. This article originally appeared in the Publications of the Astronomical Society of the Pacific. (Reproduced with permission from Babcock, H. W., *Publ Astron Soc Pac.*, 65, 386, 229–236, 1953. Copyright 2004 Astronomical Society of the Pacific.)

parabolic mirror. The schlieren image formed on the image orthicon wavefront sensor is used to modulate the intensity of an electron beam that is raster-scanned across the surface of the Eidophor, creating a charge pattern that deforms the surface of the oil, forming a spatial light modulator that corrects the wavefront aberrations. A tip–tilt corrector under feedback control from the image sensor keeps the focal point centered on the rotating knife edge.

Babcock's Eidophor-based seeing compensator idea is not too different from how modern AO systems work, the main difference being of course the much better technology now used for light modulation and wavefront sensing. A modern AO system for astronomy is shown in Figure 8.10. Here the light from a telescope with a distorted wavefront is reflected off the surface of an adaptive mirror. The shape of the mirror is the opposite of the distorted wavefront, as measured by the wavefront sensor. The wavefront sensor and adaptive mirror form a closed-loop feedback control system. After reflecting off the mirror, the corrected wavefront is recorded by a high-resolution CCD camera.

8.4 Specifications for Deformable Mirrors

The specifications for the DMs that are used in astronomical imaging are determined by the wavefront aberrations that are used to correct. Since the wavefront aberrations in astronomy are better known than the wavefront aberrations in biological imaging, which can vary widely between different biological samples, it is useful to examine how they can be corrected first.

The optical effects of the earth's atmosphere have been well characterized. The turbulence in the atmosphere is usually described by the Kolmogorov model for the velocity of motion in a fluid medium

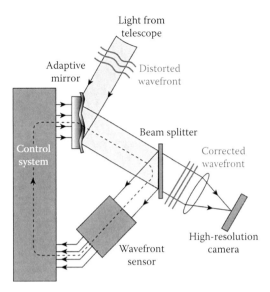

FIGURE 8.10 Astonomical AO system. Light with a distorted wavefront enters a telescope and is reflected off of an adaptive mirror. Part of the light is reflected from a beam splitter and enters a wavefront sensor that generates correction signals for a control system for the adaptive mirror. The other part of the light with a corrected wavefront passes through the beam splitter and is imaged onto a high-resolution camera. (Credit: Lawrence Livermore National Laboratory and NSF Center for Adaptive Optics.)

(Kolmogorov 1941). The structure function that relates the mean-square velocity difference between two points in space that are separated by a displacement vector $\mathbf{r}$ is given by

$$D_v = \left[v_r(r_1 + r) - v_r(r_1) \right]^2$$

The coherence length is called the Fried's parameter r_0, or the "seeing cell size" over which the overall wavefront distortion is limited to a uniform tilt, and is defined as the maximum diameter of a collector that is allowed before atmospheric distortion seriously limits performance (Fried March 1965a, 1965b, 1966). The phase structure function, which describes the expected variance in the refractive index between two points, for Kolmogorov turbulence for plane waves is given as

$$D_\emptyset = 6.88 \left(\frac{r}{r_0} \right)^{5/3}$$

The coherence length r_0 sets the number of degrees of freedom of an astronomical AO system (Figure 8.11). The primary mirror of the telescope, with diameter D, is divided into "sub-apertures" of diameter r_0. The number of subapertures is approximately $(D/r_0)^2$, where r_0 is evaluated at the desired observing wavelength. A typical range for the coherence length might be 10–100 cm. The largest telescope to date, the Keck Telescope on the peak of Mauna Kea, has a diameter $D = 10$ m, so that the number of degrees of freedom required for the DM is on the order of 100–10,000 depending on the coherence length.

A DM for correcting wavefront aberrations will have residual wavefront fitting error depending on the type of DM that is used (Hardy 1998). The mean square "fitting error" for Kolmogorov turbulence is given by

$$\sigma_F^2 = a_F \left(\frac{d}{r_0} \right)^{5/3} \text{rad}^2$$

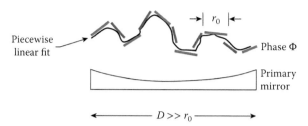

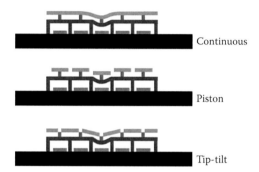

FIGURE 8.11   Degrees of freedom for a deformable mirror used in an astronomical adaptive optics system. The piecewise linear fit of the phase is fit for each coherence length r_0. (Credit: Claire Max, Astro 289C, UCSC.)

FIGURE 8.12   Different types of deformable mirrors. *Top*: Continuous facesheet mirror. *Middle*: Segmented mirror with piston-only mode. *Bottom*: Segmented mirror with piston and tip-tilt. (Credit: Don Gavel.)

where d is the subaperture size for the mirror, r_0 is the turbulence coherence length, and a_F is the fitting error coefficient, which depends on the type of DM that is being used. A physical interpretation of the fitting error is that the mirror functions as a high-pass filter, correcting the low spatial frequency components of the wavefront aberration and passing the high-frequency components. The spatial bandwidth of the filter is $1/r_0$.

Some of the different types of DMs include segmented mirrors, where each segment can be actuated independently, and continuous facesheet mirrors, where a thin continuous mirror is bonded to an array of actuators (Figure 8.12). For the segmented mirrors, the segments can be actuated in piston-only mode, where each mirror segment can be displaced in the direction normal to the segment only (i.e., up and down like a piston), or with the addition of tilting the mirror in addition to the piston motion.

The segments can also be of various shapes (e.g., square, circular). The fitting coefficients for these various configurations are given in Table 8.1 (Hardy 1998).

In addition to different fitting coefficients, segmented mirror can lose optical intensity through the gaps between segments if they are large and reduction of the Strehl ratio from diffraction by the edges of the segments. The number of actuators N that are required to make the same degree of correction will also vary depending on the configuration:

$$\frac{N_1}{N_2} = \left(\frac{d_2}{d_1}\right)^2 = \left(\frac{a_{F_1}}{a_{F_2}}\right)^{6/5}$$

So that a piston-only DM with square segments will require $(1.26/0.28)^{6/5} = 6.2$ more actuators than a continuous facesheet mirror to achieve the same amount of correction. A piston plus tilt mirror with

TABLE 8.1 Fitting Error Coefficient for Various Types of Deformable Mirrors

Configuration	Coefficient	Actuators/Segment
Piston only (square segments)	1.26	1
Piston only (circular segments)	1.07	1
Piston plus tilt (square segments)	0.18	3
Piston plus tilt (circular segments)	0.14	3
Continuous facesheet	0.28	1

Source: Hardy 1998.

square segments will require $3(0.18/0.28)^{6/5} = 1.8$ times more actuators. The factor of 3 is because a piston plus tilt mirror requires three actuators per segment to achieve the tilting motion.

In addition to fitting error, other considerations for DMs include the stroke and pitch of the mirror, the influence function between actuators, and the response time that sets the temporal bandwidth. The stroke of mirror is the magnitude of displacement of the mirror surface that can be obtained. The required stroke will depend on the magnitude of the wavefront aberrations that are required. In astronomy, several micrometers are required for the 10 m Keck telescope. As the telescope diameter increases, and it images more of the sky, the magnitude of correction that is required increases. The required mirror stroke is given by

$$\text{Stroke} = (0.5)(5)(1.2)\sqrt{\mu}\left(\frac{D}{r_0}\right)^{5/6}$$

The factor of 0.5 arises because the wavefront phase is twice the surface phase of the mirror. The factor of 5 is from the number of RMS standard deviations of the wavefront error that must be corrected. The factor of 1.2 is to account for ≈20% of the mirror stroke that must be used to flatten the mirror. The fitting coefficient, μ, depends on the type of mirror that is being used and is listed in Table 8.1. The stroke specifications for the 30 m telescope that is currently being designed calls for 10–15 μm of stroke. Vision science applications also require 10–20 μm of stroke, where the stroke requirements depend on the population of the subjects that are corrected.

The pitch of the mirror, or the spacing between actuators d, is determined by the root-mean-square fitting error σ for light with a wavelength λ:

$$\sigma_f = \left(\frac{\lambda}{2\pi}\right)\sqrt{\mu}\left(\frac{d}{r_0}\right)^{5/6}$$

The early DMs used in astronomy had piezoelectric actuators with an inter-actuator spacing of 5–7 mm, resulting in large mirror when a large number of actuators are required. The trend is toward a smaller pitch of 1 mm and a higher actuator count. The 30 m telescope requires an array of 100×100 actuators (10,000).

The influence function between actuators is determined by how much the mirror surface moves at a neighboring site when an actuator is "poked." A segmented mirror will have no coupling between actuators since the mirror segments are independent, while a continuous facesheet mirror will have on the order of 20% influence since the actuators are coupled through the facesheet. This coupling needs to be comprehended in the software that is used to control the mirror. Finally, the response time of the mirror will determine the temporal bandwidth that the mirror is capable of correcting. In astronomy, this is determined by the temporal coherence time τ_0, the timescale over which atmospheric variations can be considered to be static. This is given by the coherence length, r_0, divided by the average wind velocity, $<v>$, in a layer of the atmosphere, $\tau_0 = r_0/<v>$. For $r_0 = 10$ cm, $<v> = 10$ m/s, the coherence time τ_0 would be 10 ms.

8.4.1 Figure of Merit for Adaptive Optics System Performance

To properly specify a DM, it is important to understand the quantitative performance objectives of the AO system. For this, the traditional approach is to characterize the point-spread function of the system relative to that which would be attained if there were no aberration. The Strehl ratio is a handy scalar metric.

Strehl ratio is defined as the ratio of the peak of the aberrated point spread function (PSF) to the peak of the theoretical PSF with no wavefront aberrations ($\varphi = 0$):

$$S = \frac{\mathrm{PSF}(\theta)}{\mathrm{PSF}(0,0)\big|_{\varphi=0}}$$

A Strehl ratio of above 0.8 is a typical threshold for what is termed diffraction limited imaging. Under such a condition, points in the object form very sharp images at the focal plane, where the width of the PSF is limited by diffraction as opposed to the aberration blur (see Figures 8.13 and 8.14).

The Strehl ratio is related to variance of the wavefront's phase departure from perfectly flat or perfectly spherical. Using Marechal's approximation, for small σ_P

$$S = \frac{\mathrm{PSF}(0,0)}{\mathrm{PSF}(0,0)} \approx e^{-(\sigma_P)^2}$$

$$S = e^{-(\sigma_P)^2}$$

where σ_P is the standard deviation of the phase, in radians.

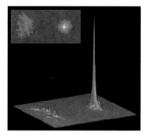

FIGURE 8.13 Images taken with the IRCAL camera and adaptive optics at the Lick Observatory Shane Telescope. *Left*: The image is with the adaptive optics system off. *Right*: The imaging is with the adaptive optics system on. (Credit: James Graham, UC Berkeley.)

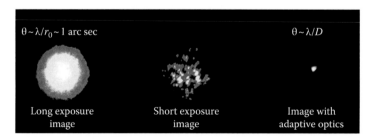

FIGURE 8.14 Image of a bright star, Arcturus, taken with the 1 m telescope at Lick Observatory. The speckles in the short-term exposure are each at the diffraction limit of the telescope. (Credit: Lawrence Livermore National Laboratory and NSF Center for Adaptive Optics.)

8.4.2 Deformable Mirror Requirements

To specify a DM for a particular application, a number of requirements must be determined. The primary considerations for the DM requirements include the following:

- *The maximum displacement or dynamic range (stroke).* This maximum stroke will determine the magnitude of the wavefront error that can be corrected. The optical stroke for a mirror is twice the mechanical stroke. In astronomy, the magnitude of wavefront error depends on the degree of atmospheric turbulence and increases as the diameter of the telescope increases, since it is imaging a larger area of the sky. In biological imaging, the magnitude of wavefront error will depend on the degree of refractive index variations and the thickness of the specimen that is being imaged.
- *The maximum inter-actuator stroke.* The inter-actuator stroke will determine the maximum gradient (slope) in wavefront error that can be corrected. This will depend on the coupling between adjacent actuators. A segmented mirror will have no coupling between the actuators, whereas a continuous facesheet may have on the order of 20% coupling between the actuators since they are mechanically linked through the facesheet.
- *The number of actuators (order).* The order determines the number of degrees of freedom of correction that can be obtained for correction of the wavefront error.
- *The spacing between the actuators.* The pitch, or distance between actuators, sets the highest spatial frequency that can be corrected by the mirror. As mentioned previously, the DM acts as a high-pass filter, correcting the low-spatial-frequency components of the wavefront aberration and passing the high-frequency components. The highest spatial frequency that can be corrected is where every other actuator is up and every intermediate actuator is down.
- *Range of Zernike coefficients that can be corrected.* The Zernike polynomials (Zernike 1934), as a set of functions that are orthogonal on a unit circle, can be used to characterize the spatial response of the mirror. Higher-order Zernike polynomials correspond to higher spatial frequencies. These polynomials are convenient to use since they can correspond to common aberrations such as astigmatism, focus, coma, spherical, and trefoil aberrations.
- *The pupil diameter (aperture).* The pupil diameter corresponds to the aperture of the DM. Typically the limiting aperture of the optical system that is being corrected is projected onto the DM, so the ratio of the mirror's aperture to the limiting aperture of the optical system will determine how much magnification is required. Typically the edge actuators of a continuous facesheet mirror will behave differently than the inner actuators since they have different boundary conditions, so they might be excluded from the pupil of the mirror.
- *The mirror surface quality (RMS roughness).* High-performance wavefront correction requires a high-quality mirror surface. Some DMs that are being used for high-contrast imaging of dim planets around bright stars call for a surface figure of a few nanometers. Some surface topography resulting from mirror manufacturing may have a repeat distance of the actuator pitch. Examples include the etch release holes and support posts for microelectromechanical systems (MEMS) DMs. Although this topography may be sizeable, since it has a well defined pitch, it acts as a diffraction grating and higher-order modes can be filtered out with an order sorting filter.
- *Initial surface bow that must be flattened (lost stroke).* Because of manufacturing defects such as residual stress, stress gradients, and other thin film defects, the initial mirror surface might not be flat. Some of the mirror's stroke may have to be used to flatten the mirror. On MEMS continuous facesheet mirrors, this can be as large as 0.5 μm, which can be a significant fraction of the total mirror stroke. Of course surface topography at a spatial frequency that is higher than the inter-actuator spacing cannot be corrected by the mirror, since the mirror functions as a high-pass filter, with the corner frequency determined by the mirror's pitch.
- *Mirror thickness.* The thickness of the mirror will determine how stiff it is. The stiffness varies as the third power of the thickness. For a continuous facesheet mirror, the stiffer the mirror, the

more coupling there will be between actuators. A stiffer mirror will remain flatter in the presence of thin film stress and stress gradients from thin film depositions such as mirror coatings and adhesion layers for mirror coatings.

- *The surface coating (reflectivity at different wavelengths (visible [VIS]/infrared [IR]), maximum optical power).* The surface coating of the mirror must be chosen to have high reflectivity at the wavelength that it will be used at. Some of the common mirror coatings include the following:
  - Gold for operation in the IR region
  - Protected aluminum for the VIS wavelength region
  - Protected silver for the VIS wavelength region
  - Dielectrics (Bragg mirror for high-power laser)
- *Protective window (transmittance at different wavelengths [VIS/IR]).* If a protective window is to be included in the packaging of the DM, it must allow light from the operating spectrum to pass through to the mirror. It must include an antireflection coating to eliminate reflections at the window interface. Ideally a DM would have no window, but many DMs need to be protected from the ambient environment and dust.
- *Mirror environment (exposed to atmosphere, hermetic seal, vacuum,* or *inert gas).* The mirror will typically be operated in an environment that can damage the mirror, such as water that can condense on the mirror below the dew point in a humid environment and dirt that can land on the mirror causing an optical defect. Moisture can also lead to electrochemical corrosion, which is accelerated in the presence of high-operating drive voltages. To avoid corrosion and dust, the mirror can be hermetically sealed in vacuum or in an inert gas environment behind an optically clear window.
- *Mirror type.* There are a number of different types of DMs that are best suited for different applications.
  - *Continuous facesheet mirrors.* These mirrors are the typical choice for astronomical applications to avoid light diffraction that can occur at the sharp edges of a segmented mirror. They are also the typical mirror of choice for microscopy. The downside of continuous facesheet mirrors is the inter-actuator coupling due to the mechanical linkage through the facesheet.
  - *Segmented mirrors.* These mirrors have cuts in the facesheet between actuators. This eliminates coupling between the mirror segments, but it also has an impact on the fitting error, since the wavefront will have to be approximated by straight line segments rather than a continuous smooth curve. The segment size and shape will have an impact on the optical performance of the mirror, and the fill factor between mirror segments will have an impact on optical losses. Typical fill factors are now approaching 99%, so optical losses are minimal. The mirror segments can be operated in piston-only (i.e., straight up and down in the surface normal direction) with a single actuator/segment or in a piston tip–tilt mode with three actuators/segment. Segmented MEMS DMs have been commonly used in vision science applications since they have been able to attain a relatively larger stroke than continuous facesheet mirrors. The larger the stroke in vision science, the wider the population that can be corrected. Mirror strokes of 10–15 μm are desired to correct the larger aberrations at the tail ends of the population. One potential problem with segmented mirrors is the large phase jump that can occur between segments if they are not properly controlled to assure continuity between segments.
  - *Mirror edge support:* There are different possible boundary conditions for the support of the mirror surface.
    - The edge of the membrane can be supported around the periphery of the mirror, either rigidly attached to a frame or connected to a support frame through springs. This spring support geometry is similar to a trampoline.

- Alternatively, the edge of the mirror can be supported by actuators. The actuators can either be flexible, like springs, or rigid, like posts. The spring support geometry is similar to box springs on a mattress.
- *Actuator type:* There are different means for actuating the mirror surface.
  - *Piezoelectric actuators (lead zirconium titanate, $PbTiO_3$ [PZT])*. An applied voltage causes a mechanical strain that causes a linear change in the length of the crystal. The piezoelectric *d* coefficients relate the strain produced to the electric field that is applied. A typical value would be $d = 390 \times 10^{-12}$ m/V to $d = 650 \times 10^{-12}$ m/V, or approximately a few angstroms/volt. Since the motion is so small, piezoelectric crystals are typically connected electrically in parallel and stacked together so that their individual displacements accumulate. In this manner, it is possible to get ±10 μm for an applied voltage of ±100–200 V. These materials are temperature sensitive. The piezoelectric *d* coefficients decrease at low temperature and they can become "depoled" above the Curie temperature.
  - *Electrostrictive actuators (lead magnesium niobate [PMN])*. An applied voltage causes a quadratic increase in the length of the crystal. If the voltage is reversed, the crystal still lengthens. To increase and decrease the length of the crystal—to push and pull on a facesheet—the actuator can be operated around a bias voltage.
  - *Electrostatic actuators.* An applied voltage induces charge between two conducting plates, giving rise to attractive Coulomb force between the negative charges on one plate and the positive charges on the other plate. If one plate is fixed and the other one is free to move, the released plate will be drawn toward the fixed plate. The attractive force between the plates is nonlinear. It varies with the square of the applied voltage and inversely with the square of the plate separation. To keep the plates from being drawn together and touching, a linear spring mechanical restoring force is typically used. Nonetheless, the nonlinear attractive electrostatic force will eventually overcome the linear mechanical restoring force, and the plates will be pulled in or drawn together. The typical operating range is one-third of the original gap before pull-in occurs. The range can be increased by using nonlinear springs or leveraging effects.
  - *Magnetic actuators.* A magnetic field from a current passing through a coil (i.e., a voice coil) interacts with the magnetic field of a permanent magnet, generating an attractive force when the poles are opposite and a repulsive force when they are aligned. The alignment of the electromagnetically generated field can be reversed by reversing the direction of the current flow. Since current must flow through the coil to generate the electromagnetic field, there is constant power dissipation when actuated.
  - *Thermal actuators.* Thermal expansion of a material when it is heated is used for actuation. Typically thermal actuators are able to generate a large force, but they dissipate a lot of power owing to the heating.
  - *Bimorph (piezo, thermal).* A bimorph actuator uses the differential expansion between two materials. The differential expansion can be caused by heating, where the difference in thermal expansion coefficients of the two materials causes them to expand at different rates. The differential expansion can also be caused by the piezoelectric effect, where one piezoelectric material is caused to expand and the other piezoelectric material, bonded to the first, is caused to contract with an applied voltage.
- *Actuator characteristics.* There are a number of characteristics required by the actuators.
  - *Repeatability (hysteresis, go-to command).* Piezoelectric actuators exhibit a hysteresis effect. Where they go depends on the history of where they have been. Typically they exhibit a hysteresis loop that describes the displacement as a function of applied voltage. Hysteresis makes open-loop, go-to command positioning difficult, since the position the actuator goes to depends on where the actuator has been previously. This problem can be overcome

by using closed-loop, feedback control of the actuator to reach a command position, but the actuator will be slowed down by the need to "hunt" for the command position based on the feedback signal. This decreases the bandwidth of systems that use these actuators.

- *Parameter calibration, drift over time.* For go-to operation, it would be helpful to have calibrated parameters that describe the response of the actuator to an applied stimulus (voltage, current, temperature) that do not drift over time. Unfortunately, the parameters that relate stimulus and response, such as the piezoelectric d coefficients, can change over time with changes in temperature and depoling.
- *Feedback.* To control an actuator in a closed-loop feedback system, an error signal must be generated that can be fed back to the control system. Positions of the actuators can be measured with capacitive sensors or strain gauges and fed back to the control system. For wavefront correction, the wavefront error can be measured with a wavefront sensor such as the Shack–Hartmann sensor and the deviance of the wavefront from a reference wavefront fed back to control the actuators on a DM to drive the error to zero.
- *Linearity.* A linear system is desirable to avoid effects such as pull-in for electrostatic actuators, where the nonlinear electrostatic force overwhelms the linear mechanical restoring force, causing the actuator plates to pull together. When the plates touch, they become stuck together. For actuators with microscopic dimensions, surface adhesion forces such as van der Waals attraction can dominate the mechanical restoring force, leading to "stiction."
- *Push–pull (bidirectional actuation).* Generally, it is helpful to have bidirectional actuation that can both push and pull the mirror facesheet. Some actuation mechanisms such as magnetic forces can generate bidirectional forces by reversing the current in a coil, whereas other actuation mechanisms can generate only attractive forces such as electrostatic actuation under voltage control. When only unidirectional forces can be generated, bidirectional actuation can be accomplished by operation around a bias condition. For electrostatic actuation, a bias voltage is applied to the actuators and bidirectional actuation is achieved by increasing or decreasing the applied voltage around the bias point.
- *Temperature sensitivity.* Some actuation mechanisms, such as piezoelectric actuation, are sensitive to temperature. At low temperatures, the piezoelectric gain d decreases, requiring a higher voltage to achieve the same displacement. At high temperatures, the piezoelectric material become depoled, again decreasing the piezoelectric gain coefficient. Once a piezoelectric material has become depoled at a high temperature (i.e., above the Curie temperature), it can be repoled by applying a poling bias.

- *High voltage or drive current.* If the required actuation voltage for an electrostatic actuator is too high, it can exceed the breakdown voltage in air or vacuum, leading to catastrophic failure for the actuator. Typically the actuation voltage will be limited by the drive electronics to a safe value. Similarly, if the drive current in a magnetic or thermal actuator is too high, it can lead to failure mechanisms associated with high power dissipation.
- *Weight.* For DMs that are mounted on actuated mounts such as tip–tilt stages, the weight of the mirror must be comprehended in the tip–tilt mount. Inertial effects can limit the actuation bandwidth for massive mirrors.
- *The response time.* The response time of the mirror determines the bandwidth, the highest temporal frequency that can be controlled.
- *Environment.* Environmental effects such as temperature and relative humidity can be important to the lifetime of a DM. Failure mechanisms such as electrical breakdown can be accelerated under extreme environmental conditions. Corrosion effects can also be accelerated by high applied voltages in high relative humidity where electrochemical effects can occur at interfaces between different materials such as gold and doped polysilicon, which are commonly used materials in

MEMS fabricated using surface micromachining processing. To protect mirrors from environmental effects, they are typically packaged in a hermetically sealed container, with optical access to the mirror surface through an appropriate window that is optically clear and treated with anti-reflection coatings for the operating wavelengths of interest.

- *Lifetime (cycles).* The lifetime of the mirror can be limited by environmental effects if the mirror is exposed to a harsh environment such as high temperature or humidity. It can also be limited by failure mechanisms such as stiction, where an actuator has been displaced sufficiently far that it touches down to the substrate and becomes stuck to it because of surface forces such as van der Waals attraction. DMs fabricated using MEMS processes are particularly vulnerable to this failure mechanism since surface forces can dominate spring restoring forces in the microdomain.

8.5 Conventional Deformable Mirrors Using Piezoelectric and Electrostrictive Actuators

There are a number of different options for DMs that can be used for wavefront correction. They can be broadly categorized as segmented and continuous facesheet mirrors and in terms of the amplitude (stroke) and spatial frequency (order) of wavefront correction they are able to make. The facesheet is the front surface mirror that reflects the wavefront. It can have various metal surface coatings (e.g., aluminum, gold, protected silver) or dielectric coatings to optimize reflection in the wavelength range of interest. Aluminum coatings have high reflectivity in the range of 400–700 nm ($R > 85\%$). Gold coatings have high reflectivity from 700–800 nm ($R > 94\%$) to 800–10,000 nm ($R > 97\%$), and protected silver has high reflectivity in the range from 500–800 nm ($R > 98\%$) to 2,000–10,000 nm ($R > 98\%$). Mirrors for high-power laser applications typically use a Bragg stack mirror coating that consists of alternating layers of high and low index of refraction dielectric layers that are each a quarter wavelength thick.

Although the method of wavefront correction using an Eidophor, as described earlier, works in principle, the astronomical community initially used adaptive mirrors rather than spatial light modulators for making wavefront corrections. The first mirrors used piezoelectric actuators to deform a thin glass mirror facesheet. The actuators were large, resulting in a large pitch (5–7 mm) between the actuators when they were assembled into an array, and the resulting mirrors were expensive since they were hand assembled into an array. An example of the DM that was used in the AO system for the Keck 10 m telescope is shown in Figure 8.15. The mirror, manufactured by Xinetics Inc., has 349 individual actuators

146 mm clear aperture   349 actuators on 7 mm spacing

FIGURE 8.15 Deformable mirror used in the Keck Observatory AO system. (Credit: Peter Wizinowich, W.M. Keck Observatory.)

arranged on a square array with 7 mm spacing. The mirror has a 146 mm clear aperture. Since the mirrors were hand assembled, the cost for the mirrors was approximately $1000 per actuator (Ealey 1994), resulting in DM that costs in the hundreds of thousands of dollars.

The first AO DMs for astronomy were hand assembled using piezoelectric or electrorestrictive actuators (PZT [piezoelectric] or PMN [electrorestrictive]) that were bonded to either segmented mirrors or a thin glass facesheet mirror. Since piezoelectrics have a typical displacement of a few angstroms/volt; they are usually fabricated as discs and stacked in series to obtain a displacement of few micrometers for 100 V. One challenge for piezoelectrics are that they exhibit a hysteresis phenomenon, where the displacement depends on the past history of actuator displacements. This tends to slow down a mirror under closed-loop feedback control since it must hunt for the demanded position and makes go-to open-loop positioning a challenge. Piezoelectric mirrors with 37–941 actuators at 7 and 5 mm spacings have been manufactured by the Xinetics corporation. These provide 3–8 µm of stroke. Northrop Grumman/AOA Xinetics has recently designed a "photonics module" approach where the grid of the actuators is machined out of a single piezoelectric ceramic substrate. This enables a smaller actuator pitch of 2.5 and 1.0 mm and, therefore, high-density actuation over a given size mirror. A continuous glass facesheet is bonded on top of the module and subsequently polished and coated to complete the high-density DM. Xinetics has delivered DMs based on this technology containing 1000 and 4000 channels at a 1 mm spacing and 37 and 349 channels at a 2.5 mm spacing. A photo of a high-density module with a 1 mm pitch is shown in Figure 8.16. The Cilas corporation in France is also developing a high-stroke, high-order DM. A 9 × 9 prototype developed under a design study for the 30-meter telescope (TMT) is shown in Figure 8.17. The TMT AO system baseline design requires two DMs with 63 × 63 and 75 × 75 arrays of actuators.

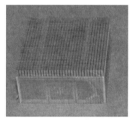

FIGURE 8.16 Xinetics high-density (32 × 32) photonic module. The actuator pitch in this image is 1 mm. The array size is scalable to 9216 channels. (Credit: John A. Wellman, David D. Pearson, Jeffrey L. Cavaco, Northrop Grumman/AOA Xinetics.)

FIGURE 8.17 Cilas high-stroke, high-order deformable mirror prototype being developed for 30 m telescope. Shown here is a 9 × 9 subscale prototype DM with a stroke of 11 µm, surface flatness of 13 nm RMS, and hysteresis of 5%–6%. (Credit: Compagnie Industrielle des Lasers [CILAS].)

8.6 Microelectromechanical System Deformable Mirrors

The use of MEMS DMs in astronomy and vision science has been reviewed by Olivier et al. (2005). The application of AO in biological imaging requires high-performance wavefront correctors at low cost. These requirements favor MEMS DMs over the piezoelectric DMs that have historically been used in astronomy. These mirrors are hand assembled from arrays of piezoelectric actuators, which limits the pitch to several millimeters and the cost to approximately $1000 per actuator. MEMS technology uses semiconductor batch fabrication of silicon wafers, or similar substrates and processing, to lower the cost to approximately $150 per actuator and to increase the actuator number, currently as high as 4000 actuators, and density, currently on the order of 6 actuators per mm^2. A typical semiconductor process is shown in Figure 8.18. The process is cyclic. First, a thin film is deposited on the wafer surface using thin film deposition techniques. A uniform photosensitive polymer (photoresist) is then deposited and exposed to light from a mask that contains the pattern that is desired on the thin film. The photoresist is developed to obtain the desired pattern. The pattern in the photoresist is then transferred to the thin film using an etching technique for removal of the unwanted material and then the photoresist is removed. The difference between semiconductor processing and MEMS processing is that MEMS processes usually require fewer processing cycles and use thicker films and deeper etches for patterning since mechanical rather than electrical components are being fabricated. Another difference is that in some MEMS processes, such as surface micromachining as described below, the final processing step involves the removal of an underlying "sacrificial" material, such as an oxide layer, to release mechanical structures that have been formed in the overlying structural layers, such as polysilicon. After the fabrication processing cycles, the wafers are sectioned into individual die and assembled into a package. Electrical connections between the die and the package are made using wire bonding. Finally, the package is hermetically sealed to protect the device from the external environment (e.g., humidity, oxidation).

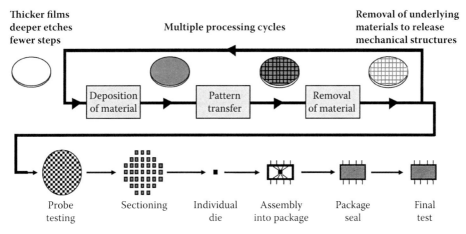

FIGURE 8.18 Semiconductor fabrication cycles for microelectronics and micromechanics. A material is deposited onto a wafer using thin-film deposition techniques. For microelectronic processing, only the electrical properties of the material need to be optimized. For micromechanical processing, both the electrical and the mechanical properties need to be optimized. Micromechanical fabrication can also include an underlying sacrificial material that is removed to release mechanical structures so that they can move. A typical micromechanical process might involve eight deposition and patterning cycles, whereas a micromechanical process could require 28 cycles. After deposition and patterning of all the layers, the wafer is tested, diced (sectioned), and packaged. The sacrificial layers can be removed at either the wafer or die level. (Reprinted from Karen W. Markus and Kaigham J. Gabriel, IEEE Trans. Electron Devices, MEMS, The Systems Function Revolution, ©1967 IEEE. With permission.)

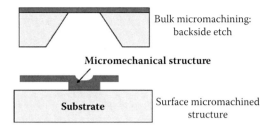

FIGURE 8.19 *Top*: Bulk micromachined wafer etched from the backside with an anisotropic etch that stops on a gray surface layer. The gray surface layer can be a dopant layer diffused in from the front side of the wafer that causes the anisotropic etch to stop, or it can be a separate layer, such as the device layer of an SOI wafer. The backside etch can then stop on the buried oxide layer. *Bottom*: Surface micromachined structure. A micromechanical structure is formed from a structural layer, such as polysilicon, that is deposited on a substrate layer. To release the micromechanical structure, it can be deposited on a patterned sacrificial layer, such as an oxide, that is removed in a chemical etch. The micromechanical structure is attached to the substrate by patterning a hole in the sacrificial layer, where the structural layer can come into contact with the substrate.

Two micromechanical processes that have been used for fabrication of MEMS DMs are bulk and surface micromachining. Cross sections of these two processes are shown in Figure 8.19. In bulk micromachining, the material is removed from the bulk of the wafer to leave behind the desired structure. An example is a thin silicon membrane that can be used as a mirror in a membrane DM. To obtain a thin supported membrane, the wafer can be etched using an anisotropic etch that stops on a doped layer, as shown by the gray layer in Figure 8.19 (top). Alternatively, a silicon-on-insulator (SOI) wafer can be used and the etch stopped on the buried oxide layer, releasing a thin membrane in the SOI device layer. The wafer with the clamped membrane can be bonded with a thin spacer layer to a second wafer that has patterned electrodes.

8.6.1 Microelectromechanical System Polysilicon Surface Micromachining Fabrication Process

The development of MEMS DMs was a major goal for the NSF Science and Technology Center for Adaptive Optics (CfAO) that spanned a decade of research efforts from 1999 to 2009 (Krolevitch 2003). A number of designs and fabrication methods were investigated including both continuous and segmented facesheet mirrors. Both designs used a form of the surface micromachining fabrication process, so we consider that process in detail here. By understanding the process, the limitations of mirrors fabricated using this process can be better understood. A schematic diagram of the surface micromachining process is shown in Figure 8.20.

The process begins by deposition of a surface insulating material, enabling different structures that are deposited on the substrate to be electrically insulating. In Figure 8.20a, silicon nitride is used as the surface material. Next a spacer layer is deposited as shown in Figure 8.20b. This layer is removed at the end of the process to release mechanical components defined in a structural material (polysilicon). In this case, a phosphosilicate glass is used because it etches quickly in wet chemical etch (hydrofluoric acid) that does not etch the structural material. The spacer layer is patterned to define an anchor area as shown in Figure 8.20c. This area will allow contact between the structural material and the surface material on the substrate, enabling parts defined in the structural layer to remain attached to the substrate following the sacrificial etch. In Figure 8.20d, polysilicon is deposited for the structural material. The part of the structural material that fills in the anchor area etched through the spacer material will keep the released structural material attached to the substrate. The part of the structural material that is deposited on the spacer material will be released from the substrate during

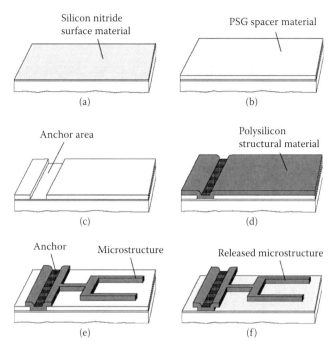

FIGURE 8.20 Polysilicon surface micromachining. (a) An insulating surface layer, such as silicon nitride, is deposited on the substrate. (b) A sacrificial spacer layer such as phosphosilicate glass (PSG), which can be etched quickly, is deposited. (c) A hole is cut through the sacrificial layer to the insulating layer, where the structure is to be anchored to the underlying surface material. (d) The structural layer, in this example polysilicon, is then deposited. The deposition is conformal and fills in the hole that had previously been cut through the sacrificial layer. The structural layer is (e) patterned and (f) then released. (Reprinted from Gary K. Fedder, ITC International Test Conference, MEMS Fabrication, ©1967 IEEE. With permission.)

the sacrificial etch. The structural material is then patterned to define a microstructure, as shown in Figure 8.20e. Finally, the spacer material is etched to realize a released microstructure, as shown in Figure 8.20f. This process can be repeated a number of times to fabricate a number of structural layers. A cross section of the process that uses three structural layers and two sacrificial layers is shown in Figure 8.21 (Cowen 2011).

One of the limitations of the polysilicon surface micromachining process is the thickness of the sacrificial space layer, as shown in Figure 8.22. For practical considerations, it is limited to several micrometers owing to the thin film deposition processing that is used in the fabrication process. This spacer layer sets limits on the maximum stroke that can be used unless the structural layer is lifted out of plane, as described below for the Iris AO segmented DM. Alternatively, bulk micromachining and wafer bonding can be used to fabricate a large gap, as described below for bulk etched membrane mirrors.

8.6.2 Electrostatic Actuation

Electrostatic actuators are commonly used in MEMS devices as they scale well in the microdomain, use very little power, and are straightforward to fabricate in a number of different processes. Two common forms are parallel plate actuators and comb drive actuators. The parallel plate actuator is a capacitor

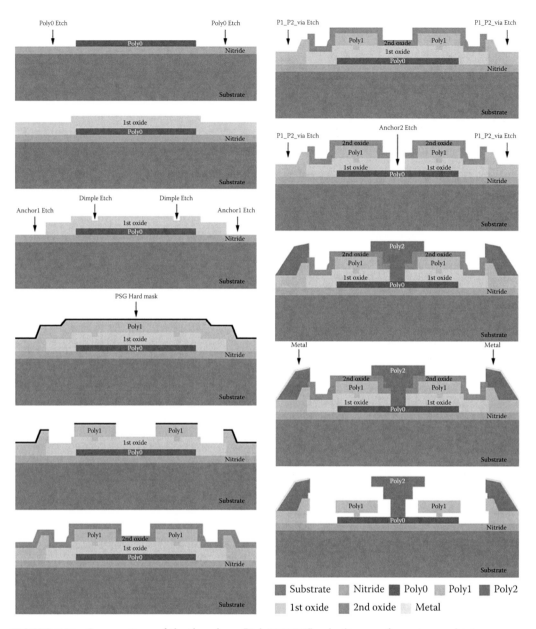

FIGURE 8.21 Cross sections of the three-layer "PolyMUMPS" polysilicon surface micromachining process offered by MEMSCAP. Polysilicon and oxide layers are deposited and patterned in a cyclic process, with anneal steps of the doped sacrificial oxide between polysilicon depositions. POLY0 is an electrical layer that is not released. POLY1 and POLY2 are structural layers that can be released. The deposition and patterning steps shown here result in a polysilicon wheel, defined in POLY1, that is constrained by a hub, defined in POLY2. Dimples defined in POLY1 keep the wheel from becoming stuck to the POLY0 layer. (Reprinted with permission from MEMSCAP Inc.)

with one of the plates released so that it is able to move, as shown in Figure 8.23. The relationship between the capacitance C, voltage V, and charge Q for a capacitor is given by

$$C = \frac{Q}{V} \qquad (8.6.2.1)$$

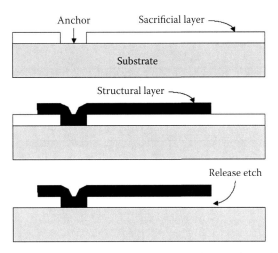

FIGURE 8.22 Cross section of the surface micromachining process showing the gap between the structural layer and the substrate that is defined by the sacrificial layer after the release etch. The thin-film deposition processing that is used in surface micromachining limits the sacrificial layer to several micrometers in thickness.

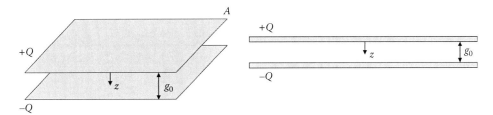

FIGURE 8.23 Parallel plate capacitor with an area A, charge Q, and an initial gap g_0. When connected to a voltage source, one plate acquires a negative charge $(-Q)$, and the other plate acquires a positive charge $(+Q)$, leading to an attractive force between the plates. The top plate moves in the z direction and the gap is decreased from its initial value of g_0.

where the capacitance of a parallel plate capacitor is given by

$$C = \varepsilon_0 \frac{A}{g} = \varepsilon_0 \frac{A}{g_0 - z} \tag{8.6.2.2}$$

ε_0 is the dielectric permittivity of free space (8.85×10^{-12} F/m), g is the distance between the plates (m), and A is the area of the plates (m²).

The incremental work dU done in charging the capacitor by transferring an incremental charge dQ from one plate to the other through a voltage V is given by

$$dU = V \, dQ \tag{8.6.2.3}$$

Substituting for V,

$$dU = V \, dQ = \frac{Q \, dQ}{C} \tag{8.6.2.4}$$

Integrating dU for the total work U,

$$U = \int \frac{Q\,dQ}{C} = \frac{1}{2}\frac{Q^2}{C} = \frac{1}{2}CV^2 \qquad (8.6.2.5)$$

To find the force generated by a parallel plate actuator, we can use the principle of virtual work by considering the work done when the plates of a capacitor are moved a small distance Δz closer together when a constant voltage V, set by a battery, is applied between the plates. Since the plates have opposite charge, the force between the plates is attractive. The decrease in the gap causes the capacitance of the capacitor to increase by ΔC and an amount of charge ΔQ to be transferred from the battery to the capacitor, increasing its stored energy. We can then balance the work done by the battery to transfer the charge to the work done by the actuator and the potential energy stored in the capacitor when the plates move closer together:

$$\Delta W_{\text{battery}} = \Delta W_{\text{capacitor}} + \Delta U_{\text{capacitor}} \qquad (8.6.2.6)$$

$$V\Delta Q = F\Delta z + \frac{1}{2}V^2\Delta C \qquad (8.6.2.7)$$

Using Equation xx to substitute for ΔQ at constant V,

$$Q = CV \rightarrow \Delta Q = V\Delta C|_V \rightarrow V\Delta Q = V^2\Delta C \qquad (8.6.2.8)$$

$$V^2\Delta C = F\Delta z + \frac{1}{2}V^2\Delta C \qquad (8.6.2.9)$$

$$F\Delta z = \frac{1}{2}V^2\Delta C \qquad (8.6.2.10)$$

$$F = \frac{1}{2}V^2\frac{\Delta C}{\Delta z} \qquad (8.6.2.11)$$

We can calculate the force by taking the derivative of the capacitance with respect to the separation between the plates:

$$\frac{\partial C}{\partial z}\bigg|_V = \frac{\partial}{\partial z}\left(\varepsilon_0 \frac{A}{g_0 - z}\right) = \varepsilon_0 \frac{A}{(g_0 - z)^2} \qquad (8.6.2.12)$$

So that the electrostatic force is given by

$$F_e = \frac{1}{2}V^2\frac{\Delta C}{\Delta z} = \frac{\varepsilon_0 A}{2}\frac{V^2}{(g_0 - z)^2} \qquad (8.6.2.13)$$

8.6.3 Mechanical Restoring Force

A spring is typically used to apply a mechanical restoring force F_{m} for electrostatic actuators, as shown in Figure 8.24. The spring can be linear, following Hooke's Law,

$$F_{\text{m}} = -kz \qquad (8.6.2.14)$$

where k is the spring constant and z is the distance to which the spring is either stretched or compressed. As described below, in some situations it can be useful to use a nonlinear spring where the restoring force does not vary linearly with the displacement.

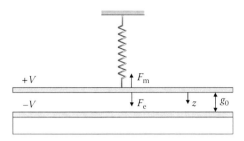

FIGURE 8.24   Parallel plate electrostatic actuator with a mechanical spring that provides a restoring force F_m in opposition to the attractive electrostatic force F_e. The initial gap is g_0.

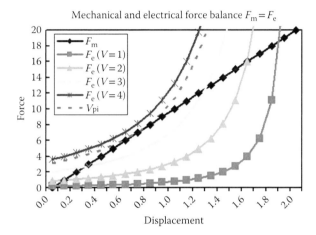

FIGURE 8.25   Graphical solution for balancing the mechanical and electrical forces. The capacitor has an initial gap g_0 equal to 2.1 μm.

The force balance between the electrostatic force that pulls the released plate down toward the fixed counter electrode and the mechanical restoring force that pulls holds them apart can be determined graphically as shown in Figure 8.25. The electrostatic force is shown for a few different voltages. There are two solutions for low voltages, and no solutions for the highest voltage shown. At a critical voltage, called the pull-in voltage, there is only a single solution. The electrostatic force for this single solution is shown as a dashed line. If the voltage is increased further, the nonlinear electrostatic force is greater than the mechanical spring force and the two plates pull-in and touch. At the critical voltage, the electrical and mechanical forces are equal:

$$F_m = F_e \tag{8.6.2.15}$$

$$kz = \frac{\varepsilon_0 A}{2} \frac{V^2}{\left(g_0 - z\right)^2} \tag{8.6.2.16}$$

The slopes of the electrical and mechanical forces are also equal:

$$\frac{\mathrm{d}F_m}{\mathrm{d}z} = \frac{\mathrm{d}F_e}{\mathrm{d}z} \tag{8.6.2.17}$$

$$k = \varepsilon_0 A \frac{V^2}{\left(g_0 - z\right)^3} \tag{8.6.2.18}$$

Substituting for *k* and solving for *z*,

$$\varepsilon_0 A \frac{V^2}{(g_0 - z)^3} z = \frac{\varepsilon_0 A}{2} \frac{V^2}{(g_0 - z)^2} \tag{8.6.2.19}$$

$$z = \frac{g_0}{3} \tag{8.6.2.20}$$

When one-third of the initial gap has been closed, the plates snap together or pull-in. This pull-in instability limits the useful range of parallel plate electrostatic actuators with linear springs. For parallel plate electrostatic actuators formed in surface micromachining processes, the initial gap is defined by the sacrificial layer thickness, which is practically limited to a few micrometers, so that the useful actuation range is typically less than a micrometer, unless a nonlinear spring or leveraged bending is used (Hung 1999). To find the pull-in voltage, the gap at pull-in, $g_0/3$, can be substituted into Equation 8.6.2.18 and solved for the voltage:

$$k = \varepsilon_0 A \frac{V^2}{(g_0 - z)^3} \bigg|_{z = g_0/3} = \varepsilon_0 A \frac{V^2}{(2g_0/3)^3} = \frac{27\varepsilon_0 A}{8g_0^3} V^2 \tag{8.6.2.21}$$

$$V_{\text{pull-in}} = \sqrt{\frac{8kg_0^3}{27\varepsilon_0 A}} \tag{8.6.2.22}$$

8.6.4 Electrostatically Actuated Membrane Mirrors

Example of an MEMS DM that uses bulk micromachining is the membrane mirror developed at Delft University, which was later commercialized by Flexible Optical B.V., now known as Flexible Optical. An aluminum-coated membrane is defined in a silicon chip using bulk micromachining. The silicon chip is then bonded to a substrate with control electrodes through a spacer layer. The spacer layer defines a gap across which a bias voltage is placed, which causes the membrane to deflect under electrostatic forces (Figure 8.26).

A similar approach is used by Agile Optics in their membrane mirror, as shown in Figure 8.27. Here a silicon frame supports a silicon nitride membrane that is approximately 1 μm thick. Silicon nitride is an insulating layer that is typically used as masking material for wet anisotropic bulk micromachining

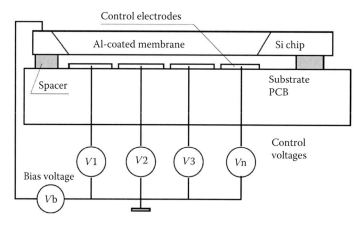

FIGURE 8.26 MEMS membrane mirror fabricated with bulk micromachining. A silicon chip with an aluminum-coated mirror membrane supported along the edges is bonded to a substrate with a spacer layer. The thickness of the spacer layer defines the gap across which an applied bias voltage induces an electrostatic field, causing the mirror membrane to deflect. The control electrodes can be individually biased to control the shape of the mirror. (Credit: Flexible Optical B.V., The Netherlands.)

of silicon. This membrane is coated with either an aluminum or a multilayer dielectric film. The bottom of the mirror layer is metalized with conductive gold to define the counter electrode for actuation of the mirror. The mirror is actuated by applying voltage to the conductive gold actuator pads defined on the silicon pad array substrate. Figure 8.27 (bottom) shows a cross section of the membrane structure, including the top aluminum reflective coating, the middle structural silicon nitride membrane, and the bottom gold conductive coating. A thin chromium layer is sometimes used to increase adhesion for the aluminum reflective layer and the gold conductive coating. The total thickness of the metalized membrane is approximately 1.25 μm thick. The silicon frame chip is bonded to the silicon pad array substrate using flip-chip solder bonding with precision spacer beads to define the gap.

Push–Pull Membrane Mirror: Adaptica has designed and fabricated a push–pull membrane DM by adding additional transparent electrodes over the top side of the mirror membrane, in addition to the usual backside electrodes that are used in the membrane mirrors from Flexible Optical B.V. and Agile Optics, as described earlier. A cross section of the Adaptica mirror is shown in Figure 8.28. By using

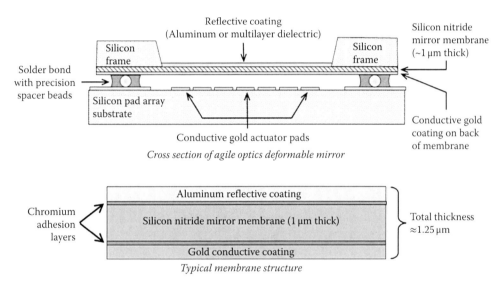

Cross section of agile optics deformable mirror

Typical membrane structure

FIGURE 8.27 MEMS deformable membrane mirror from Agile Optics. *Top*: A bulk micromachined chip provides a silicon frame that supports a silicon nitride mirror membrane. The silicon nitride has a reflective coating (aluminum or a multilayer dielectric) on the front side of the mirror and a gold counter electrode on the back side of the mirror. The silicon frame chip is flip-chip bonded to a silicon pad array substrate using precision spacer beads to define the gap between them. The silicon pad array substrate has conductive gold actuator pads that can be biased to deflect the mirror membrane. *Bottom*: A cross section of the metalized silicon nitride mirror membrane. (Credit: From Justin Mansell, Active Optical Systems, LLC. With permission.)

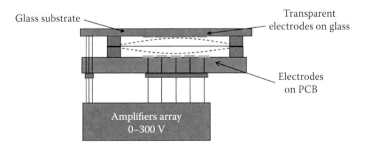

FIGURE 8.28 Push–pull mirror from Adaptica. In addition to backside electrodes that are defined on the bottom side PCB, there are also upper electrodes on the front side defined in transparent conducting indium tin oxide (ITO). (Credit: Adaptica, Italy.)

actuation in both directions, the stroke of the mirror can be increased and it can be actuated from its neutral position (i.e., no applied voltage) rather than from an intermediate position as is typically obtained by the application of an offset bias for unidirectional actuation (Bonora 2006, 2010).

8.6.5 Magnetically Actuated Membrane Mirrors

The membrane mirrors from Imagine Optics and ALPAO use electromagnetic actuators to deform a mirror membrane that is clamped around the edge. The membrane has small magnets bonded to it that can be attracted or repelled by a current induced in opposing induction coils, as shown in Figure 8.29.

The magnetically actuated DM52 (52 actuator) deformable membrane mirror from ALPAO is able to generate the Zernike polynomials up through the fourth order with over 10 μm of surface deformation, as shown in Table 8.2. Higher-order mirrors with additional actuators (88, 97, 241) are able to make higher-accuracy corrections and/or larger amplitude Zernike corrections, as shown in Figure 8.30.

ALPAO also offers a high-speed version of their DM. The ALPAO 97 high-speed DM has a settling time of less than 1 ms, as shown in Figure 8.31, offering high-bandwidth (>900 Hz) correction for making fast corrections of rapidly changing phenomenon.

8.6.6 Electrostatically Actuated Continuous Facesheet Microelectromechanical System Mirrors

MEMS DMs are so named by the MEMS fabrication processes that are used to fabricate them. Rather than the bulk micromachining and bonding process that is used for membrane mirrors, MEMS mirrors typically use the surface micromachining process. This allows for fabrication of complex structures, enabling more localized actuation for high-order correction.

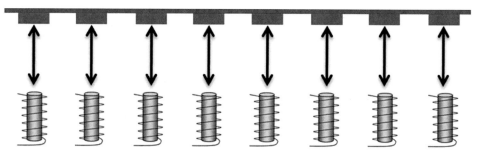

FIGURE 8.29 Magnetically actuated deformable membrane mirror. An array of magnetic "voice" coils each creates a localized magnetic field when a current is passed through them. The direction of the field can be controlled by the direction of the current. An array of permanent magnets is bonded to a flexible membrane. When the electromagnets are powered, they induce a force, either attractive or repulsive, between the coil and the permanent magnet, pushing or pulling on the membrane.

TABLE 8.2   ALPAO Hi-Speed Deformable Mirror Characteristics

Array Size (Order)	37	52	69	97	88	241
Stroke (μm) Tip–Tilt		±60			±20	
Stroke (μm) 3 × 3 Array			>30		>14	
Stroke (μm) Inter-Actuator		>3			>3	
Aperture Diameter (mm)	7.5	9.0	10.5	13.5	15	20
Pitch (mm)		1.5			2.5	
Bandwidth (Hz)		>750			>750	

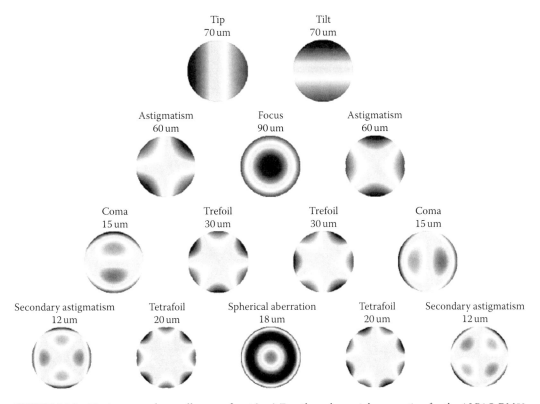

FIGURE 8.30 Maximum peak-to-valley wavefront (μm) Zernike polynomials generation for the ALPAO DM52 deformable mirror. (Credit: ALPAO, France.)

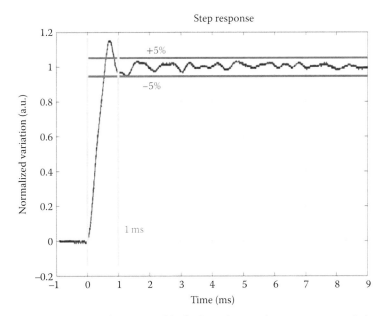

FIGURE 8.31 The ALPAO Hi-Speed DM97 enables high-stroke wavefront correction with fast stabilization as needed for correcting aberrations for rapidly changing phenomenon. (Credit: ALPAO, France.)

Boston Micromachines commercialized continuous and segmented MEMS DMs based on the prototypes developed at Boston University that used a three-layer polysilicon surface micromachining process (Bifano 1996, Krishnamoorthy 1999, Bifano 2002, Perreault 2002, and Cornelissen 2006 and 2009). The prototypes used a standard process and the commercialization involved revisions to the standard process to make commercial products. Cross sections of the Boston Micromachines DMs are shown in Figure 8.32.

Boston Micromachines currently offers MEMS continuous facesheet DMs with 32 actuators in a 6×6 array (i.e., without the four corner actuators), 140 actuators in a 12×12 array, and 1020 actuators in a 32×32 array. These are currently the highest-order MEMS continuous facesheet mirrors on the market. Nonetheless, they have a stroke that is limited to 1.5–5.5 µm, as shown in Table 8.3, and approximately 0.5 µm of stroke must be used to flatten the initial curvature in the mirror. In addition, a custom-built 4096 actuator array with a 25 mm aperture and 4 µm of stroke is under development for the Gemini Planet Imager (GPI) (Cornelissen 2006, 2009, 2012, Poyneer 2011).

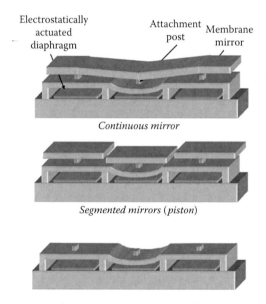

FIGURE 8.32 *Top*: A continuous membrane mirror that is attached by posts to an electrostatically actuated diaphragm that is clamped to the substrate. Below the electrostatically actuated diaphragm, isolated counter electrodes are used to selectively actuate portions of the mirror. *Middle*: The cross section shows a segmented mirror that has piston only (up-down) actuation. *Bottom*: The cross section shows just the actuators. The middle actuator is deflected in each of the cross sections. (Credit: Boston Micromachines, MA.)

TABLE 8.3 Specifications for Boston Micromachines Continuous Facesheet Deformable Mirrors

Array Size (Order)	32 (Mini-DM)			140 (Multi-DM)			1020 (Kilo-DM)
Stroke (µm)	1.5	3.5	5.5	1.5	3.5	5.5	1.5
Aperture (mm)	1.5	2.0	2.25	3.3	4.4	4.95	9.3
Pitch (µm)	300	400	450	300	400	450	300
Response Time (µm)	20	100	500	<20	<100	<500	<20
Inter-actuator Coupling (%)	15	13	22	15	13	22	15

The 32×32 array (kilo-DM) is also offered as a segmented mirror with piston (vertical) actuation on each segment for use as a spatial light modulator.

8.6.7 Electrostatically Actuated Segmented Facesheet Microelectromechanical System Mirrors

Segmented mirrors have small segments that can be independently actuated, typically in a piston mode, with motion limited to translation along the normal to the mirror surface, or in piston tip–tilt mode, where the displacement along both the normal (piston) and the slope (tip–tilt) within the plane of the segment are controlled. Typically segmented mirrors require higher order (i.e., more actuators) to make the same amount of wavefront correction as a continuous facesheet mirror, as shown by the fitting error coefficients in Table 8.1 (Hardy 1998). Since the mirror segments are independent, they do not influence their neighbors when they are actuated. On the other hand, since the segments can move independently, there can be a large jump in the phase between independent actuators. An additional challenge is the space between the mirror segments. Unless they are closely spaced, the fill factor can be limited. Typical fill factors of leading segmented mirrors are from 98 to 99%.

There are currently two manufactures of segmented MEMS mirrors. Boston Micromachines offers a segmented, piston-only MEMS DM with 1020 actuators, 1.5 μm of stroke with actuators on a pitch of 300 μm. The segments have a response time that is less than 20 μs. They also offer a low-latency driver for this mirror with a 34 kHz frame rate. Iris AO offers segmented mirrors with 163 hexagonal piston-tip segments, 350 μm on a side, with a pitch of 600 μm from center to center. Each tip–tilt segment has three actuators for a total of 489 actuators. The stroke is 5–8 μm. They also offer smaller DMs as described in Table 8.4. The larger stroke is made possible by lifting the mirror segments out of plane on actuator platforms using temperature-insensitive bimorph flexures, as shown in Figure 8.33.

TABLE 8.4 Iris AO Segmented Deformable Mirror Specifications

Model	Number of Actuators	Number of Segments	Maximum Stroke (μm)	Maximum Tilt Angle (mrad)	Minimum Frequency Response (kHz)	Aperture (mm)
PTT111-5	111	37	5	±5	2	3.5
PTT111-8	111	37	8	±8	2	3.5
PTT489-5	489	163	5	±5	2	7.7
PTT489-8	489	163	8	±8	2	7.7

Each mirror segment has three actuators to provide tip–tilt actuation.

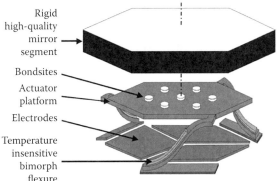

Rigid high-quality mirror segment

Bondsites

Actuator platform

Electrodes

Temperature insensitive bimorph flexure

FIGURE 8.33 Segmented deformable mirror from Iris AO. (Credit: Michael Helmbrecht, Iris AO, CA.)

The high-quality mirror segments themselves are made out of thick single-crystal silicon that is transferred and bonded to the mirror platforms following the polysilicon surface micromachining process that forms the electrodes, bimorphs, and actuator platforms. The thick single-crystal mirror segments allow the deposition of protected silver or dielectric coatings without significant deformations (<20 nm RMS typical) due to stress or stress gradients in the deposited coatings. The mirror surfaces remain flat under temperature changes (0.56 nm/°C peak-to-valley). The heights of the bimorphs above the substrate are also temperature insensitive (14 nm/°C, σ = 0.8 nm/°C). The mirrors have a response time (20–80%) that is less than 150 μs. The fill factor is greater than 98% (6 μm gaps between segments). The mirror driver comes precalibrated, enabling go-to open-loop positioning of the mirror segments.

8.7 High-Stroke, High-Order "Woofer-Tweeter" Two-Mirror System

As described earlier, membrane DMs that have high stroke, such as the mirrors from Flexible Optical B.V., ALPAO, and Imagine Optics, usually provide only low-order correction. Since each actuator is not supported individually, they are strongly coupled, so actuation of one leads to movement of the neighbors, limiting the ability of membrane mirrors to perform high-spatial-frequency corrections. The membrane is like a trampoline that is supported along the edges. If you step on the trampoline, the deformation is extended, as the support comes only from the edges. On the other hand, MEMS continuous facesheet mirrors, such as the Boston Micromachines DM, can provide high-order, but only low-stroke, correction. Since each actuator is supported individually by two nearby posts that surround the electrode, as shown in Figure 8.32, the coupling between actuators is much lower than the membrane DMs. The Boston Micromachines continuous facesheet MEMS DM is support in a way that is similar to a mattress with box springs. Deformation of one area of the mattress does not strongly influence the nearby neighbors. Typically the coupling is around 10–20%. Although these mirrors provide high-order correction, they typically have only low-stroke actuation capability, since the gap is defined by a thin-film sacrificial oxide that has a limited thickness, and only one-third of the initial gap can be used for electrostatic actuation before the pull-in instability is reached. If the mirror actuator pulls-in, it can come into contact with the substrate and remain stuck down, since surface forces such as van der Waals interactions are dominant at small dimensions. This phenomenon is so prevalent in MEMS devices that it even has a name: stiction. Once an actuator has been stuck down, it is usually hard to release it without destroying the mirror. While there are some approaches to overcome the limitation to using only one-third of the initial gap, such as the use of nonlinear springs or leveraged bending (Hung 1999), these approaches can drive up the actuation voltage until electrical breakdown occurs. Since applications often call for highstroke, high-order correction, which is not presently available in membrane or MEMS DMs (Morzinski 2009). This combination is shown in Figure 8.34.

The correction of wavefront aberrations is shown in Figure 8.35. The first mirror, a bimorph mirror, corrects the low-order, high-stroke aberrations. In analogy to acoustic speakers it is called a woofer. The light is then relayed to a second MEMS mirror that corrects the high-order aberrations, which typically has a lower magnitude, so that they can be corrected with a smaller-stroke "tweeter" mirror (Hu 2006, Lavigne 2008, Zou 2008). While this combination can provide high-order, high-stroke correction, it requires reflection off two mirror surfaces and requires two mirror drivers, decreasing optical performance while increasing the complexity of the drive electronics.

8.8 High-Stroke, High-Order Microelectromechanical System Mirrors

To obtain high-stroke and high-order correction using a single MEMS DM, a high-aspect ratio micromachining (HARM) process has been used for mirror fabrication. By using a high-aspect ratio process, the stroke limitations that arise from the use of thin sacrificial films in the surface micromachining

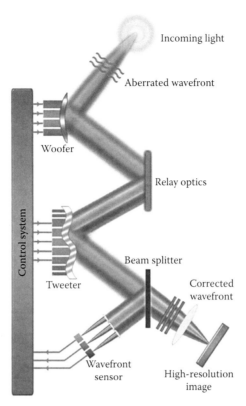

FIGURE 8.34 A woofer–tweeter system of deformable mirrors consisting of a high-stroke, low-order woofer membrane mirror with a low-stroke, high-order tweeter MEMS mirror. The function of the relay optics is to make images of the pupil at each deformable mirror. (Credit: Boston Micromachines.)

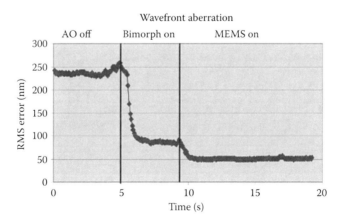

FIGURE 8.35 Measured wavefront aberration over a 6 mm pupil before and after AO compensation with a woofer–tweeter dual deformable mirror. Initially the adaptive optics system is off. Then the woofer (bimorph) mirror is activated. Finally the tweeter (MEMS) mirror is activated. (From Chen, D., et al., *J Opt Soc Am. A*, 24, 1305–1312, 2007a; Chen, D. C., et al., High-resolution adaptive optics scanning laser ophthalmoscope with dual deformable mirrors for large aberration correction, SPIE Photonics West 2007. San Jose, CA, United States, January 20–25, 2007, Vol. 6426, *Ophthalmic Technologies XVII*, edited by Fabrice Manns, Per G. Soederberg, Arthur Ho, Bruce E. Stuck, Michael Belkin, p. 64261L, 2007b.)

process described above can be avoided. In a HARM process, thick sacrificial layers can be obtained, enabling larger gaps and longer mirror strokes. To increase the vertical dimensions, thick film polymer photoresists are used as molds for electrodeposition of structural and sacrificial layers. One process, called LIGA (Becker 1986), a German acronym for *Lithographie, Galvanoformung, Abformung* (lithography, electroplating, and molding), uses synchrotron radiation to expose a thick resist. The synchrotron radiation is highly collimated and forms vertical sidewalls in polymethyl methacrylate (PMMA) photoresists that can be hundreds of micrometers thick. Using this process, feature sizes as small as 0.1 μm can be replicated using the developed PMMA layer as a mold for electrodeposition of metals such as copper, gold, and nickel. Since there are etches with high selectivity between these metals, one metal can be used as the structural layer and one layer used as the sacrificial layer. An overview of a LIGA process is shown in Figure 8.36, where a small metal gear has been fabricated.

The LIGA process can also be used to electroform two different metals, one for a structural layer and one for a sacrificial layer, and the process can be repeated after the planarization step to build up more complex-layered structures. An example is the electrochecmical fabrication (EFAB) process where one metal layer, such as copper, is used for the sacrificial material and a different metal layer, such as nickel, is used for the structural layer (Cohen 2001). A schematic diagram of the EFAB process is shown in Figure 8.37.

A high-aspect-ratio micromachining process has been used to fabricate large-stroke DMs for AO (Fernandez 2010). By using a HARM fabrication process, a thick sacrificial layer (copper) could be electrodeposited on top of gold structural layers to enable a large electrostatic gap for high-stroke actuation.

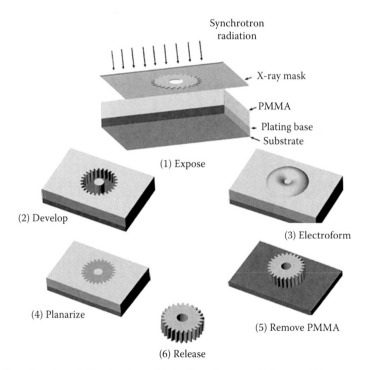

FIGURE 8.36 Overview of the LIGA (*Lithographie, Galvanoformung, Abformung*) high-aspect-ratio micromachining process. (1) A highly collimated synchrotron radiation is used to expose a thick polymethyl methacrylate (PMMA) photoresist on a plating base through an X-ray mask. (2) The PMMA is then developed to form a mold for electroplating. (3) Using electroforming, this mold is then filled with a metal. (4) Using mechanical polishing techniques, the PMMA and fill metal are then planarized. (5) The PMMA is then removed, leaving the electroformed metal part attached to the plating base. (6) The metal part is then released from the plating base.

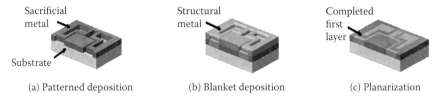

(a) Patterned deposition   (b) Blanket deposition   (c) Planarization

FIGURE 8.37 Overview of the EFAB process. (a) A LIGA process is used to electrodeposit a patterned sacrificial metal, such as copper, on top of a substrate. (b) A blanket electrodeposition of a structural metal, such as nickel, occurs on top of the sacrificial metal. (c) Using polishing techniques, the structural metal is then polished back to the sacrificial layer, completing the first layer. The process can be repeated in cycles to build up multiple sacrificial and structural layers to form more-complex structures. (Credit: EFAB process.)

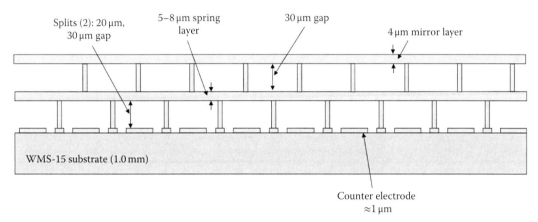

FIGURE 8.38 Cross section of a high-stroke deformable mirror for fabrication in a HARM process. The mirror was fabricated in two splits. One split had a 20-μm-thick gap between the counter electrode and spring layers. The other split had a 30-μm-thick gap. The structural material was gold, and the sacrificial material was copper. The mirror was formed on top of a 1-mm-thick glass–ceramic substrate (WMS-15) that was thermally matched to the gold structural layers. Not shown in this figure are small etch release holes that were included to allow the etchant to remove the sacrificial layers of copper. (From Fernandez, B. R. and J. Kubby, *J Micro/Nanolitho MEMS MOEMS.*, 9, 041106–1, 2010.)

A cross section of the final structure is shown in Figure 8.38. The mirror is formed on top of a glass–ceramic substrate (WMS-15, Ohoro) that has a coefficient of thermal expansion (CTE) closely matched to the CTE of gold. The gold layers include a counter electrode layer for the electrostatic actuators that is ≈1 μm thick. This layer, which was not released, was deposited onto a chrome adhesion layer on the WMS-15 substrate. The next structural layer was a spring layer, 5–8 μm thick that provided a mechanical restoring force for the electrostatic actuator formed between this layer and the counter electrode layer. A copper sacrificial layer provided a gap 20–30 μm thick between the counter electrode and the spring layers. From the aforementioned discussion of electrostatic actuators, this should allow one-third of the initial gap or 7–10 μm of stroke to be used before reaching the pull-in instability. Slightly more than one-third of the initial gap can be used since the spring layer stretches and provides a nonlinear restoring force. A 4-μm-thick mirror layer is deposited on top of a post layer and is separated from the spring layer by a 30 μm gap.

A number of different electrostatic actuator structures were fabricated using their process to avoid tilting of the actuator as it was displaced. Solid models of these different actuator structures are shown in Figure 8.39. The square and circular actuators supported by folded springs were found to tilt when they were actuated, which led to premature pull-in (Fernandez 2010). The X-beam actuators used fixed-guided springs, which are stiffer than the folded springs. The fixed-guided springs provide a nonlinear restoring force when they are displaced more than the thickness of the spring layer, and if the actuator starts to tilt, the corner that tilts down the most further stretches the spring, increasing the nonlinear spring force. This provides a feedback mechanism for vertical displacement rather than tilting. The trade-off for the stiffer nonlinear spring is an increased actuation voltage. The actuator shown in Figure 8.40 required 377 V to obtain a displacement of 9.7 µm. To decrease the actuation voltage, either a smaller gap or more-flexible springs must be used. The gap on this actuator was 28.5 µm and the average thickness of the spring layer was ~3.5 µm. The gap can be decreased to 20 µm and the spring layer to ~2 µm to decrease the actuation voltage.

Gold mirror layers were electrodeposited on top of the actuator arrays to fabricate DMs. Gold is a good reflector in the IR spectrum, as used in astronomy and two-photon microscopy, and it does not tend to oxidize, so it can be packaged without a mirror to protect it from the environment. This eliminates two passes through a glass surface as would be required for DMs packaged behind a protective window. The small deformations that can be seen on the mirror surface shown in Figure 8.41 result from residual tensile stress in the electrodeposited gold layer. These deformations result from torques placed

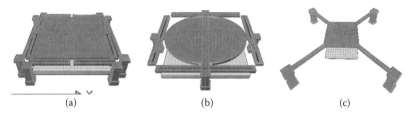

(a)   (b)   (c)

FIGURE 8.39 Solid models of different electrostatic actuator designs. (a) Square actuators supported by eight folded springs at the corners. (b) Circular actuators supported by four folded springs. (c) X-beam actuators supported by four fixed-guided beams. (From Fernandez, B. R. and J. Kubby, *J Micro/Nanolitho MEMS MOEMS.*, 9, 041106–1, 2010.)

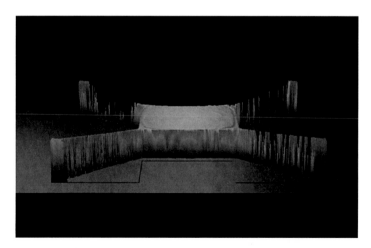

FIGURE 8.40 White light interferogram showing an X-beam actuator displaced by 9.3 µm at 377 V. (From Fernandez, B. R. and J. Kubby, *J Micro/Nanolitho MEMS MOEMS.*, 9, 041106–1, 2010.)

FIGURE 8.41   High-stroke, high-order MEMS deformable mirror fabricated in a high-aspect-ratio microma-chining process. The gold mirror surface is deposited on top of a 16 × 16 array of X-beam actuators fabricated in gold on a WMS-15 glass–ceramic substrate. The effect of the tensile stress in the electrodeposited gold mirror layer can be seen along the edges where the edge of the mirror appears to be serrated. Rows of test X-beam actuators can be seen outside the mirror area. (From Fernandez, B. R. and J. Kubby, *J Micro/Nanolitho MEMS MOEMS.*, 9, 041106–1, 2010.)

on the mirror surface by the support posts as the mirror shrinks to relieve tensile stress after release. These deformations can be decreased by decreasing the residual tensile stress or by applying more-rigid boundary conditions along the edges of the mirror.

8.9   Comparison of Microelectromechanical System Mirrors

An optical system designer has a considerable number of DMs to choose between that are now available. There have been some reviews comparing different MEMS mirror options in vision science (Dalimier 2005; Daly 2006) and which have also been extended to applications in astronomy (Devaney 2008). Both segmented and continuous facesheet MEMS DMs have been used in vision science, but most imaging applications in astronomy have used continuous facesheet mirrors. Nonetheless, segmented MEMS DMs have been used for correcting the laser guide-star uplink in astronomical AO. The mirrors used in vision science tend to require a large-stroke to fit a wide population distribution, but only require low-order correction since most of the common vision science aberrations such as astigmatism and focus can be corrected with low-order Zernike modes. Applications in astronomy typically require high order for high-contrast imaging (e.g., Gemini Planet Imager) (Poyneer 2011) but not high stroke. Nonetheless, the need for high stroke in astronomy will develop as larger diameter telescopes are built since they will view more of the sky and thus require higher-amplitude wavefront correction.

So far no standards have been developed for the application of MEMS DMs in biological imaging. Similar to vision science, most of the aberration in biological imaging occurs at lower order, so low-order MEMS DMs can be used to correct most wavefronts. However, there are no standard models for biological aberrations as there are now for astronomy, where the Kolmogorov spectrum is typically used (Kolmogorov 1941) or the population distributions that have been determined in vision science

(Porter 2006). The aberrations in biological specimens vary from specimen to specimen, as discussed in Chapter 4. Here phase-stepping interferometry has been used to measure wavefront aberrations in different biological specimens. The stroke requirements for biological imaging are currently met by low-stroke MEMS DMs; however, larger stroke will become necessary for imaging more deeply into tissue samples for correction of aberrations that increase with the optical path length, such as spherical aberration that arises due to mismatches between the lens, the immersion media (e.g., water or index matching oil), and the sample. Here either large-stroke DMs, or woofer–tweeter mirror combinations, will be required. A significant fraction of the stroke in an MEMS continuous facesheet DM can also be used up for correction of the microscope optics and optical alignment, even before using stroke for correction of specimen-induced wavefront aberrations. An additional portion of the stroke will also be used to flatten the DM if it has residual deformations from stress and stress-gradients that arise during manufacture and packaging of the mirror.

8.10 Microelectromechanical System Mirror Solutions

A number of the MEMS mirror component suppliers have either offered their own MEMS AO optical systems or partnered with larger optical companies to offer full AO systems including both MEMS DMs and wavefront sensors.

Boston Micromachines has partnered with Thorlabs to offer an AO kit that includes a DM, 32 actuators or 140 actuators, a Shack–Hartmann wavefront sensor, intervening imaging optics (collimation and relay optics), and stand-alone control software. The software is capable of minimizing wavefront aberrations by analyzing the signals from the Shack–Hartmann wavefront sensor and using those signals to determine the appropriate drive signals to send to the DM actuators. The mirror can then compensate for the measured wavefront aberrations. The control software allows the user to monitor the wavefront corrections and intensity distribution in real time. In addition, user-defined aberrations can

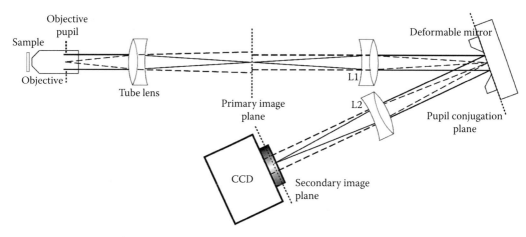

FIGURE 8.42 The MicAO system from Imagine Optics. The system is an afocal system, which consists of two identical lenses (L1, L2). The distances between the deformable mirror (DM) and the lenses are equal to the focal length of the lenses (f). In the same way, distances between the primary and secondary image planes and L1 and L2, respectively, are also equal to (f). The DM is positioned in the focal plane of L1, which corresponds to the conjugate plane of the objective lens pupil. The DM (mirao™ 52-e from Imagine Eyes) has 52 actuators controlled electromagnetically, which provides high deformation dynamics (up to ±35 μm of focus on a 15 mm mirror aperture). (From Andilla, J. and X. Levecq, MICAO: first universal all-in-the-box adaptive optics plug-in accessory for standard high resolution microscopy. *Imaging, Manipulation, and Analysis of Biomolecules, Cells, and Tissues VIII*, edited by Daniel L. Farkas, Dan V. Nicolau, Robert C. Leif, *Proc SPIE*. Vol. 7568, 75680U1–8, 2010.)

be introduced via the software, and the wavefront deviations can be compared to the new user-defined reference. For a research group that is just starting out and is not affiliated with an AO astronomy or vision science group that has already developed this software, this is a way to get off to a quick start in biological imaging.

Imagine optics offers a MicAO system, an add-on for commercial biological microscopes, including super-resolution approaches, such as stochastic optical reconstruction microscopy and photoactivated localization microscopy. A schematic diagram of the optical system is shown in Figure 8.42. The AO system is designed to be plugged into the camera port of a microscope to correct for the wavefront aberrations introduced by the system. It is compatible with most commercial microscopes, and the design of the system allows working with almost all the commercial biological objectives that are available. It uses a mirao™ 52-d electromagnetic DM and a sensorless correction algorithm, as described in Chapter 10, to improve the image quality. Since the system does not require a wavefront sensor, the cost can be substantially lower than an AO system that does use a wavefront sensor.

References

Andilla, J. and X. Levecq. MICAO: First universal all-in-the-box adaptive optics plug-in accessory for standard high resolution microscopy. *Imaging, Manipulation, and Analysis of Biomolecules, Cells, and Tissues VIII*, edited by Daniel L. Farkas, Dan V. Nicolau, Robert C. Leif. *Proc SPIE*. Vol. **7568**, pp. 75680U1–8 (2010).

Babcock, H. W. The possibility of compensating astronomical imaging. *Publ Astron Soc Pac.* **65**, 386, 229–236 (1953).

Becker, E. W., W. Ehrfeld, P. Hagmann, A. Maner, and D. Münchmeyer. Fabrication of microstructures with high aspect ratios and great structural heights by synchrotron radiation lithography, galvanoforming, and plastic moulding (LIGA process). *Microelectron Eng.* **4**, 35–56 (1986).

Bifano, T. G. and R. Krishnamoorthy. Surface micromachined deformable mirrors. Proceedings of the 5th IEEE International Conference on Emerging Technologies and Factory Automation, Kauai, Hawaii, Nov. 18–21, 1996, Vol. **2**, pp. 393–399 (1996).

Bifano, T. G., H. T. Johnson, P. Bierden, and R. K. Mali. Elimination of stress-induced curvature in thin-film structures. *J Microelectromech Syst.* **11**, 592–597 (2002).

Bonora, S., B. Umberto, H. Jean Pierre, and R. Stefania. Optically addressable deformable membrane mirror, *MEMS Adaptive Optics IV*, edited by Scot S. Olivier, Thomas G. Bifano, Joel A. Kubby, *Proc SPIE*. Vol. **7595**, pp. 75950R-75950R-8 (2010).

Bonora, S. and L. Poletto. Push-pull membrane mirrors for adaptive optics. *Opt Express.* **14**, 11935–11944 (2006).

Chen, D., M. Steven, J. Dennis, A. Silva, and S. S. Olivier. High-resolution adaptive optics scanning laser ophthalmoscope with dual deformable mirrors. *J Opt Soc Am. A* **24**, 1305–1312 (2007a).

Chen, D. C., S. M. Jones, D. A. Silva, and S. S. Olivier. High-resolution adaptive optics scanning laser ophthalmoscope with dual deformable mirrors for large aberration correction. SPIE Photonics West 2007. San Jose, CA, United States, January 20th–25th, 2007, **6426**, *Ophthalmic Technologies XVII*, edited by Fabrice Manns, Per G. Soederberg, Arthur Ho, Bruce E. Stuck, Michael Belkin, p. 64261L (2007b).

Cohen, A. Electrochemical fabrication. *The MEMS Handbook*, Chapter 19, Boca Raton, FL: CRC Press, (2001).

Cornelissen, S. A., P. A. Bierden, and T. G. Bifano. Development of a 4096 element MEMS continuous membrane deformable mirror for high contrast astronomical imaging, *Proc SPIE*. **6306**, 630606–1 (2006).

Cornelissen, S. A., P. A. Bierden, T. G. Bifano, and C. V. Lam. 4096-element continuous face-sheet MEMS deformable mirror for high-contrast imaging. *J Micro/Nanolith MEMS MOEMS*. **8**, 031308–1 (2009).

Cornelissen, S. and T. G. Bifano. Advances in MEMS deformable mirror development for astronomical adaptive optics. *Paper 8253-5 of Conference 8253*, Photonics West (2012).

Cowen, A., B. Hardy, R. Mahadevanand, S. Wilcenski. PolyMUMPs Design Handbook. *MEMSCAP Inc., Rev.* 13.0 (2011).

Dalimier, E. and C. Dainty. Comparative analysis of deformable mirrors for adaptive optics in the eye. *Opt Express.* **13**, 11, 4275–4285 (2005).

Daly, E., E. Dalimier, and C. Dainty. Requirements for MEMS mirrors for adaptive optics in the eye. *Proc SPIE.* **6113**, 611309 (2006).

Devaney, N., D. Coburn, C. Coleman, J. C. Dainty, E. Dalimier, T. Farrell, D. Lara, D. Mackey, and R. Mackey. Characterisation of MEMs mirrors for use in atmospheric and ocular wavefront correction, *MEMS Adaptive Optics II*, edited by S. S. Olivier, T. G. Bifano, J. A. Kubby, *Proc SPIE.* Vol. **6888**, p. 688802-1, (2008).

Ealy, M. A. and J. A. Wellman. Xinetics low cost deformable mirrors with actuator replacement cartridges, *SPIE.* **2201**, *Adaptive Optics in Astronomy*, 680–687 (1994).

Fernandez, B. R. and J. Kubby. High-aspect-ratio microelectromechanical systems deformable mirrors for adaptive optics. *J Micro/Nanolitho MEMS MOEMS.* Vol. **9**, p. 041106-1 (2010).

Fried, D. L. The Effect of Wave-Front Distortion on the Performance of an Ideal Optical Heterodyne Receiver and an Ideal Camera, presented at the Conference on Atmospheric Limitations to Optical Propagation at the U. S. National Bureau of Standards CRPL, 18–19 March 1965a.

———. Statistics of a geometric representation of wavefront distortion. *J Opt Soc Am.* **55**, 1427 (1965b).

———. Limiting resolution looking down through the atmosphere. *J Opt Soc Am.* **56**, 1380 (1966).

Hardy, J. W. *Adaptive Optics for Astronomical Telescopes.* Oxford Series in Optical and Imaging Sciences, Oxford University Press, Oxford (1998).

Hornbeck, L. J. From cathode rays to digital micromirrors: a history of electronic projection display technology. *Tex Instrum Tech J.* **15** (3), 7–46 (1998).

Hu, S., B. Xu, X. Zhang, J. Hou, J. Wu, and W. Jiang. Double-deformable-mirror adaptive optics system for phase compensation. *Appl Opt.* **45**, 2638–2642 (2006).

Hung, E., and S. D. Senturia. Extending the travel range of analog-tuned electrostatic actuators. *J Microelectromech Syst.* Vol. **8**, pp. 497–505 (1999).

Ji, N., D. E. Milkie, and E. Betzig. Adaptive optics via pupil segmentation for high-resolution imaging in biological tissues. *Nat Methods.* **7**, 141–147 (2009).

Kolmogorov, A. N., 1941. Dissipation of energy in locally isotropic turbulence. Doklady Akad. Nauk SSSR 32, p. 16. *Translation I Turbulence, Classic Paper on Statistical Theory*, edited by S. K. Friedlander and L. Topper, Interscience, New York, 1961.

Krishnamoorthy, R. M., T. Bifano, and D. Koester, A design-based approach to planarization in multilayer surface micromachining. *J Micromech Microeng.* **9**, 294–299 (1999).

Krulevitch, P., P. Bierden, T. Bifano, E. Carr, C. Dimas, H. Dyson, M. Helmbrecht, et al. MOEMS spatial light modulator development at the Center for Adaptive Optics. *MOEMS and Miniaturized Systems III*, edited by James H. Smith, Peter A. Krulevitch, Hubert K. Lakner. *Proc SPIE.* **4983**, 227–234 (2003).

Lavigne, J., and J. Véran. Woofer–tweeter control in an adaptive optics system using a Fourier reconstructor. *J Opt Soc Am A.* **25**, 2271–2279 (2008).

Maurer, C., A. Jesacher, S. Bernet, and M. Ritsch-Marte, What spatial light modulators can do for optical microscopy. *Laser Photonics Rev.* **5**, 81–101 (2011).

Milkie, D. E., E. Betzig, and N. J. Pupil-segmentation based adaptive optics with full-pupil illumination. *Opt Lett.* **36**, 4206–4208 (2011).

Morzinski, K., B. Macintosh, D. Gavel, and D. Dillon. Stroke saturation on a MEMS deformable mirror for woofer-tweeter adaptive optics. *Opt Express.* **17**, 5829–5844 (2009).

Olivier, S. Adaptive optics, Chapter 9, *MOEMS Micro-Opto-Electro-Mechanical Systems*, edited by Manouchehr E. Motamedi. *SPIE Press*, Bellingham, Washington, USA, p. 453 (2005).

Perreault, J. A., P. A. Bierden, M. N. Horenstein, and T. G. Bifano. Manufacturing of an optical-quality mirror system for adaptive optics. *Proc SPIE.* **4493**, 13–20 (2002).

Porter, J., H. Queener, J. Lin, K. Thorn, A. A. S. Awwal. *Adaptive Optics for Vision Science: Principles, Practices, Design and Applications*, Wiley Series in Microwave and Optical Engineering (2006).

Poyneer, L. A., B. Bauman, S. Cornelissen, S. Jones, B. Macintosh, D. Palmer, and J. Isaacs. The use of a high-order MEMS deformable mirror in the Gemini Planet Imager. MEMS Adaptive Optics V. San Francisco, CA, January 27th, Vol. 7931, *Proc SPIE.* (2011).

Vuelban, E. M., N. Bhattacharya, and J. J. M. Braat. Liquid deformable mirror for high-order wavefront correction. *Opt Lett.* **31**, 1717–1719 (2006).

Vellekoop, I. M. and A. P. Mosk. Focusing coherent light through opaque strongly scattering media. *Opt Lett.* **32**, 2309–2311 (2007).

Vellekoop, I. M., E. G. van Putten, A. Lagendijk, and A. P. Mosk. Demixing light paths inside disordered metamaterials. *Opt Express.* **16**, 67–80 (2008).

Webb, R. H., M. J. Albanese, Y. Zhou, T. Bifano, and S. A. Burns. Stroke amplifier for deformable mirrors. *Appl Opt.* **43**, 5330–5333 (2004).

Zernike, F., Beugungstheorie des Schneidenverfahrens und seiner verbesserten Form, der Phasenkontrastmethode [English translation: Diffraction theory of the cutting process and its improved form, the phase contrast method]. *Physica* **1**, 689 (1934).

Zou, W., X. Qi, and S. A. Burns. Wavefront-aberration sorting and correction for a dual-deformable-mirror adaptive-optics system. *Opt Lett.* **33**, 2602–2604 (2008).

MEMSCAP Inc. 3021 E.Cornwallis Rd PO Box 14486 Research Triangle Park, Durham, NC 27709 USA

Adaptive Optics System Alignment and Assembly

Diana C. Chen
Lawrence Livermore
National Laboratory

9.1 Introduction

This chapter is written for biologists and technicians needing fundamental knowledge and practical techniques to achieve alignment of an instrument exploiting adaptive optics (AO). It provides procedures on how to realize an optical system design from concepts to a working AO system in the laboratory environment.

Optical alignment means adjusting the positions of sources, optical components, and detectors to build an optical system as designed.[1–3]AO systems[4–6] quite often involve multiple sources, many optical components, and several optical function paths. Getting an AO system working effectively can take several iterations, where errors and misalignment are slowly worked out. Optical system designs and alignments have been discussed in many excellent books.[7–11] Readers are recommended to read the referenced books if they are interested in gaining more knowledge on the topic.

In this chapter, we will discuss the specific issues related with an AO system in terms of optical system design considerations, the fundamental imaging errors associated with optical systems, how to diagnose the fundamental imaging errors by observing the interferograms, misalignment penalties, and how alignment fits into the overall error budget. In addition, we will show the various tools available for aligning optical systems and discuss the special alignment considerations related to an AO system. More importantly, we will discuss the procedures on how to align and assemble the optics systematically, so that a complex AO system can be built quickly and effectively as designed.

9.2 Adaptive Optics Optical System Design Considerations

As with any optical system, the AO system starts with an object, with light going through a series of optics, and finally ends at the image. The object typically refers to a laser source that provides illumination for the biological samples and a detector that is placed at the image plane to record the image of the samples. Optical system design is optimized to have the clearest image possible at the detector.[12,13]

Several optical design software packages such as Zemax, CodeV, and Oslo for optical system design and system optimization are available commercially. Optical design fundamentals and principles are covered by many books and also by short courses at the conferences of the International Society for Optics and Photonics (SPIE) and Optical Society of America (OSA). Optical design software companies also offer design classes and design services if needed.

AO refers to optical systems that adapt to compensate for optical aberrations introduced by the medium between the object and its image. This is typically achieved by using a deformable mirror (DM), as described in Chapter 8. The shape of the deformable mirror can be adjusted to compensate the wavefront errors in the light path.

Figure 9.1 illustrates an AO microscope that consists of two collimated laser sources at different wavelengths, a deformable mirror for aberration compensation, a wavefront sensor to measure the wavefront error in the optical path, scanning mirrors for both x and y directions for two-dimensional viewing of the biological samples, and a photodetector for scientific image detection.[14] These components are typically connected together by relay telescopes in an AO system. These relay telescopes play important roles in the AO system.

When designing an AO system, compared with a typical optical system, special attention is needed to ensure the conjugation of two sets of planes to achieve correct phase compensation and thus a good image quality. The first set of conjugate planes includes the illumination source, specimen samples, and a detector for recording images. The second set of conjugate planes consists of a set of pupil planes, which includes the DM, the lenslet array, the scanning mirrors for both x and y directions, the pupil for eyepieces in the microscope, and the iris diaphragm in the human eye. Conventional optical microscopes require the first set of conjugation, while the second set is uniquely important to ensure that an AO instrument performs well.

The conjugation between two pupil planes is accomplished by an afocal relay telescope. The afocal telescope consists of two lenses or two curved mirrors separated by the sum of two focal lengths $F_1 + F_2$, as shown is Figure 9.2. The entrance pupil is usually placed at a focal distance F_1 away in front of the telescope. And the resulting exit pupil is created at a focal length F_2 behind the telescope.

The F-number of the system is defined by the focal length divided by the diameter of the optics.

$$\frac{F}{\#} = \frac{F_1}{D_{in}} = \frac{F_2}{D_{out}} \tag{9.1}$$

The telescope usually has collimated light in and collimated light out, and the beam size is changed by the magnification of the telescope. For an afocal system, linear magnification, which is defined by the ratio of the height of the image to the height of the object, is given by

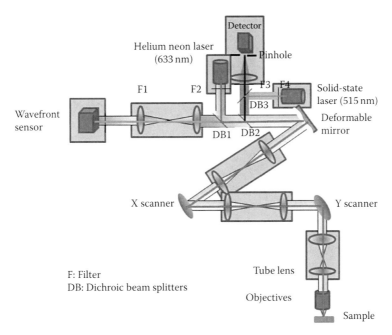

FIGURE 9.1 An example of an AO microscope. The microscope is made of functioning subassemblies such as laser input subassemblies, relay telescope subassemblies, wavefront sensor subassembly, and detector subassembly. Each subassembly is aligned independently offline and illustrated in a gray box. The figure also illustrates a complex optical system that has multiple beam paths with folds. The light delivery path and image acquisition path share common path errors, and the wavefront sensor has an additional non-common path error.

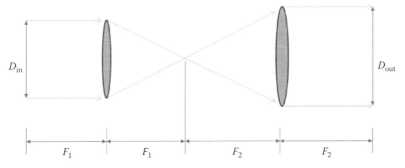

FIGURE 9.2 A transmissive relay afocal telescope.

$$M = \frac{D_{\text{out}}}{D_{\text{in}}} = -\frac{F_2}{F_1} \tag{9.2}$$

The negative sign indicates that the output image is inverted with regard to the input. In most AO systems, positive focal lengths are typically chosen for the two lenses or mirrors to create *real* pupils at the entrance and exit planes of the telescope. This would make it easier for the characterization of optical subsystem alignment and troubleshooting as opposed to virtual pupils. As the ratio between the two pupil diameters is proportional to that of the two focal lengths of an afocal relay telescope, a pair of lenses or mirrors can be conveniently chosen based on availability and space as long as the ratio remains the same.

The angular magnification is given by the reciprocal of the linear magnification:

$$M_a = \frac{1}{M} = -\frac{F_1}{F_2} \tag{9.3}$$

Relay telescopes utilize two reflective mirrors instead of two lenses following the same principles as in the transmissive-type relay telescopes. The linear magnification and the angular magnification are shown as in Equations 9.2 and 9.3.

Transmissive- and reflective-type relay telescopes have their own pros and cons. Optical designers need to decide the type of relay telescope to be used for pupil conjugation based on the system requirements and overall system performance.

Table 9.1 provides general guidelines for choosing reflective or transmissive relay telescopes for an AO biological imaging system. Typically, commercial off-the-shelf components are used to construct an AO microscope to reduce the cost in scientific experiments. Quite often, transmissive lenses and associated optical mounts tend to have more options than those of reflective elements. To have the required optical clearance for the light beam in an off-axis reflective telescope and to minimize optical aberrations, reflective mirrors with longer focal lengths are typically used compared to transmissive telescopes. When smaller focal length optics are chosen for transmissive telescopes, the optical system becomes more compact. They could be significant if the available space is limited. Another advantage of transmissive relay telescopes is that the optical alignment is more forgiving than in reflective telescopes.

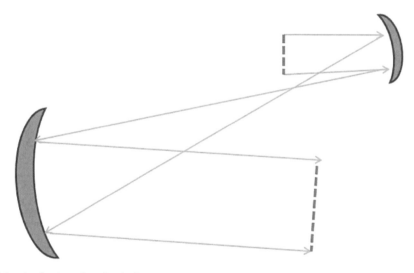

FIGURE 9.3 A reflective relay afocal telescope.

TABLE 9.1 Comparison of Transmissive and Reflective Relay Telescopes

	Transmissive	Reflective
Components availability	More choices	Fewer choices
Space constraints	Compact	Need more space
Alignment	Less sensitive	More sensitive
Light budget	More loss per component	Less loss per component
Ghost images	Yes	No
Background noise	More	Less
Multiple wavelengths	Chromatic aberrations	No chromatic aberrations
Pupil conjugate	Yes	Normal
DM compensation	Point-to-point	Quasi point-to-point

Reflecting telescopes have a number of advantages over transmissive telescopes. They are not subject to chromatic aberrations because the reflected light does not disperse according to wavelength. They do not create ghost images due to surface reflection of transmissive lenses, and because of this, the background noise is typically less.

In a transmissive relay telescope, the entrance pupil plane and the exit pupil plane are generally normal to the light beam, and thus they are exactly conjugate to each other. However, in an off-axis reflective relay telescope, the mirrors are tilted at non-normal incident angles to the incoming beam, thus making the entrance pupil plane and the exit pupil plane only "quasi" conjugate to each other. This may result in some errors in terms of phase corrections. If the field of view is small, the error is often negligible.

9.3 Optical Aberrations

Optical systems with smaller focal-length optics are more compact than those with longer focal-length optics. However, the optical system aberrations are typically higher in a compact optical system. The most common aberrations for any optical systems are spherical aberration, coma, astigmatism, field curvature, distortion, and chromatic aberration. These aberrations tend to degrade image quality as shown in Figure 9.4.

Spherical aberration is quite often the dominant error. The optical rays at the edge of the lens do not focus to the same spot as the rays at the center of the lens do. As a result, the image is blurred across the image plane. The image is uniformly degraded across the whole image by spherical aberrations as shown in Figure 9.4b. As the focus shifts, the image plane shifts with it. There is a position for which the circular image is at the minimum. This minimum image is called the circle of least confusion. The detector should

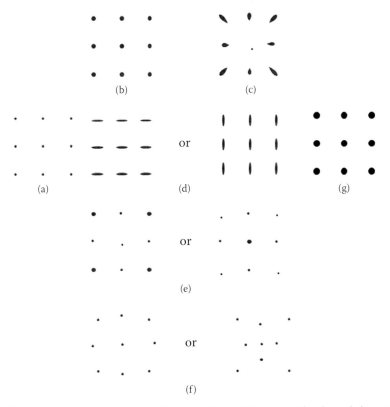

FIGURE 9.4 Illustration of images with aberrations. (a) Object. (b) Image with spherical aberration. (c) Image with coma. (d) Image with astigmatism. (e) Image with field curvature. (f) Image with distortion. (g) Image with chromatic aberration.

be placed at the plane of best focus or circle of least confusion for the best image result. For a simple thin lens or a reflective spherical mirror, the spherical aberration is inversely proportional to the cube of $F/\#$ (Equation 9.4). As the focal length gets smaller, the spherical aberration also increases rapidly:

$$S \propto \frac{K_s}{\left(F/\#\right)^3} \tag{9.4}$$

where S is the spherical aberration and K_s is a constant.

Coma is an off-axis aberration. Coma is absent on the axis and the magnitude increases with the field angle and decreases with the square of F-number. Coma limits the field of view of an optical system. For a simple thin lens or a reflective spherical mirror, the coma is given by

$$C \propto \frac{K_c}{\left(F/\#\right)^2} \tag{9.5}$$

where C is the coma and K_c is a constant. A shift in focus does not improve a comatic image unlike the case with spherical aberrations. In testing an optical system, coma may appear on axis, where it should be zero. This indicates tilted or decentered optical components in the system due to the misalignment.

Astigmatism is caused by lens asymmetry or off-axis object. Rays focus at different points for horizontal or vertical rays as shown in Figure 9.4d. In a system where astigmatism exists, a detector should be placed at the plane in between the two focal planes for horizontal and vertical rays for best focus and best image quality. For infinite conjugates, the astigmatism of a thin lens or reflective spherical mirror is given by

$$A \propto \frac{K_a}{\left(F/\#\right)} \tag{9.6}$$

where A is the astigmatic aberration and K_a is a constant.

Astigmatism is introduced by reflective relay telescopes due to off-axis reflection, and it is more severe in a compact system with shorter focal-length optics. A cylindrical lens can be inserted in the optical system to compensate the astigmatism introduced by the off-axis reflective telescopes. Often, however, when images show severe asymmetry over the entire field of view, it indicates the optical system is misaligned or badly reassembled.

Field curvature is the aberration where the sharpest focus is on a curved surface in the image space rather than a plane. Objects in the center and edges of the field are not in focus simultaneously as shown in Figure 9.4e. With a flat detector, it can never focus all the points at once. Again, the best focus is the best compromise for the best image quality. Field curvature is not directly related to the F-number of the optical system and it is mainly noticeable at the edges of the field. Most microscope objectives and eyepieces are designed to be well corrected for field curvature, especially for the wide-field microscopes.

Distortion happens when parts of the image are more magnified than others, so the optical system cannot create a rectilinear image of the object. Distortion does not modify the sharpness of the image but deforms its shape as shown in Figure 9.4f. Distortion occurs quite often in microscope eyepieces using thick lenses. It can be corrected using post-image processing if necessary. Field curvature and distortion cannot be improved by moving the image plane or by better alignment.

Chromatic aberrations are different from the preceding aberrations as they are not caused by imperfect surfaces or surface errors. Chromatic aberrations happen only when polychromatic light is used in the optical system, whereas other aberrations occur with a single color of light. When different colors of light propagate in a medium, they travel at different speeds. The refractive index is thus wavelength dependent, which is known as dispersion. Chromatic aberrations are those departures from perfect imaging that are due to dispersion.

There are two types of chromatic aberrations: longitudinal and transverse chromatic aberrations. Longitudinal chromatic aberration causes the variation of focus or image plane with wavelengths. Lateral chromatic aberration causes images to be different sizes for different colors or wavelengths. Chromatic aberrations often limit the performance of otherwise well-corrected optical systems. Longitudinal chromatic aberration reduces the image sharpness whereas lateral chromatic aberration gives rise to colored fringes.

9.4 Optomechanics

In a complex optical system, three-dimensional presentation of the system is very useful for visualizing how the components fit together. Many visualization tools have been developed in the modern optical design software packages to assist scientists and engineers to understand the overall optical system layout. More complicated computer-aided design (CAD) mechanical engineering tools provide detailed layout of the complex system—for example, AutoCAD, SolidWorks, and Pro-E. Optical design software can usually export design data in a format that can be translated into the format that is readable by most mechanical CAD software packages. Commercial suppliers of components often provide detailed optical and optomechanical CAD models that can be downloaded from their websites in the mechanical CAD software format to assist design integration and visualization.

In general, optomechanical design starts from the exported version of the optical design and it usually displays ray traces and element surfaces. Design modification is typically needed to display the full three-dimensional models with details. The three-dimensional computer models of the system hardware integrated with the beam paths (shown in Figure 9.5) are very useful for quickly identifying the conflicts between any components and light paths. When designing a compact AO system, CAD tools are particularly helpful to identify the space constraint limitation, especially in a situation where optical paths are folded by mirrors.

A good optomechanical design for an AO system will require serious considerations of several aspects of the system such as functionality, tolerances, alignment aids, characterization, and adaptability. Functionality in biological imaging typically means the CCD camera records the imaging at high resolution. This is achieved by optimizing the optical design and picking high-quality optical components

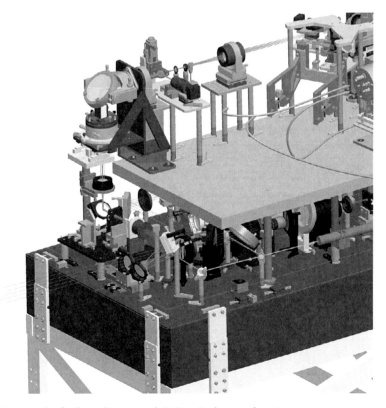

FIGURE 9.5 An example of a three-dimensional CAD optical system layout.

that satisfy the specifications. Tolerance analysis in optical design provides guidance for sensitivity in positioning the optical components to achieve high-resolution imaging. The sensitivity of the individual component determines the exact requirement of the particular fixture. In this way, optomechanical fixtures are designed to suit needs without being over- or underspecified.

In addition, characterization and alignment procedures should be carefully planned during the optical and optomechanical design process so that enough space is available for additional characterization equipment or alignment aids. Quite often in a scientific research environment, the optical system needs to be adaptable so that additional testing can be performed without rebuilding the entire optical system from the beginning. This can be achieved by inserting fold mirrors or beam splitters in the optical paths. Typically, off-the-shelf optical components introduce rms wavefront errors up to a quarter of wavelength, so care must be taken to choose high-quality mirrors and beam splitters to minimize effects on the total wavefront error in the optical system.

9.4.1 Optical Mechanical Hardware

Depending on overall system requirements, optical mechanical hardware may be custom designed or specified from vendor catalogs. Custom mounts are typically more compact, but more expensive and not easy to adapt to modification or upgrades. This section focuses mainly on commercial optomechanical mounts and fixtures. The advantages are that they are typically low-cost and readily available. Particularly in a scientific laboratory environment, new ideas and inspirations are quite often generated while performing the experiments no matter how well the initial design is planned. It is much quicker and cost-effective to make changes using off-the-shelf commercial mounts.

9.4.2 Optical Tables and Breadboards

Many companies such as Newport, Thorlabs, and CVI Melles Griot, offer a broad range of optical tables, optical breadboards, honeycomb structures, granite structures, base platforms, pneumatic vibration isolators, elastomeric vibration isolators, and microscope workstations. These platforms provide not only ways to mount optical components stably but also isolation in sensitive laboratory measurements. The legs are designed to provide isolation using active, passive, elastomeric, and rigid support methods depending on the system requirements. A simple optical system can be assembled directly onto an optical breadboard if vibration is not a serious issue.

Three important parameters should be considered for the selection of an optical table or breadboard. First of all, the table must satisfy the isolation requirements of all sources of mechanical, thermal, electromagnetic, and acoustic noises. Second, the tabletop must meet the standards of flatness and stiffness. Third, the ergonomic of the whole setup must fit into the workspace and allow easy access to other testing equipments. Equipment and component manufactures offer selection guides on their websites. Users are recommended to define their specific requirements and visit the manufacturers' web pages for details.

9.4.3 Mounts, Rods, Posts, and Holders

Commercial vendors offer a full range of mounts, posts, rods, and holders to meet various experimental requirements. Optical components are stably fixed to optical mounts as shown in Figure 9.6. The mounts are mounted on top of rods or posts that are held tightly to the optical table or breadboard by the holders.

9.4.4 Optical Subsystem, Cages, Rails, and Stages

The cage, rail, and stage assembly systems (Figure 9.7) provide convenient ways to construct large optomechanical systems with established lines of precision-machined building blocks designed for high flexibility and accurate alignment. Cage, rail, and stage assembly systems achieve optical alignment by tight tolerance of mechanical fixtures.

FIGURE 9.6 An illustration of an optical setup with optical breadboards, mounts, posts, and holders.

FIGURE 9.7 An example of a rail system with several components aligned along one direction by mechanical fixtures.

Rail and stage assembly systems center optical components along one direction. Cage systems use four rigid steel rods on which optical components are mounted along one common optical axis along both x and y directions. The distance between optical elements can be adjusted easily by sliding back and forth and then locked down with a locking setscrew. These optical subsystems are very useful tools for simplifying optical alignment processes as discussed in Section 9.5.

9.5 Optical Alignment

Alignment of complex AO systems can be quite challenging. Multiple-beam paths with folds, several sources at different wavelengths, off-axis elements, cylindrical optics, aspheric lenses, and tight tolerances all can cause potential alignment issues. AO components such as lens arrays, sensors, and DMs deserve special attention in the alignment process.

The AO microscope in Figure 9.1 shows an example of a complex optical system including multiple-beam paths with folds. Two laser input sources with two different wavelengths are coupled into the AO system by beam splitters. There are mainly three functional optical paths in this example. One path delivers the light to the biological sample; one path is for measuring the wavefront errors using a wavefront sensor; and one path is for image detection with a detector. All three paths share the DM, scanners, microscope objective, and pupil-relay telescopes between them.

It can be difficult to decide where to start and which methods to use when dealing with a complex optical system. But alignment does not need to be an overwhelming task even if the system is complex. When planning the alignment and assembly processes, a number of general strategies as well as utilization of well-known alignment tools and optical-instrument feedback can be very helpful.

9.5.1 Using a Full-Size Mechanical Layout

It is very useful to print out a full-size drawing of the optical design and place it on top of the optical breadboard or optical table. If the mechanical design has been done using a CAD tool, print out a full-size drawing of the system layout with the bases of the optomechanical fixtures and mounts as well. This will provide a quick idea on how the system fits together and if it is what the designers or scientists have intended.

Tape the full-size drawing to the optical table and then place the optical components and optomechanical mounts on the table over their footprint on the drawing. This will provide a direct view on whether there is enough work space available to allow hands in between the components for alignment adjustment in a compact system. If there are any issues, modifying the design and layout before aligning the complete optical system will save a lot of time and effort.

This method also provides an effective way to quickly place the components very close to the nominal design and enable one to tell if the integration is starting to deviate from the nominal design. Fine optical alignment and adjustment are still needed based on optical feedback. The paper can be removed when necessary to access the table's threaded holes and eventually can be removed entirely.

9.5.2 Use Mechanical Constraints

As discussed in Section 9.4.4, a cage, a rail, or a stage assembly system is one of the most effective and inexpensive alignment tools to prealign optical components along the optical axis in either one direction or two directions. By putting the complexity into the metal, the assembly process becomes much simpler; all that is needed is to place the optical component in its appropriate location. When there are multiple paths and folds in a system, these metal mounts may not be able to integrate all the components together. Nonetheless, these tools can be used to build subassemblies of the more-complicated system and partially simplify the system integration process.

When tolerances are too tight for drop-in assembly, this basic approach is still valuable for getting components close to their optimum position and limiting their deviation from their nominal positions during the alignment process. When mechanical constraints limit the deviation of an optic's position from the ideal location, component adjustments are easier and time-consuming gross misalignments are prevented. Commercial vendors are increasingly expanding their production lines so that fine adjustments can be made in all six degrees of freedom with these precision building blocks.

9.5.3 Divide into Functioning Segments

In a complex AO system with multiple optical paths, each path in the system should be divided into segments that are delimiting by a beam splitter or folding mirror. A strategy should be devised for each segment to ensure that it can be aligned independently without having to iterate through the sequence multiple times.

When there are multiple powered elements in a segment, typically it is most convenient to align the lenses in a separate subassembly housing and then mount the completed subassembly into the segment. In systems in which there are multiple paths that share segments, first align the shared segments to keep the final adjustments as independent as possible (Figure 9.8).

For example, phase conjugation of the pupil planes is very important in an AO system. The conjugation between two pupil planes is accomplished by an afocal relay telescope.

Several telescopes relay conjugation of the DM, the lenslet array, the scanning mirrors for both x and y directions, the pupil for the objective lens in the microscope, and the specimen. Each relay telescope should be built and tested independently using a collimated source and built into a subassembly. These subassemblies later on can be directly "dropped-in" as independent modules. In addition, laser input assemblies, the wavefront sensor subassembly, and the detector subassembly can each be independently built and aligned offline and then integrated together with other components or subassemblies. Figure 9.1 shows an AO system consisting of several functioning subassemblies such as laser input subassemblies, relay telescope subassemblies, a wavefront sensor subassembly, and a detector subassembly. Each subassembly is aligned independently offline. When system function changes are needed, an individual module may be modified to satisfy the requirements, thus avoiding the work of realigning the entire optical system.

9.5.4 Establish Optical Axis and Line of Sight

Light travels from the object to the image, establishing the optical axis of the optical system and the line of sight following the optical path. It is very important to center all optics along the optical axis. If an

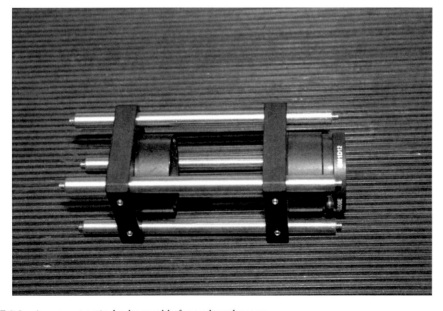

FIGURE 9.8 A compact optical subassembly for a relay telescope.

optical component is off-center, additional aberrations such as astigmatism and coma may be created and the system performance will be degraded.

Centering optical components using the line-of-sight concept is straightforward. Establish the line of sight using an alignment laser and a target without any optical components in between and recording where the line of sight intersects the target. Put the transmission lens in the optical beam path and adjust the lens transversely until the beam hits the same mark on the target. Typically based on the scattered light from the reflection of the lens surface, one can position the lens roughly in the center of the optical path before starting fine adjustment. Place the target as far as practically possible from the alignment laser so that the measurement is more sensitive and the placement of the optical component is more accurate.

9.5.5 Align the Science Camera Last

The science camera should be aligned and integrated into the system last. This approach is important, because it provides a final opportunity to optimize the alignment using performance feedback that is measured through the complete optical system. This is particularly important if multiple wavelengths are involved in the optical system. The position of the CCD camera should be adjusted to the optimized position for all wavelengths. Whenever possible, an effort should be made to avoid allowing all six degrees of freedom in the final alignment, as this can greatly complicate the mount and increase material and labor costs.

9.5.6 Multiple Wavelengths

Multiple wavelengths are typically built in the biological imaging tools for flexibility in imaging different types of biological samples for best resolution and contrast. It is best to start optical alignment using a single source and finish the entire system alignment from end to end. If multiple sources need to operate simultaneously, separate sources can be inserted in to the system by using beam splitters, gratings, or folding mirrors. If necessary, more adjustment freedom is given to the mounts of beam splitters, gratings, or folding mirrors to avoid the situation where the alignment of shared optical paths needs to be readjusted. In imaging tools where an optical source with a single wavelength is used one at a time instead of being used simultaneously, laser sources can be coupled into optical fibers. Wavelength switching can be easily achieved by switching the fiber connector attached to the fiber-coupled laser sources. As discussed in Section 9.5.5, the CCD camera should be placed on a translation stage with freedom of motion along the optical axis to compensate the defocusing error caused by different wavelengths.

9.6 Alignment Tools

9.6.1 Point Source with a Single-Mode Optical Fiber

The core diameter of a single-mode fiber is typically about 3–5 μm, making it an ideal point source for illumination, characterization, and alignment. An optical fiber is a very versatile tool in alignment. The through-hole in the fiber optics connector can be used to provide a reference for quick and coarse alignment of the line of sight in an AO system. The reflection coupled back to the single-mode fiber source is a perfect indicator that the optical surface measured is at the conjugate plane of the fiber source. For example, if the AO system is well aligned, the back reflection from the CCD camera of the wavefront sensor should be coupled back to the fiber source.

Figure 9.9 shows a compact point source with a single-mode fiber. The fiber connector is inserted to a cage plate mounted on a stand-alone post. The compact design makes it easy to be placed virtually anywhere in the optical system where a point source is needed for illumination, characterization, and alignment.

FIGURE 9.9   A compact, versatile point source with a single-mode fiber.

9.6.2 Alignment Sources

Whenever possible, start the optical system alignment using a laser with visible wavelength from 400 to 700 nm. Coarse alignment of optical subassemblies and the entire optical system can be done much more effectively with a visible laser rather than an infrared laser. Fine adjustment, especially along the optical axis, must be performed if the wavelength used for biological imaging is not the same as the alignment laser.

A helium-neon (HeNe) laser is widely used as an alignment laser for its low cost and ease of operation. It is primarily used at a wavelength of 632.8 nm, which is in the red portion of the visible spectrum. The output beam quality of a HeNe laser is typically not good enough to be used directly for alignment. Additional beam cleaning and collimation are needed.

Light coming out from a HeNe laser is focused down by a spherical lens to a pinhole. The pinhole filters out the high-order aberrations in the beam and passes the good-quality lower-order modes of the beam through. Adjust the pinhole so that most of the light passes through to avoid light loss. A good collimation spherical optics is placed behind the pinhole to re-collimate the beam.

With the recent development of semiconductor laser diodes, low-cost laser diodes are used more and more in optical alignment and illumination in biological imaging instruments. The light that comes out of the laser diode is highly diverging and has an elliptical cross section. The divergence is typically specified in both the x and y axes. The beam is collimated by passing it through a pair of cylindrical lenses that are perpendicular to each other.

Collimated light can be focused down to a point source by a good focusing lens or coupled into a single-mode fiber. Based on the light-reversal principle, light coming out of a single-mode fiber can be collimated well if the lens is placed one focal length away. A fiber-coupled laser beam is exploited in many biological instruments for its significant benefits to the system. Once the fiber-coupled laser is collimated, minimum optical realignment is needed. By using a connectorized fiber and a sturdy connector, the position of the fiber can be easily controlled and repeated at the micrometer level. Optomechanical alignment considerations of the laser beam to the application are thus much simplified. It allows the laser source to be separated from the microscopes and isolates the heat and vibration generated by laser sources and power supplies. The small footprint also makes it much easier to insert

FIGURE 9.10   A collimator setup with fiber-coupled light input.

into optical-beam paths for the purpose of checking alignment and characterizing deformable mirrors or wavefront sensors.

An optical fiber-coupled source is often conveniently used as a collimating source in an optical sub-assembly process for afocal relay microscopes and wavefront sensors. It is also an ideal source for the characterization of deformable mirrors and wavefront sensors. Once the AO system is in operation, recalibration may be needed. To achieve the optimized overall performance, it is best to verify the performance of the individual subassemblies, adjust the subassemblies if necessary, and finally tune up the entire optical system together. The compactness of the fiber-coupled point source allows the possibility of verifying the individual performance of subassemblies without disturbing the rest of optical system. In troubleshooting, this method provides an effective way to figure out the issues while avoiding the need for realignment of all the components.

How well is the beam collimated (Figure 9.10)? There are several tools for characterizing the beam collimation such as shearing interferometers and autocollimators.

9.6.3 Shearing Interferometer

Conjugate planes are relayed by afocal telescopes in most AO imaging systems.[15] Both the incident beam and the output beam of an afocal telescope are collimated. Afocal telescopes are typically built into an optical subassembly module and tested independently to minimize alignment errors.

The shearing interferometer is a convenient tool to observe interference and to use this phenomenon to test the collimation of light beams as well as some low-order aberrations. The laser source must have a coherence length several times longer than the thickness of the shear plate to satisfy the basic requirements for interference.

A collimated beam would create equally spaced straight interference fringes parallel to a reference line that is determined by the direction in which the two wavefronts are shifted or sheered. When a defocused beam is incident on the shear plate, the screen shows tilted fringes. Figure 9.11 shows the interference pattern observed for a diverging, collimated, and converging beam.

FIGURE 9.11 Interference patterns for a divering, collimated, and converging beam.

If the interference fringes are not straight lines, but distorted with wiggles, it indicates higher-order aberrations in the beam. Sometimes, straight lines are observed but not equally spaced. Again, this is the indication of higher-order aberrations. Depending on the extent of the wiggles or the straight-line offset, the optical components may need to be replaced to achieve a higher-quality collimated beam.

If the beam is astigmatic, it is possible to collimate it in one direction, but not in the orthogonal direction. As a good practice, the beam should be tested and sheared both horizontally and vertically.

9.6.4 Autocollimator

An autocollimator is an optical instrument that is used to measure small angles with very high sensitivity. The autocollimator projects a beam of collimated light. An external reflector reflects all or part of the beam back into the instrument where the beam is focused and detected by a photodetector. The autocollimator measures the deviation between the emitted beam and the reflected beam. It is a very useful tool in precision alignment and laser collimation.

9.6.5 Viewing Card

It is helpful to have a small (1″ × 1″) piece of white paper or an infrared card that can be used as a viewing card. Placing the card in the optical beam will help to track the optical paths, reflected light, scattered light, and diffused light. Placing the card close to the center of an optical component provides information on the rough centering of the beam on the optical component. If the beam is in the collimate space, the spot size on the card along the beam path should stay the same size. Otherwise, the optical component needs to be realigned so that the beam size does not change along the beam path. This is a quick check in troubleshooting the optical system alignment. In addition, the card can help track unwanted light in the system so that a beam block can be used to terminate any undesired beam or an iris used to filter out scattered light to reduce the background optical noise.

9.6.6 Iris (Diaphragm)

An iris is a very useful component in constructing and aligning an optical system. It is a thin disk with an adjustable opening at its center. The center of the iris coincides with the optical axis of the optical path. It may serve several functions in a complex optical system. First, the size of the opening regulates the amount of light that passes through the optical path. Second, it stops unwanted light, especially randomly scattered light, from reaching the imaging camera. This decreases the background noise and increases the imaging contrast. Third, it is an effective visual tool for alignment. If the iris opening is adjusted to be slightly smaller than the beam size of the alignment laser, a ring of light can be viewed on the iris. If the ring is not uniform on the iris, it indicates that the iris is not centered on the optical path. Adjustment in either the x or y direction is needed until the circular ring is uniform. The other way to use an iris to assist alignment is to close the opening to a very small circle. Check the circular dots on each optical element in the subsequent downstream optical path after the iris to see if all of the optics are centered. If not, adjust the positions of optical components so that the dot falls on the center of the elements. An iris provides a quick and effective way to carry out coarse alignment, and visualization makes the process easy.

9.6.7 Machine Target/Fiducially

Many components are difficult to align based on their features. In such cases, adding reference features can greatly simplify the alignment. For example, microlens arrays with secondary alignment features etched into the substrate are very useful for minimizing alignment errors in rotation. An easy-to-align, nonfunctional reference feature that is precisely registered to the functioning part of the optic can be an essential part of an optical alignment process.

9.7 AO System–Specific Considerations

9.7.1 Deformable Mirrors on Kinematic Mounts

A DM in general is not optically flat in its natural state. It is recommended to use a flat mirror in the position of the DM to perform optical alignment of the AO system. Otherwise, the optical alignment may not be done properly, as DMs typically have some curvature, and the curvature might even be off-centered or irregular.

Furthermore, deformable mirror manufacturing takes substantial engineering and research effort to deliver high-quality DMs, depending on the specifications. The lead time is typically fairly long. Quite often, DMs may not be available when the system alignment and integration are performed. In addition, DMs need to be changed or fixed if there are electrical, mechanical, or reliability issues when doing scientific imaging of biological samples after the imaging system is complete and totally operational.

The DM should be mounted and aligned using a removable kinematic mount with a permanent kinematic base with interchangeable tops. The DM and the flat mirror should each be mounted on separate kinematic top plates and coaligned in both position and angle, so that they can be interchanged offline. This way, one can integrate and troubleshoot the AO system without worrying that previous work in alignment will be lost.

Obviously, repeatability is essential to ensure the performance of AO systems. One needs to use high-precision kinematic mounts for DMs and flat mirrors to achieve the targeted positions over many attempts and over time. Figure 9.12 shows an example of kinematic mounts. Figure 9.12a shows the separate top and bottom plates with optics mounted on the top plate and Figure 9.12b shows how the top plate, bottom plate, and optics are mounted together. The top mounting plate and the bottom base plate of the kinematic base are magnetically coupled, and the top plate can be inserted or removed typically with 10 µrad precision and repeatability. Repeatability of the kinematic mounts should be measured and verified before integrating the subassembly together with the alignment of the whole AO system.

9.7.2 Shack–Hartmann Wavefront Sensor Subassembly

The wavefront sensor (WFS) measures the aberration in the optical path to provide the input signal to the DM, so that the DM can be shaped by the control electrodes to cancel out the wavefront error in the beam path. It is one of the most important elements in an AO system. Its measurement sensitivity and accuracy determines the overall performance of the AO system and thus the image quality of biological samples.

The Shack–Hartmann WFSs consist of an optical source, a lenslet array, and a CCD camera. The lenslet array is a two-dimensional array of small lenses with the same diameters, curvatures, and

(a) (b)

FIGURE 9.12 An example of a kinematic swappable mount. (a) Separate top and bottom plates with optics mounted on the top plate. (b) Top plate, bottom plate, and optics mounted together.

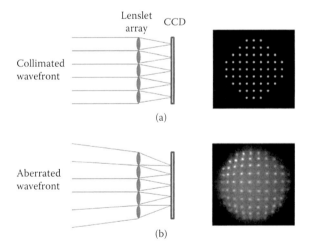

FIGURE 9.13 Wavefront sensor measurement results of (a) a collimated wavefront and (b) an aberrated wavefront.

focal lengths. The CCD camera is placed at the focal plane of the lenslet array and records the pattern of images formed by the lenslets in the array. Figure 9.13 shows the images on the CCD camera for a collimated wavefront and an aberrated wavefront. The spots are sharp and clear when there are no wavefront errors, and they are fuzzy and distorted for a beam with wavefront errors. The locations of the focal spots of the aberrated beam are also shifted relative to those of the collimated beam.

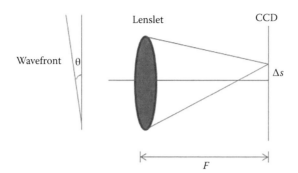

FIGURE 9.14   Wavefront measurement of a single lenslet.

The Shack–Hartmann WFS reconstructs the wave aberration based on the displacements of the focused Shack–Hartmann spots from the reference-spot positions for a perfect wavefront (Figure 9.14). The slope of the wavefront (θ) is given by spot displacement Δs as in Equation 9.7:

$$\theta = \frac{\Delta s}{F} \tag{9.7}$$

where F is the focal length of the lenslet. Measurement accuracy of the wavefront sensor is directly related to the precision of the measurement of Δs. The lenslet and the CCD camera should be mounted and aligned permanently as an optical subassembly. Align the lenslet array and the CCD camera with a well-collimated wavefront. The incoming alignment beam should be perpendicular to the CCD camera and the tilt of the camera with respect to the alignment beam must be minimized to achieve the maximum dynamic range for the wavefront measurement.

Adjust the tip and tilt angles so that the retroreflection from the CCD is reflected back through the optics to the source of the alignment beam. Then insert the lenslet array in the beam and adjust the lenslet array positions laterally and longitudinally so that the array is centered and the focused spots on the CCD camera are as sharp as possible. Adjust the tip and tilt angle so that the retroreflection from the lenslet array and the CCD camera are going back to the incoming laser source. Fine-tune the lenslet array positions both laterally and longitudinally until the focused spots on the CCD are as sharp as possible. Finally, fix the relative position between the lenslet array and the CCD camera and integrate the two components together as one optical subassembly. The centered spots provide the reference spots of the Shack–Hartmann WFS, while the displacement of the focused spots from the reference spots can be used to reconstruct the wavefront aberrations.

9.7.3 Pupil Plane Camera

AO biological imaging systems achieve high-resolution by using AO to compensate for the aberrations in the optical beam. In a properly aligned AO optical microscope, the pupil for eyepieces in the microscope, the lenslet array in the Shack–Hartmann WFS, and the DMs should be simultaneously in focus to ensure the correct measurement and compensation of optical aberrations. In addition, if there are scanners involved, the scanner should be placed at the conjugate planes with the DMs and WFS, so that the image of the DM on the WFS is quasi-static, which means the image does not move around as a function of time.

Typically a pupil camera is placed in the optical beam to help with the alignment process. Since the pupil for eyepieces in the microscope, the lenslet array in the Shack–Hartmann WFS, the DMs, and the scanners are at conjugate planes, their images should all be in focus and superimposed on the camera.

It may be confusing to interpret what we see on the camera when several images show up together, and it is also difficult to tell if the images are in focus. One method is to place a test target or fiducial mark on the component and observe if we see a clear image of the test target on the camera. Quite often, there are some dust particles or sharp edges on the surface of the optical component. These can be helpful to judge if the component is in focus as well.

Start with the optical component closest to the pupil camera and block all other components in the optical path. If that is not possible, block half of the beam so that an image of the other half of the component is viewed. If the image is not clear on the camera, move the component longitudinally until the image is in sharp focus. Then repeat the procedure with the next optical component along the optical beam path until all the conjugate components show up clearly in the pupil plane camera. Typically, only small adjustments for the positions of the optical components are needed.

Even though the pupil camera is not a part of the AO science imaging system, it is recommended to keep it permanently in the system to use as a diagnostic tool for troubleshooting and calibration of the AO system.

9.7.4 Photomultiplier Tube Pinhole Adjustment

The photomultiplier tube (PMT) subassembly consists of a high-quality focusing lens, a pinhole, and a science camera, as shown in Figure 9.15. The pinhole is a critical element to achieve diffraction-limited images. The diameter of the pinhole d is determined by Equation 9.8:

$$d = 2.44 \frac{\lambda F}{D} \tag{9.8}$$

where λ is the wavelength, F is the focal length of the lens, and D is the diameter of the optical beam.

The pinhole adjustment may be challenging to those who are not skilled in the art. Start with a pinhole size several times larger than what is required for the best image quality. Either the focusing lens or the pinhole needs to be placed on a x, y, z translation stage with micrometer adjustment accuracy. Illuminate the PMT with a collimated optical beam; place the pinhole close to the focal plane of the lens by measuring the distance between the pinhole and the lens; and adjust the x, y positions of the pinhole with the micrometers, until the science camera receives some photons, and images show up on the computer screen. Fine-tune the x, y positions so that the image is brightest for the axial z position. Slightly increase the distance between the focusing lens and the pinhole by adjusting the micrometer for axial adjustment. Repeat the above procedures with the x, y position adjustment. Then increase the axial distance between the focusing lens and pinhole and repeat the same process. When the maximum

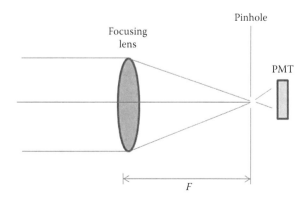

FIGURE 9.15   PMT subassembly.

brightness decreases with the increase in axial distance, adjust the micrometer back in the axial direction and find the location of the pinhole with the maximum brightness.

9.8 Assembly Procedures for an AO System

A complete AO system typically involves several optical paths and many optical and optoelectrical components. It is strongly advised to perform the system assembly in a systematic way. It is recommended to quickly go through the entire alignment procedure to verify that there are no obvious mechanical problems—such as any conflicts or narrow spaces—and there are sufficient ranges for adjustment. In addition, exploring the alignment tools helps to make sure that they can achieve their targeted functions and are easy to use. Operate the active components such as sources and detectors before integrating them into the system. If there is any issue, at this point it is best to resolve the issue, before full integration and alignment of the component. Once all major initial problems have been resolved, one may proceed with a more careful and critical fine alignment.

1. Measure the wavefront errors of the optical components, preferably mounted in their holders. Optomechanical mounting may cause stress and thus create additional aberrations. If spare parts are purchased, use the better-quality components.
2. If the wavefront error is within the tolerance of the components specifications, the optical design does not need to change. Otherwise, the initial optical design may require slight modification to compensate the wavefront errors or better-performance optical components may need to be repurchased.
3. As a good practice, label optical components on the sides or mounts to eliminate any confusion about the optic's identity.
4. Print out a 1:1 ratio optical design layout and place all the components on the optical table or breadboard according to the layout. Check if there are obvious mechanical conflict issues.
5. Center all optomechanical adjustments for the optical components initially within their travel ranges to minimize the possibility of running out of travel during alignment.
6. Assemble the building blocks—optical subassemblies such as relay telescopes, WFSs, and laser source collimators—offline.
7. Use a flat mirror in place of the DM. Build the optomechanical mounts for the DM, so that the flat mirror and DM are interchangeable without the need for realignment.
8. Establish the line of sight or optical axis with a collimated alignment laser (quite often used as an illuminator laser as well).
9. Install beam splitters and mirrors in the beam paths. Start from the mirror or beam splitter furthest upstream and work downstream. Adjust the tilt of each mirror so that the line of sight hits the center of each mirror.
10. Start installing the relay telescope furthest upstream and work downstream. The specified location on the optomechanical design printout can help to place the relay telescope close to its correct positions. Adjust the lateral position and the tilt of the subassembly so that the line of sight is still at the center for the downstream mirrors and hits the final target. Relay telescopes are used to relay conjugate planes in an AO system. The lenses in the relay telescopes are one focal length away from the pupil planes. Set up a pupil camera to adjust the position of the relay telescope so that the conjugate planes are in focus.
11. After installation of the optical elements in the major optical paths from the illumination source to the science imaging samples, install the other functioning optical paths involving the active optical elements, such as the WFS and the detector.
12. Adjust the optical subassembly of the WFS so that the optical beam incident to the WFS is well centered and perpendicular to the lenslet array. Check the image on the pupil camera to ensure

that the lenslet array plane is in focus. If not, adjust the position of WFS subassembly longitudinally or axially. Typically the alignment marks or fiducials on the lenslet array are very helpful to achieve this goal.

13. The detector path should be installed last. In the case of a PMT assembly, center the detector subassembly and make sure that image of the focusing lens for the PMT shows up sharply in the pupil plane camera. Pinhole positions are usually optimized for a single wavelength and typically need to be readjusted depending on the fluorescence wavelength from the science samples, especially when several wavelengths are used in imaging. The pinhole size and location should be optimized for each individual wavelength for overall performance.

9.9 Optical System Performance Improvement/Troubleshooting

It may take several iterations to achieve high performance of the AO system including improvement in the control algorithms for wavefront sensing and reconstruction; fine adjustment for the optical system alignment; and minimizing the electrical, mechanical, and optical noises while performing scientific measurements. In this section, we focus on how to improve the system performance related to better optical alignment. We also discuss what needs to be done if the acquired images are fuzzy or degraded images are acquired.

1. Starting with coarse alignment—making sure that spot sizes on optical components are centered and the optical beam is going through all the optical trains without being blocked or unintentionally vignetted by any optomechanical mounts.

2. If there are collimated spaces in the design, check for collimation using a shear plate and observe the shear fringes. The DM, WFS, detector, and scanners are all typically in the collimated space. Check all relay telescopes to see if the incoming collimated beam that passes through the relay telescopes is still well collimated after the telescope.

3. Check the pupil camera to see if all the related conjugate planes can be observed with clear, sharp images so that there is a correct phase correlation in AO system. If a scanner is involved and is not positioned correctly at the conjugate plane, the image on the CCD camera will move around when the scanner is in operation. Otherwise, the image of the scanner should remain stationary.

4. If there is any scattering of light in the system, place an iris or an aperture at the focal point of the relay telescope to filter out the high-order scattered light. The size of the aperture should be adjusted to let the main beam pass through while blocking the high-order scattering. This will significantly increase the signal-to-noise ratio.

5. Each subassembly should be independently calibrated and adjusted. A compact fiber optics–coupled laser source can be especially useful. It can be inserted into a finished system to check the performance of the part of optical path either as a point source or a collimated source. This way, problems resulting from the optical subassembly or functioning path can be identified and resolved quickly. In addition, a perfect point source or well-collimated source would make the calibration or alignment much more accurate instead of using the optical beam that has accumulated aberrations from the upstream optics.

6. Place a collimated beam in front of the WFS, and measure the aberrations in the beam. The aberration measured by the WFS should be very small—close to zero. Otherwise, the WFS needs to be adjusted and recalibrated.

7. The DM should be flattened offline using an interferometer before being inserted into the AO optical system. To confirm the flatness, a point source can be inserted at the image plane and used to measure the flatness of the DM with the WFS in the system. Even though the DM is mounted on a high-precision interchangeable top plate with high repeatability, in sensitive measurements, the AO system most likely still needs recalibration for optimized performance.

8. Because typical pinhole sizes range from several micrometers to 100 μm in a PMT, the alignment of the pinhole is very critical. For a new system, it is recommended to start with a pinhole size many times larger than what is required by the best image quality and then try several smaller pinhole sizes to reduce the pinhole size to what is needed. If the optical system has residual aberrations after being compensated by the DM, the diffraction-limited pinhole size may not be the best choice as it may cut off too much of the signal and reduce the intensity too much. When a different wavelength is used in the system, the pinhole axial position should be adjusted and reoptimized for the best images. It is a good practice to check the pinhole alignment every day when the AO system is operated. Finely adjust the *x*, *y* positions until the acquired images are brightest and then perform the measurement.

9.10 Understanding Alignment Penalties

When discussing the misalignment, it is important to understand their effects and penalties on performance. First of all, if the optical system is perfect without aberrations, what image quality can be expected from the system?

If the optical system is aberration-free, the resolution of the image is limited by diffraction. The point spread function (PSF) describes the response of an imaging system to a point source. For an aberration-free optical system, the response for a point source is an airy disk pattern.

The minimum resolvable details are defined by the Rayleigh criterion. Two objects can be resolved as long as the central peaks of the images do not overlap. The imaging process is said to be diffraction-limited when the central peak of the diffraction pattern of one point source coincides with the first dark ring of the diffraction pattern of the second point source.

If there are aberrations in the optical system, what criterion should be used to evaluate the image quality?

The Strehl ratio is defined as the ratio of the central peak intensities of the aberrated PSF and the diffraction-limited PSF.

$$S = \frac{I(0,0)\text{aberrated}}{I(0,0)} \tag{9.9}$$

For small aberrations, the Strehl ratio can be approximated by the wavefront error:

$$S = 1 - \left(\frac{2\pi}{\lambda}\sigma\right)^2 \tag{9.10}$$

where σ is the wavefront error.

The system is close to diffraction limited when the Strehl ratio is greater than 80%. This is the Maréchal criterion. The corresponding rms wavefront error must be better than 1/14 of a wavelength. This means that the total system aberrations must be less than 1/14 of wavelength after compensation by the DM.

In an AO system, the total wavefront error comes from several sources, including common path errors, non-common path errors, and phase registration errors.

Common path wavefront errors refer to aberrations that are common to the optical paths of both the science camera and the WFS. Since the WFS can measure the common path aberrations, DMs can be used to compensate the aberrations in both the science and WFS optical paths. This is acceptable as long as correcting the optical system aberrations does not take away too much phase compensation capability of the DM for compensating specific sample aberrations. Commercially available DMs typically have modulation capability from several micrometers up to 100 μm, so using half a micrometer to compensate for common path errors is acceptable. However, these errors should be minimized.

Non-common path errors refer to aberrations in only one functional path of the system. If the aberration is only in the detector path, then the WFS will not be able to measure it and therefore the AO system will not correct the aberration for the final image that is acquired. Thus, the image quality is degraded. Conversely, if the aberration is only in the WFS portion, the AO system will correct for an aberration that does not exist in the image acquisition path, which will also degrade the image quality.

In the example of an AO microscope as shown in Figure 9.1, the light delivery path and the image acquisition path share common path errors and the WFS has an additional non-common path error that is created by the relay telescope and the filter in front of the WFS.

Non-common path aberrations can, in principle, be negated by calibrating the WFS. The WFS uses the measured slightly aberrated wavefront as the "perfect wavefront" in calculating the compensation needed by the DM, so that a perfect diffraction-limited image is produced on the detector camera.

To achieve a diffraction-limited image, the wavefront measured by the WFS, the wavefront generated at the DM, and the wavefront at the pupil plane at the science camera have to be correctly correlated and interpreted. Any misinterpretation will create additional phase registration error, which will degrade the image quality. The strong emphasis of correct pupil plane relay in an AO system is to ensure the correct phase relationship. For example, if the scanner is not at the conjugate plane of the WFS, the pupil projected onto the WFS is moving with time instead of being stationary. As the speed of the scanner is usually much faster than the response time of the WFS and DM, the Shack–Hartmann WFS can measure only the time-averaged wavefront and any high spatial frequency aberration may be averaged out and uncompensated.[16]

References

1. W. J. Smith. *Modern Optical Engineering*. McGraw-Hill (2008).
2. P. Hobbs. *Building Electro-Optical System: Making It All Work*. John Wiley (2009).
3. D. Malacara. *Optical Shop Testing*. John Wiley (2007).
4. R. Tyson. *Introduction to Adaptive Optics*. SPIE, 2000.
5. J. Porter, H. Queener, J. Lin, K. Thron, and A. Awwal. *Adaptive Optics for Vision Science*. John Wiley (2006).
6. J. Hardy. *Adaptive Optics for Astronomical Telescopes*. Oxford University Press (1998).
7. R. Liang. *Optical Design for Biomedical Imaging*. SPIE Vol. PM **203** (2010).
8. R. E. Fischer, B. Tadic-Galeb, and P. Yoder. *Optical System Design*. McGraw-Hill (2008).
9. R. Kingslake and R.B. Johnson. *Lens Design Fundamentals*. Academic Express (2009).
10. M. J. Kidger. *Fundamental Optical Design*. SPIE Vol. PM **92** (2002).
11. G. H. Smith. *Practical Computer-Aided Lens Design*. Willmann-Bell (1998).
12. V. N. Mahajan. *Aberration Theory Made Simple*. SPIE (1991).
13. W. T. Welford. *Aberration of Optical Systems*. Adam Hilger (1991).
14. X. Tao, B. Fernadez, O. Azucena, M. Fu, D. Garsia, Y. Zuo, D. Chen, and J. Kubby. "Adaptive Optics Confocal Microscopy Using Direct Wavefront Sensing." *Opt. Lett.*, **36**, 7, 1062–1064 (2011).
15. M. P. Rimmer. "Method for Evaluating Lateral Shearing Interferograms." *Appl. Opt.*, **13**, 3, 263 (1974).
16. D. Chen, S. Oliver, S. Jones, R. Zawadzki, J. Evans, S. Choi, and J. Werner. "Compact MEMS-based adaptive optics-optical coherence tomography for clinical use," *Proc. of SPIE*, **6888**, 68880F-1 (2008).

III

Applications

Part 1: Indirect Wavefront Sensing

10

Sensorless Adaptive Optics for Microscopy

Martin J. Booth
University of Oxford

Alexander Jesacher
Innsbruck Medical University

10.1 Introduction

The imaging properties of a microscope suffer from the presence of aberrations, and optimum performance is obtained only when aberrations are zero. As aberrations are introduced by imperfections in the optical system or by the inhomogeneous refractive index distribution of the specimen, most microscope systems are in some way affected. To ensure optimum image quality, optical systems are designed so that the overall aberration is set below an acceptable tolerance. However, if the system is used outside of its design specifications, including when the specimen refractive index differs from the objective immersion medium, then aberrations can be significant. Correction of these aberrations through adaptive optics (AO) is essential if optimum imaging performance is to be ensured.

In AO, an adaptive element, such as a deformable mirror (DM), is used to introduce additional wavefront distortions that cancel out other aberrations in the system. Traditional AO systems, such as those developed for astronomical telescopes, also use a wavefront sensor to measure aberrations. Common methods for direct wavefront sensing are the Shack–Hartmann sensor (Hardy 1998) or devices based on interferometry (Hariharan 2003). An advantage of these sensors is their speed, which makes them the method of choice for astronomical or free-space optical communication applications, where corrections have to be performed at the fast timescale given by air turbulence.

Although AO techniques have been applied to microscopy, most of these systems have not used wavefront sensors but have instead relied on indirect aberration measurements (Booth 2007a) (Figure 10.1). One reason for this is that, for many microscopy applications, it is not straightforward to apply direct wavefront sensing because of the three-dimensional nature of the specimen. The sensor would ideally detect light emanating from the focal region, but light is instead emitted from a larger, three-dimensional

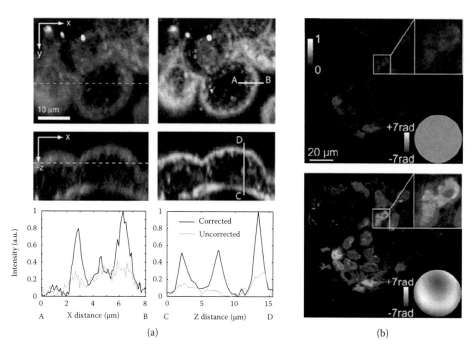

(a)   (b)

FIGURE 10.1 Examples of the correction of aberrations in multiphoton microscopes using sensorless adaptive optics. (a) Third harmonic generation images of a 5.5-day-old live mouse embryo with correction of only system aberrations (*left*) and after additional correction of specimen-induced aberrations (*right*). The dashed lines show where the *xy* and *xz* planes intersect. Also shown are intensity profiles along the solid lines AB and CD as drawn in the images. (Reproduced from Jesacher et al., *Opt. Lett.*, 34, 20, 3154–3156, 2009.) (b) Two-photon excitation fluorescence images of a mouse embryo: upper, before correction; lower, after correction). (Reproduced from Débarre et al., *Opt. Lett.*, 34, 16, 2495–2497, 2009.)

region of the specimen. As most wavefront sensors have no mechanism for discriminating between in-focus and out-of-focus light, the sensor readings can be ambiguous. For this reason, most adaptive microscope systems have "sensorless" AO systems, in which the correction aberration is inferred from a collection of image measurements, where each image is acquired with a different aberration introduced by the adaptive element. We refer to these additional aberrations as bias aberrations. As in any AO system, the goal of a sensorless AO system is to find a DM shape that cancels all other aberrations in the system or at least reduces them to an acceptable level. The sensorless AO scheme must, therefore, be able to determine from a set of aberrated images what the optimal correction aberration is or, equivalently, what the system aberration is. We explore the principle behind this approach. We also describe the methods that can be used to design efficient sensorless adaptive schemes by appropriate choice of aberration modes.

10.2 Indirect Wavefront Sensing

As aberrations affect the imaging quality of a microscope, some information about the nature of the aberrations must be encoded in the images. Therefore, one should expect that some of this information could be extracted indirectly from an aberrated image or, more readily, from a collection of aberrated images. This is the principle behind sensorless AO.

Mathematically, an image is a function of both the object structure and the imaging properties of the microscope. For example, in the case of a confocal fluorescence microscope (an incoherent imaging system), the image is the convolution of the fluorophore distribution in the object and the intensity point spread function (PSF) of the microscope. As the aberrations affect only the imaging properties, the sensorless AO scheme must be able to separate this information from the effects of the object structure. Phase retrieval methods have been used to extract aberration information from a single image, but these methods are effective only if the object structure is known (e.g., if the specimen is a single point-like bead) (Kner et al. 2010). For unknown specimen structures, the aberrations can be determined only by acquiring more images, each with a different bias aberration introduced by the adaptive element. Assuming that the object structure is unknown but does not change measurements between images, then comparisons between the qualities of the biased images should reveal the effects of the aberrations on the imaging properties of the microscope. This information about image quality forms the basis for the determination of the aberration in the system and consequently the optimum setting for the correction element.

The definition of "image quality" is somewhat subjective. However, there are many ways in which quality can be reasonably represented by a mathematical quantity. For example, in the confocal fluorescence microscope, the magnitude of the intensity PSF is reduced as the size of aberration increases; consequently, the total image intensity (the sum of all pixel values) is also reduced. It is straightforward to show that for most object structures, the total image intensity is maximum when aberrations are zero. Hence, the total image intensity is an appropriate metric for the representation of the image quality in this microscope.

Once we have defined an appropriate image quality metric, we can consider the sensorless AO scheme as a mathematical optimization problem, whose objective is to maximize the metric function, which we denote as M. The maximum of M corresponds to the optimum DM setting that minimizes the total aberration in the system. The metric M is a function of input parameters that determine the aberration applied by the adaptive element. These input parameters can be represented by the set of P control signals $\{c_1,...,c_P\}$ driving the actuators of the DM. The calculation of the image quality for each applied aberration is, therefore, equivalent to the evaluation of the metric function at the coordinates given by the DM control signals. Therefore, the objective of the optimization is to find the values of $c_1, ..., c_P$ that maximize $M(\{c_1, ...,c_P\})$.

The general outline of this optimization procedure is shown in Figure 10.2. There exist many different strategies for the implementation of this procedure, including the various heuristics used in convex optimization theory. One possible approach is to use model-free stochastic methods, for which we need to assume only that M has a well-defined maximum value. For example, one could test all possible shapes of the DM and select the setting with optimum image quality—this would be equivalent to an exhaustive search through all possible aberrations represented by all possible combinations of control signals. Although from a mathematical point of view this is guaranteed to find the optimum correction, the number of image measurements would be impractically large (Booth 2006). Other more-developed model-free methods, such as genetic or hill-climbing algorithms, have been demonstrated (Sherman et al. 2002; Wright et al. 2005). Although they are more efficient than an exhaustive search, they typically require a large number of image measurements. Although, in principle, the acquisition of these images is possible, there are several reasons why this is incompatible with practical microscopy. Specimens are often sensitive to cumulative light exposure—live specimens suffer from phototoxic effects and even fixed fluorescent specimens undergo photobleaching. Many multiple exposures are clearly undesirable in each of these cases. Specimens may also move during the imaging sequence, hence the specimen structure will not stay constant. It is, therefore, desirable to minimize the measurement time by ensuring that the number of exposures is as small as possible.

Using the mathematical analogy, each specimen exposure (and quality metric calculation) corresponds to a function evaluation in the optimization problem. It follows that reducing the required number of exposures is equivalent to deriving an efficient optimization algorithm that minimizes the number of function evaluations. This can be achieved using a priori knowledge of how aberrations

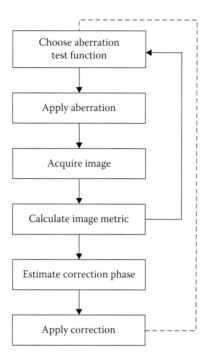

FIGURE 10.2   Flow chart of the indirect wavefront sensing scheme. A sequence of images is acquired, each with a different aberration applied. For each image, the quality metric is calculated. From the obtained metric values, the correction phase is estimated. The dashed line represents an optional repeat of the process, if full correction was not achieved after the first cycle.

influence the metric through, for example, a functional description of the microscope's image formation process in the presence of aberrations.

In Sections 10.3 to 10.6, we show how the appropriate mathematical formulation of the optimization problem informs the design of the sensorless AO scheme. In particular, we show how the choice of the modal expansion used to represent aberrations plays an important role in the efficiency of the process.

10.3 Modal Representation of Aberrations

In Section 10.2, we explained how the optimization metric M can be represented as a function of the control signals that drive the actuators of the DM. However, it is usually more convenient to express a wavefront as a sequence of basis functions or aberration modes. For example, an arbitrary wavefront $\Phi(r,\theta)$ is expressed as

$$\Phi(r,\theta) = \sum_i a_i X_i(r,\theta) \tag{10.1}$$

with a_i representing the mode coefficients and X_i the basis functions. If the DM is used in a linear operating regime, then the modal coefficients are related to the DM control signals by a linear transformation.

The aim of modal wavefront sensing is to obtain the coefficients a_i directly. This contrasts with zonal sensing methods, such as the Shack–Hartmann sensor, where the measured wavefront is constructed from separate measurements from different regions of the wavefront (Hardy 1998). The choice of the basis modes X_i can be influenced by various factors, both mathematical and practical. For example,

Zernike polynomials are often used for systems with circular apertures as they form a complete, orthogonal set of functions defined over a unit circle (Zernike 1934; Born and Wolf 1983). The expansion may also be based on the deformation modes of a DM or the statistics of the induced aberrations. For an efficient sensorless AO scheme, the modal expansion is chosen to ensure that the metric function has certain mathematical properties. These are discussed in Section 10.4.

10.4 Sensorless Adaptive Optics Using Modal Wavefront Sensing

In the design of an efficient sensorless AO scheme, we should use a priori knowledge of the form of the optimization metric M as a function of the mode coefficients a_i. In most practical situations, this form is approximately parabolic in the vicinity of the maximum of M. This observation assists us in designing an efficient optimization algorithm that uses minimal function evaluations and hence fewest specimen exposures. We will first show the principle of this approach in a system where only one aberration mode is present. In the subsequent discussion, we extend this description to the case of multiple aberration modes.

10.5 Measurement of a Single Mode

When only one aberration mode X_i is present in the system, then the aberrated wavefront can be expressed as $\Phi(r,\theta) = aX_i(r,\theta)$, and the optimization problem is equivalent to a one-dimensional maximization of a parabolic function. Close to its maximum, the metric M can be expressed as

$$M \approx M_0 - \alpha\, a^2 \tag{10.2}$$

where a denotes the aberration mode coefficient and α is a constant. The metric has the maximum value M_0 when the aberration coefficient a is zero. The values of M_0 and α are usually unknown as they depend on factors such as specimen structure, illumination intensity, and detection efficiency. Let us now include the effect of an adaptive element that adds the aberration $cX_i(r,\theta)$. Equation 10.2 then becomes

$$M \approx M_0 - \alpha\, (a+c)^2 \tag{10.3}$$

where it is clear that M is maximum when $c = -a$, and the correction aberration fully compensates the system aberration. As we are free to choose the value of c, the right-hand side of Equation 10.2 contains three unknown variables, one of which, a, we wish to determine. It follows that we can extract the value of a from three measurements of M that we obtain using different values of c. To achieve this practically, bias aberrations corresponding to three different values of c are intentionally introduced to the wavefront and the corresponding metric values are measured. This process proceeds as follows. The first step requires taking an image with zero additional aberration applied by the adaptive element ($c = 0$). The metric value M_Z is calculated from the resulting image. Subsequently, two further measurements are taken, with small amounts ($c = \pm b$) of the bias aberration introduced to the system. The corresponding images provide the metric values M_+ and M_-. We then use a simple quadratic maximization algorithm to find the aberration magnitude, a, from the three metric measurements (Press et al. 1992):

$$a = -\frac{b(M_+ - M_-)}{2M_+ - 4M_Z + 2M_-} \tag{10.4}$$

From this, the necessary correction amplitude is obtained as $c = -a$. This measurement principle is illustrated in Figure 10.3.

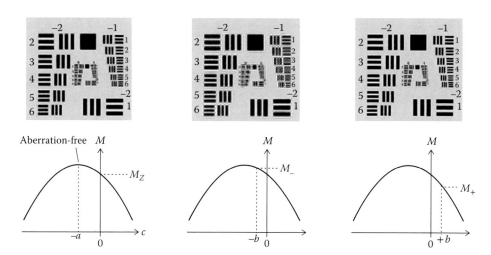

FIGURE 10.3 Principle of modal wavefront sensing of a single aberration mode. Three images are acquired with intentionally applied bias aberrations $-b$, 0, and $+b$, and the corresponding metric values M_-, M_Z, and M_- are calculated from the images. The aberration magnitude, a, is then derived from the three metric values using a quadratic maximization calculation.

10.6 Measurement of Multiple Modes

When multiple aberration modes are present in the system, then $\Phi(r,\theta) = \sum_i a_i X_i(r,\theta)$ and the optimization problem is equivalent to a multidimensional maximization of a paraboloidal function. Using a Taylor expansion around the maximum point, where $a_i = 0$ for all i, the metric M can be expressed in a general form as

$$M \approx M_0 - \sum_i \sum_j \alpha_{i,j}\, a_i\, a_j \qquad (10.5)$$

where M_0 and $\alpha_{i,j}$ are unknown constants. The one-dimensional quadratic maximization approach of Section 10.5 is not obviously extendable to this more-complicated multidimensional function. In particular, we see that the coefficients for each mode do not appear as separable terms in the Taylor expansion. If we consider the shape of the metric function, this means that the primary axes of the paraboloid do not align with the coordinate axes formed by the coefficients a_i. This is illustrated in Figure 10.4a using the simple case where two aberration modes are present. Performing an optimization in one aberration coefficient will find the maximum along a section parallel to the coordinate axes. Repetition of this process for the second aberration coefficient leads to a combined correction aberration for both modes. However, this does not find the overall maximum of the paraboloid (although further repetition of the process would move the correction closer to the optimum value).

In practical terms, this effect means that a standard set of basis modes, such as the Zernike polynomials, will not, in general, be the best choice for the control of a sensorless adaptive microscope. Indeed, it has been found that each different type of microscope would have its own ideal set of modes, which do permit independent optimization of each mode using the quadratic optimization method.

The nonideal basis modes can be converted by recasting Equation 10.5 through a coordinate transformation so that the coordinate system is aligned with the paraboloid. The metric can then be expressed as

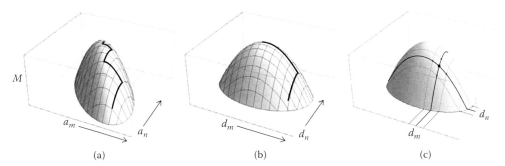

M a_m a_n d_m d_n d_m d_n

(a)   (b)   (c)

FIGURE 10.4 (a) Principle of modal wavefront sensing for a system with two basis modes, X_m and X_n. Maximization with respect to these modes is not optimal and several cycles are required to reach the peak. (b) The equivalent process for the derived modes Y_m and Y_n. As the axes of the paraboloid are aligned with the coordinate axes, the maximization of each mode is independent and the peak is obtained after a single measurement cycle. (c) Schematic illustration of the biased aberration measurements for the two-mode system. The positions of the five dots represent the aberration biases required for the measurements. This set of measurements is sufficient to reconstruct the whole paraboloid and hence to determine the correction aberration.

$$M \approx M_0 - \sum_i \beta_i \ d_i^2 \tag{10.6}$$

where the variables d_i are new coordinates derived from the original coordinates a_i and the constants β_i are related to the constants $\alpha_{i,j}$ of Equation 10.5. This coordinate transformation is equivalent to the derivation of a new set of aberration modes $Y_i(r,\theta)$, each of which is a linear combination of the original modes $X_i(r,\theta)$. The total aberration can, therefore, be expressed in terms of either set of modes as $\Phi(r,\theta) = \sum_i a_i X_i(r,\theta) = \sum_i d_i Y_i(r,\theta)$. Methods for derivation of the new modes are discussed in Section 10.8.

As each aberration coefficient d_i appears in a separate term on the right-hand side of Equation 10.6, it is possible to employ the one-dimensional quadratic maximization in sequence to each separate coefficient. As these calculations are mutually independent, only one maximization per mode is necessary to find the peak of the metric function. This is illustrated for the two-mode system in Figure 10.4b. The orientation of the coordinate axes with the axes of the paraboloid ensures that the correction aberration is found efficiently with the smallest number of maximization cycles. We can, therefore, consider the choice of modes Y_i as being optimum for this system.

The single-mode maximization procedure presented in Section 10.5 required three measurements with bias aberrations corresponding to $c = -b, 0, b$. This procedure can be applied, in turn, to each of the multiple aberration modes. It appears, therefore, that maximization of M when N aberration modes are present would require $3N$ measurements. However, as the bias aberration corresponding to $c = 0$ is common to all modes, the total number of measurements can be reduced to $2N + 1$. This is illustrated in Figure 10.4c.

10.7 Example of a Sensorless Adaptive System

In this section, we illustrate the principles of sensorless modal wavefront measurement using a simple adaptive optical system, as shown in Figure 10.5. The measurement system consists of an adaptive element, a convex lens, a subwavelength-sized pinhole, and a photodetector. The intensity of the light passing through the pinhole is chosen as the optimization metric. This is an appropriate choice as the measured intensity will be maximal for an aberration-free beam. We derive analytic expressions showing how the metric depends on the aberrations in the system and we demonstrate how choosing specific properties of the aberration modes allow its simplification to the form given in Equation 10.6.

FIGURE 10.5   Simple sensorless adaptive system consists of an aberrated input beam, a focusing lens, and a pin-hole detector.

The intensity at the pinhole plane is given by the modulus squared of the Fourier transform of the pupil field:

$$I(v,\xi) = \left| \frac{1}{\pi} \iint P(r,\theta) \exp\left[\iota r v \cos(\theta - \xi) \right] r \, dr \, d\theta \right|^2 \tag{10.7}$$

where ι is the imaginary unit, (v,ξ) are the polar coordinates in the pinhole plane, and $P(r,\theta)$ is the pupil function. It is assumed that the pupil is circular, with unit radius so that $P(r,\theta) = 0$ for $r > 1$. Assuming that the intensity of the input beam is uniform, we define $P(r,\theta) = \exp[\iota\Phi(r,\theta)]$. In the limit of a point-like pinhole, the metric (the signal measured by the photodetector) is proportional to the on-axis intensity, which is found by setting $v = 0$ in Equation 10.7. The metric takes a relatively simple form:

$$M = \left| \frac{1}{\pi} \int\limits_{r}^{1} \int\limits_{\theta}^{2\pi} \exp\left[\iota\Phi(r,\theta) \right] r \, dr \, d\theta \right|^2 \tag{10.8}$$

If the aberration amplitude is small, the exponent can be expanded to give

$$M = \left| \frac{1}{\pi} \int\limits_{r}^{1} \int\limits_{\theta}^{2\pi} \left[1 + \iota\Phi(r,\theta) - \frac{1}{2}\Phi(r,\theta)^2 + \ldots \right] r \, dr \, d\theta \right|^2 \tag{10.9}$$

Using the aberration expansion of Equation 10.1 in Equation 10.9 yields

$$M = \left| 1 + \frac{\iota}{\pi} \sum_i a_i \iint X_i \ r \, dr \, d\theta - \frac{1}{2\pi} \sum_i \sum_j a_i a_j \iint X_i X_j \ r \, dr \, d\theta + \ldots \right|^2 \tag{10.10}$$

where we have omitted the explicit dependence of X_i on (r,θ) for notational brevity. The second term on the right-hand side of the equation can be set to zero if we assume that the aberration modes X_i have zero mean value. This assumption is reasonable in practice as it is equivalent to choosing modes that do not contain any piston component. As piston has no effect on the metric measurement, this assumption does not restrict us in any practical way. The metric can then be written as

$$M \approx 1 - \sum_i \sum_j \alpha_{i,j} a_i a_j \tag{10.11}$$

where we define

$$\alpha_{i,j} = \frac{1}{\pi} \iint X_i X_j \ r \, dr \, d\theta \tag{10.12}$$

Equation 10.11 is now in the same form as Equation 10.5. The integral on the right-hand side of Equation 10.12 has the properties of an inner product between the modes X_i and X_j. Indeed, this inner product is identical to that defining the orthogonality of the Zernike polynomials. We can also note that the value of $\alpha_{i,j}$ in Equation 10.12 can equivalently be obtained by calculating the partial derivative of the metric:

$$\alpha_{i,j} = \frac{\partial M}{\partial a_i \, \partial a_j} = \frac{1}{\pi} \iint X_i X_j \; r \, dr \, d\theta \qquad (10.13)$$

If the modes X_i are chosen appropriately, then one can ensure that the modes are orthonormal, and $\alpha_{i,j} = \delta_{i,j}$ with $\delta_{i,j}$ being the Kronecker delta. Equation 10.11 simplifies to

$$M \approx 1 - \sum_i a_i^2 \qquad (10.14)$$

which has the desired form of Equation 10.6, where each coefficient can be maximized independently of the others.

We have seen that, for this sensorless adaptive system, an efficient scheme can be derived if the modes are orthogonal according to the definition of the inner product in Equation 10.12. The Zernike polynomials form a set of modes that fulfils this property. However, one is free to use other sets of modes that have similar mathematical properties but are better suited to a specific application. This approach is outlined in Section 10.8.

10.8 Derivation of Optimal Modes

The example described in Section 10.7 illustrates how the required modal properties can be extracted from a mathematical expression for the optimization metric. In principle, this approach can be applied to more-complex sensorless adaptive systems, such as adaptive microscopes using image-based quality metrics. However, the increased complexity of the mathematics describing the imaging process means that simple expressions for the inner product are not readily obtained. Similarly, conventional analytic modal sets, such as the Zernike polynomials, may not have the required mathematical properties. Generally, it is necessary to derive new sets of modes for a particular application to ensure optimal performance of the sensorless system. Three different methods that have been used to obtain optimum modes in these systems—analytical, numerical, and empirical—are explained below.

1. Analytical: In some situations, it is possible to define a set of analytic functions that are orthogonal with respect to the inner product. This method has been employed, for example, using Zernike modes in a focusing system (Booth 2006) (as described in Section 10.7) or Lukosz modes in a focusing system (Booth 2007b) or in an incoherent microscope using image low spatial frequency content as the metric (Débarre et al. 2007). This analytical approach is of relatively limited application, as there are few sets of known analytic modes that could be matched to any particular adaptive system.

2. Numerical: If the functional form of the inner product is known—for example, from the Taylor expansion or by direct evaluation of $\partial M/\partial a_i \, \partial a_j$—then the optimum orthogonal modes can be obtained numerically. As a starting point, one selects a suitable set of modes as a basis set (e.g., a subset of low-order Zernike polynomials). The optimum modes are then constructed from the basis set using an orthogonalization process based around the inner product. This has been shown in structured illumination and two-photon fluorescence microscopes (Débarre et al. 2008, 2009).

3. Empirical: If the functional description of the inner product is not available, the basis modes can be orthogonalized using an empirical process in which the form of the metric function is determined from image measurements. A sequence of images is acquired with different bias aberrations applied by the adaptive element. The corresponding metric measurements map out the shape of the metric

function in the vicinity of the paraboloidal peak. The new modes are obtained as a linear combination of the basis modes by finding the coordinate system that aligns the primary axes of the paraboloidal maximum with the coordinate axes. This approach has been demonstrated in structured illumination, two-photon, and harmonic generation microscopes (Débarre et al. 2008, 2009; Olivier et al. 2009). One possible implementation of this empirical method is outlined in Section 10.9.

The analytical approach is applicable to a very limited range of systems. The numerical approach has wider use and could be used with nonanalytic basis functions, such as the deformation modes of a DM. However, this requires an accurate numerical model of the mirror properties and full specification of other system parameters. The empirical approach is more widely applicable, as it could be applied to any sensorless system, as long as the metric has a paraboloidal maximum. This method neither requires full specification of the system nor an accurate model of the DM properties.

10.9 An Empirical Approach for the Derivation of Optimal Modes

It is possible to obtain optimal modes through experimental measurements, avoiding the need for a mathematical description of the inner product. One method for achieving this is presented in this section. In principle, this involves analyzing a set of measurements of the metric M with different applied bias aberrations. From these measurements, one can obtain the shape of the metric function and determine the orientation of the paraboloid encoded in the coefficients $\alpha_{i,j}$ of Equation 10.5. In turn, this enables the derivation of the optimal modes.

We choose a set of basis modes X_i, the definition of which can be arbitrary at this point—typically, one would choose either Zernike modes or a set derived from the properties of the adaptive element, such as mirror deformation modes. We assume that the modes have zero mean value and are all normalized to have a root-mean-square phase of 1 rad. We also assume that the basis modes do not contain components of tip, tilt, and defocus; as these components correspond to three-dimensional images shifts, they should be excluded from AO correction in a three-dimensionally resolved microscope system.

For any pair of basis modes with $i \neq j$, the metric value is measured for a number of test aberrations $\Phi_{(i,j),n}$, which contain certain combinations of the modes X_i and X_j. The mode combinations are chosen to have a constant total magnitude γ so that the nth test aberration is

$$\Phi_{(i,j),n} = \gamma \left[\cos\left(\frac{2\pi}{N} n \right) X_i + \sin\left(\frac{2\pi}{N} n \right) X_j \right] \tag{10.15}$$

where N is the total number of test aberrations. The constant γ should be chosen small enough such that the assumption of small aberration magnitude is valid; a root-mean-square phase value of <0.5 rad is usually appropriate. In terms of the shape of the metric function, the N measurements represent the height of the function along a circular path, a fixed distance away from the peak. This is illustrated in Figure 10.6. The collection of N metric measurements forms a polar plot, where the measured metric values are interpreted as radial coordinates and the numbers $2\pi n/N$ as angular coordinates. The resulting elliptical plot is a "section" through the multidimensional paraboloidal peak of M that shows the orientation of the paraboloid to the axes. Several of these plots, acquired for different pairs of modes, provide sufficient information to reconstruct M and obtain the principle axes, which correspond to the optimum modes. One can obtain the coefficients $\alpha_{i,j}$ by fitting a multidimensional ellipsoid $\sum_{i,j} \alpha_{i,j} \, a_i \, a_j = C$ into the data points, where C is a constant. An example of this process was initially presented by Débarre et al. (2008) as applied to an adaptive structured illumination microscope.

The selection of test aberrations is not restricted to the pairs of modes, as explained in the above illustration. It should be possible, for example, to use random combinations of all basis modes to provide random samples of the metric function. Similarly, one would obtain the coefficients $\alpha_{i,j}$ by fitting a multidimensional ellipsoid to the test points.

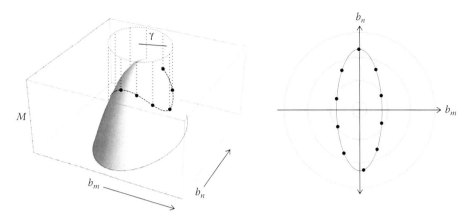

FIGURE 10.6 The shape of the multidimensional metric ellipsoid can be determined by performing measurements with different test aberrations applied (*left* image).

10.10 Orthogonalization of Modes

In principle, the set of new aberration modes is derived through orthogonalization of the basis modes. This process uses a particular inner product that depends upon the nature of the imaging system and the choice of optimization metric. In practice, one obtains values of the coefficients $\alpha_{i,j}$ of Equation 10.5 either through direct calculation of the inner product or indirectly through empirical measurements, as explained in Section 10.9. Once the coefficients $\alpha_{i,j}$ have been obtained, the orthogonalization process is facilitated though a matrix formulation. We outline this process as follows. First, we define a matrix **A** whose elements are the values of $\alpha_{i,j}$. Equation 10.5 can then be expressed in the convenient form

$$M \approx M_0 - \mathbf{a}^T \mathbf{A} \mathbf{a}. \tag{10.16}$$

where the vector **a** consists of the aberration coefficients a_i. We then derive an alternative representation by diagonalizing **A** using standard eigenvector/eigenvalue decomposition:

$$\mathbf{A} = \mathbf{V} \mathbf{B} \mathbf{V}^T \tag{10.17}$$

where **B** is a diagonal matrix whose on-diagonal entries β_i are the eigenvalues of **A**. The columns of **V** are the corresponding eigenvectors. The optimization metric then becomes:

$$M \approx M_0 - \mathbf{a}^T \mathbf{V} \mathbf{B} \mathbf{V}^T \mathbf{a} = M_0 - \mathbf{d}^T \mathbf{B} \mathbf{d} \tag{10.18}$$

where $\mathbf{d} = \mathbf{V}^T \mathbf{a}$. Denoting the elements of the vector **d** as d_i, we can show that Equation 10.18 is now equivalent to the desired form of Equation 10.6, where

$$M \approx M_0 - \sum_i \beta_i d_i^2 \tag{10.19}$$

The values of d_i are the coefficients of the aberration expansion in terms of the optimal modes $Y_i(r,\theta)$, where $\Phi(r,\theta) = \sum_i a_i X_i(r,\theta) = \sum_i d_i Y_i(r,\theta)$. The new modes Y_i can themselves be calculated from the basis modes as

$$Y_i(r,\theta) = \sum_j V_{i,j} X_j(r,\theta) \tag{10.20}$$

Once the values of $\alpha_{i,j}$ have been found for a particular sensorless AO system, it is a simple mathematical procedure to obtain a new set of orthogonal modes that are optimal for that system.

10.11 Conclusion

We have explained the motivation behind sensorless AO based on modal wavefront sensing, described the basic principles of the approach, and provided an outline of its mathematical basis. We presented, in particular, a method using sequential quadratic maximization, which has already proven to be effective in a range of adaptive microscopes. Central to this approach is the choice of aberration expansion in terms of modes that are orthogonal, where the definition of orthogonality depends on the imaging properties of the microscope and the optimization metric. This means that, in general, different sets of optimum modes are required for different adaptive microscopes.

The sensorless adaptive schemes are designed to be specimen-independent. Although the value of the optimization metric depends on both the aberrations and the specimen structure, the orthogonal form of the metric function does not change if the aberration modes are appropriately chosen. As explained in Section 10.5, the coefficients of the metric function depend on factors such as specimen structure, illumination intensity, and detection efficiency. This means that the width of the parabolic function can change between different specimens and image settings. However, the metric function retains the necessary form that permits independent optimization of each aberration mode.

It should be emphasized that once the optimum modes have been determined, the aberration measurement and correction process is fast, limited only by the rate at which images can be acquired. Calculations are not usually the speed-limiting factor, as neither the evaluation of the quality metric nor the maximization calculation of Equation 10.4 is time-consuming.

For the purposes of live specimen imaging, it is also important to consider the cumulative exposure during the correction process. When using an image quality metric such as total intensity, the value is averaged across the large number of image pixels. It follows that the image quality required for this measurement could be considerably lower than that required in the final image. The sequence of images used for the aberration correction process can be acquired at higher frame rates and lower illumination intensity than would normally be required for imaging. As an example, it was shown by Débarre et al. (2009) in an adaptive two-photon microscope that the total cumulative energy dose during correction was only twice the energy dose used in the final image acquisition.

Efficient optimization algorithms are not limited to the sequential quadratic maximization approach explained here. In some systems, it is possible to determine the aberrations using fewer measurements. For example, it has been shown that the system in Section 10.7 actually requires only $N + 1$ measurements to find N modal coefficients (Booth 2006). As it is based on the approximation of Equation 10.11, the method is limited in the range of aberration amplitudes that can be corrected. When the aberration amplitude lies near but outside of this range of approximation, the method still provides rapid convergence, but extra iterations may be required before acceptable correction is obtained. Further iterations may also be required if the modes used are not optimum, according to the definition we have used here. This has been shown in confocal and harmonic generation microscopes, where Zernike modes were used in an efficient but nonoptimum sensorless scheme (Booth et al. 2002; Jesacher et al. 2009). The methods described here could also be combined with more advanced control strategies to refine performance in practical adaptive systems (Song et al. 2010).

Acknowledgments

M. J. Booth was an Engineering and Physical Sciences Research Council (EPSRC) Advanced Research Fellow supported by the grant EP/E055818/1.

References

Booth, M. J. (2006). Wave front sensor-less adaptive optics: a model-based approach using sphere packings. *Opt. Express. 14*, 1339–1352.

Booth, M. J. (2007a, Dec). Adaptive optics in microscopy. *Philos. Transact. A Math. Phys. Eng. Sci. 365* (1861), 2829–2843.

Booth, M. J. (2007b). Wavefront sensorless adaptive optics for large aberrations. *Opt. Lett. 32* (1), 5–7.

Booth, M. J., M. A. A. Neil, R. Juškaitis, and T. Wilson. (2002). Adaptive aberration correction in a confocal microscope. *Proc. Natl. Acad. Sci. USA 99* (9), 5788–5792.

Born, M. and E. Wolf. (1983). *Principles of Optics* (6th ed.). Pergamon Press, Oxford.

Débarre, D., M. J. Booth, and T. Wilson. (2007). Image based adaptive optics through optimisation of low spatial frequencies. *Opt. Express. 15*, 8176–8190.

Débarre, D., E. J. Botcherby, M. J. Booth, and T. Wilson. (2008, Jun). Adaptive optics for structured illumination microscopy. *Opt. Express. 16* (13), 9290–9305.

Débarre, D., E. J. Botcherby, T. Watanabe, S. Srinivas, M. J. Booth, and T. Wilson. (2009, Aug). Image-based adaptive optics for two-photon microscopy. *Opt. Lett. 34* (16), 2495–2497.

Hardy, J. W. (1998). *Adaptive Optics for Astronomical Telescopes*. Oxford University Press, Oxford.

Hariharan, P. (2003). *Optical Interferometry*. Academic Press.

Jesacher, A., A. Thayil, K. Grieve, D. Dèbarre, T. Watanabe, T. Wilson, S. Srinivas, and M. Booth. (2009, Oct). Adaptive harmonic generation microscopy of mammalian embryos. *Opt. Lett. 34* (20), 3154–3156.

Kner, P., J. W. Sedat, D. A. Agard, and Z. Kam (2010, Feb). High-resolution wide-field microscopy with adaptive optics for spherical aberration correction and motionless focusing. *J. Microsc. 237* (2), 136–147.

Olivier, N., D. Débarre, and E. Beaurepaire (2009, Oct). Dynamic aberration correction for multiharmonic microscopy. *Opt. Lett. 34* (20), 3145–3147.

Press, W., S. Teukolsky, W. Vetterling, and B. Flannery. (1992). *Numerical Recipes in C* (2nd ed.). Cambridge University Press, Cambridge.

Sherman, L., J. Y. Ye, O. Albert, and T. B. Norris. (2002). Adaptive correction of depth-induced aberrations in multiphoton scanning microscopy using a deformable mirror. *J. Microsc. 206* (1), 65–71.

Song, H., R. Fraanje, G. Schitter, H. Kroese, G. Vdovin, and M. Verhaegen. (2010, Nov). Model-based aberration correction in a closed-loop wavefront-sensor-less adaptive optics system. *Opt. Express. 18* (23), 24070–24084.

Wright, A. J., D. Burns, B. A. Patterson, S. P. Poland, G. J. Valentine, and J. M. Girkin. (2005). Exploration of the optimisation algorithms used in the implementation of adaptive optics in confocal and multiphoton microscopy. *Microsc. Res. Tech. 67*, 36–44.

Zernike, F. (1934). Beugungstheorie des Schneidenverfahrens und seiner verbesserten Form, der Phasenkontrastmethode "Diffraction theory of the knife edge method and its improved version, the phase contrast method". *Physica (Utrecht). 1*, 689–704.

11

Implementation of Adaptive Optics in Nonlinear Microscopy for Biological Samples Using Optimization Algorithms

John M. Girkin
Durham University

11.1 Introduction

The desire of humankind to image the very large (stars and galaxies) and the very small (cells and microbes) has lead to significant advances in optics and improvements in the fundamental understanding of the imaging process. The influence on the two fields has been bidirectional with Abbé's diffraction theory, developed for optical microscopes, being applied to astronomical observations and the development of all modern instruments. The application of adaptive optics (AO) in optical microscopy is, thus, just a continuation of this historical record. A stimulus behind the introduction and development of low-cost AO into biology has been partly the most recent advances in microscopy, that of nonlinear imaging.

In this chapter, the practical synthesis of these two emerging methods will be explored both from a hardware and a software perspective. I will initially look at the physical considerations behind nonlinear imaging and then move to provide a link to the optical challenge of deep imaging that is covered in detail in Chapters 4 and 5 and in journal publications (Girkin et al. 2009). This chapter will then look at the practical considerations that should be made when designing AO into a nonlinear, beam-scanning microscope before examining optimization algorithms as a means of determining the shape of the mirror to be used. This has recently been termed "wavefront sensorless" AO, and the specific details on how they apply to nonlinear imaging are described in the following text.

11.1.1 Basics of Nonlinear Imaging

Before considering the exact role that AO now plays in nonlinear optical microscopy, it is worth considering the basic physics behind the nonlinear imaging methodology as this really sets the requirements for the adaptive system. The original underpinning theoretical physics behind multiphoton fluorescence microscopy was developed nearly 80 years ago by Prof. Göppert-Mayer (1931), as she worked in the then rapidly emerging new field of quantum mechanics. The concept that two photons could be absorbed simultaneously to cause an electron transition within an atom or a molecule was at the time very much a curiosity of theoretical physics. Practical use of this predication was not possible until the laser was invented in 1960 leading to the development of nonlinear optics, a technique also now used in microscopy and multiphoton absorption for high-resolution laser spectroscopy. The concept of nonlinear microscopy, where the returned signal is not linearly dependent on the excitation intensity, and more than one photon is required for the excitation process, was demonstrated in the 1970s, but the main complication was the laser power required to generate sufficient signal without destroying the sample. Q-switched lasers had the energy required but the length of pulse tended to lead to smoke and dust rather than an intact sample after imaging.

As laser physicists developed ever shorter pulse lasers, the possibility of a source with very high peak power but low average power became a reality, leading to the first article on two-photon fluorescence microscopy by Denk et al. (1990). The use of femtosecond pulses, albeit from a colliding pulse dye laser, meant that the average power on the sample was low, leading to minimal thermal damage; the energy in each pulse was also low minimizing explosive damage, but the photon density was high leading to sufficient two-photon excitation events. As this nonlinear absorption happens only at the focus of the excitation beam, the method provides inherent optical sectioning, enabling three-dimensional images to be built up, although clearly the size of that focal volume is crucial to the resolution and number of excitation events for a given laser power.

Such three-dimensional optical images were, however, already available through confocal microscopy. The crucial advantage of the nonlinear approach was the use of longer wavelength light that could penetrate further into samples because of the reduced scattering and also, in general, was not directly absorbed by endogenous fluorophores. The longer wavelength light also significantly reduced the risk of phototoxic damage that was always present if ultraviolet light was used. At this stage in the development of nonlinear microscopy, the major downside to the method was the requirement for a laser producing ultrashort pulses in the near infrared. The Ti:Sapphire laser provided a solution to this problem and the source issue was further helped by the significant improvements that were being made at the time in laser diodes for pumping Nd-based laser sources, which when frequency doubled were rapidly replacing the argon ion laser pump source (Girkin and McConnell 2005).

With such sources becoming routinely available, the concept of using nonlinear optical methods for imaging, with their inherent sectioning, grew rapidly with the movement toward imaging without adding exogenous fluorophores and second harmonic, third harmonic imaging of live biological tissue arrived along with coherent anti–Stokes Raman scattering (CARS), in which direct molecular vibration could be excited within molecules already present in the tissue (Zumbusch et al. 1999). One other crucial point that should be noted is that in parallel with the technological developments came molecular and genetic advances in biology, most specifically the development of fluorescent proteins that could be expressed very specifically within living cells with minimal perturbation to the biology (except for the presence of a large nonendogenous protein), producing a desire for the life scientists to image long term, in vivo, and at depth.

Now I will consider the basics of all nonlinear imaging. For a given number of molecules capable of excitation (N), the number actually excited is given by N times the probability of an excitation event taking place, P_{excite}. This term depends on several factors: the instantaneous peak power P_{peak}, which is a function of the average power P_{cw}; the pulse length τ; repetition rate f; the multiphoton absorption cross section at the given wavelength σ; and from our perspective inversely with the excitation volume V^2. In total, this gives

$$\text{Signal} \propto \frac{N\sigma\left(P_{cw}\right)^2}{\left(f\tau V\right)^2}$$

Thus, any increase in the focal volume, from whatever source, will significantly decrease the expected signal to the sixth power of the radius, assuming a spherical excitation volume. The squared terms come from two-photon fluorescence imaging; in the case of three-photon or high-order excitation methods, this becomes cubed or higher. It is thus clear that minimizing the excitation volume plays a significant role in maximizing the signal to noise in nonlinear microscopy and determining the maximum possible resolution.

11.1.2 Sources of Aberration

It is worth briefly considering the root cause of increases in the excitation volume. In general, there are two main sources of aberrations in nonlinear microscopy and the level of contribution is not exactly the same as in the case of linear-excitation-based systems. The first and most obvious source of aberrations is sample-related because of the local changes in refractive index causing different parts of the excitation wavefront to be delayed relative to other areas, leading to a significantly enlarged or nondiffraction-limited spot at the focal point of the system. These aberrations become larger with depth and are also very sample-dependent. The use of longer wavelength excitation sources required for nonlinear imaging methods tends to mean this effect is slightly lower than that for single-photon excitation, but the difference is small. The main advantage of longer wavelength in this context is that scattering, which is very wavelength-dependent, is reduced in the sample. Full details on these effects can be found in Chapter 5 of this book or summarized in a recent review (Girkin et al. 2009).

The other significant sources of aberrations in a nonlinear microscope are those present in the system. Although all beam-scanned microscopes are optimized for performance, some compromise has to be made in the design process, perhaps trading off resolution for improved light throughput or balancing out chromatic aberration effects because of the large difference in excitation and emission light in such microscopes. In addition, over time even in the best-maintained microscope components drift slightly, causing aberrations that may not be obvious to facility-based users. The use of different objectives can also increase the source of such aberrations and AO provides one method to ensure that systems are kept optimized even before sample aberrations add further complications in the imaging path.

There is one further source of aberration that is present in nonlinear imaging systems, which is not normally seen in more-conventional single-photon excitation microscopes: beam distortions present in the laser source itself. In conventional microscopes, the excitation light is delivered either directly from the laser or, more commonly now, via a single-mode optical fiber. This means that the light entering the scanning system has an intensity profile and wavefront determined by the optical fiber, which will in general be round, nonastigmatic, and typically have a Gaussian intensity profile. The spectral bandwidth of the laser source will also be small, and even when several laser sources with different wavelengths are delivered simultaneously, the wavelength range is frequently less than 100 nm. But this is not the case with nonlinear systems.

To achieve femtosecond laser pulses, the laser cavity of Ti:Sapphire lasers contains a number of extra optical elements (Girkin 2008). These will not be described in detail here but crucially they add distortion to the output beam of the laser, frequently making it astigmatic. When this beam is then expanded and manipulated through the beam-scanning optics, the optical train can increase these aberrations leading to large focal volumes. In addition, the broad spectrum of the laser pulse (typically around 50 nm for a center wavelength of 800 nm), even before the laser tuning is considered (680–1060 nm in a typical modern Ti:Sapphire laser), adds to this complication. Thus, the light source is adding to the aberrations that need to be corrected. Recently, the use of AO to remove these effects has been demonstrated in a two-photon microscope used for imaging ocular tissue (Gualda et al. 2010).

From this brief introduction, it is clear that AO potentially has a significant role in optimizing and improving the performance of the beam-scanned nonlinear microscopes, and with their inherent ability to image at depth in live samples, the practical implementation of such systems will now be considered.

11.2 Instrumentation

In this section, we will consider the basic practical requirements needed to include an AO element into the system. Initially, the core elements of such a microscope will be examined before moving on to the specific details of incorporating an AO element into the optical path. Throughout, we will consider the AO element to be reflective, although it is appreciated that liquid crystal devices are available that can operate in a transmissive mode. The basic principles of design outlined later, however, apply in all cases.

11.2.1 Basic Beam-Scanned System

The basic optical system for a beam-scanned microscope is shown in Figure 11.1, and the excitation path is common for both linear and nonlinear excitation systems. The main difference between the two excitation modes, apart from the coatings required on the dichroic elements and filters, is that in a confocal microscope, the signal light needs to be descanned before passing through the confocal pinhole and reaching the detector. In a nonlinear microscope, as will be discussed later, the signal light does not have to be descanned.

The light from the laser source is initially directed onto the scanning system, and the spot size of the beam is matched to the scanning optics. For most nonlinear microscopes, the scanning is undertaken using a pair of scanning mirrors that eventually, under computer control, raster-scan the beam over the sample. It is possible to use acousto-optic modulators to undertake the beam scanning, which does result in fast scanning, but the complication is that the material used in such devices tends to lengthen the laser pulse, and thus pulse-length-dispersion compensation is required.

In a scanning mirror system, the mirrors can be configured in two ways: close coupled or reimaged. In the close-coupled configuration, the mirrors are placed as close together as possible, with one mirror being directly reflected onto the other. The advantage of such an optical design is clearly that the scan head can be small and compact; the disadvantage is that that the beam on the second scanner moves

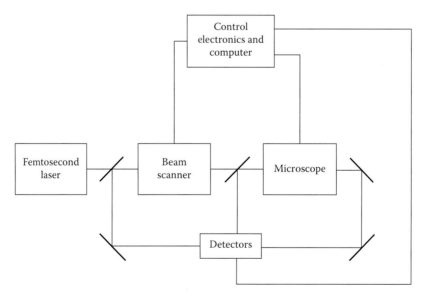

FIGURE 11.1   Basic components of a beam-scanned nonlinear microscope.

around slightly, and that the midpoint between the scanners is reimaged onto the back aperture of the microscope objective rather than onto a true stationary scan point. For a reimaging system, one scanning mirror is relayed onto the second, using either lenses or mirrors with the final mirror, then is reimaged onto the back aperture of the microscope objective, ensuring that one has a true stationary point in the back plane. Clearly, this involves more optical components but theoretically it provides a better imaging system. In practice, when real biological samples are placed on the sample stage both configurations work well and are used in accepted commercial microscopes.

As mentioned earlier, the scanned beam now has to be relayed onto the back aperture of the objective such that the beam pivots around the aperture rather than being spatially scanned. However, there is a further consideration. To exploit the full numerical aperture of the objective lens, the beam needs to fill the back aperture, which will vary from objective to objective. Typically, systems are optimized for the largest back aperture and for lenses with smaller apertures the user just accepts the light loss. The beam expansion required is normally around 10 times so that the 1 mm diameter laser beam then fills the 10 mm diameter objective. Again, this relaying and expansion can be achieved either using lenses or mirrors with the advantage of mirrors being their achromatic performance.

In nonlinear microscopes, as mentioned earlier, the detector does not have to be mounted such that the emission beam is descanned before reaching the detector as the nonlinear excitation provides inherent optical sectioning negating the need for a confocal pinhole for three-dimensional imaging. The detector can thus be placed in any of the three positions as illustrated in Figure 11.1. In the case of CARS and harmonic imaging, which are coherent processes, the natural direction of the signal light is forward rather than being isotropic in the case of fluorescence-based imaging. This means that a forward-mounted detector frequently provides the largest signal even if it is inconvenient to mount. The entire system is clearly controlled by computer, which operates the scanning mirrors, the light collection, and the Z-drive on the microscope.

11.2.2 Incorporation of the Adaptive Optic Element: Pre- and Postscanning

We now need to consider the insertion of the AO element and there are two options: before or after the scanning optics. In both configurations, the adaptive element needs to be reimaged to ensure that the AO element, scanner, and back aperture are all conjugate with each other. If this is not achieved, the beam will wander around the back aperture. It is clear why the scanners and objective need to be conjugate but perhaps this is not so obvious for the AO element. This can clearly be recognized if one considers two of the simplest aberrations: those of tip and tilt. Clearly, if the AO element needs to add a tip to the beam to compensate for a sample aberration, it will move the beam over the back aperture of the objective unless the two are conjugate. The same thinking also applies to all the complex aberrations and thus adjusting the optics to ensure all the relevant optical planes are correctly imaged onto one another is crucial and indeed the most difficult alignment task in an AO microscope system.

A typical postscanned system is shown in Figure 11.2, illustrating the key elements of previously published systems (Marsh et al. 2003) in a block diagram. The light from the laser is directed straight into the beam-scanning head. In the diagram, the typical 1 mm laser beam does not need to be expanded, or contracted, onto the 3 mm scanning mirrors. The output from the scan head is then reimaged onto the AO element using beam expansion if required. In the configuration shown, the coupling onto and off the AO element is provided via polarization optics. The incoming light passes through a polarizing beam splitter, through a quarter waveplate, and onto the mirror. On reflection, the beam passes through the waveplate a second time so that it is now reflected by the beam splitter before being directed onto the objective lens such that the AO element is conjugate with the back aperture. In these reimaging optics, the beam size can be adjusted to ensure optimal throughput of excitation light as described earlier.

An alternative configuration is to place the AO element before the scanning element (Wright et al. 2007), with the block diagram in Figure 11.3 showing the key components. Typically, the beam now needs to be expanded onto the AO element and then reduced in size as it is reimaged onto the scanning

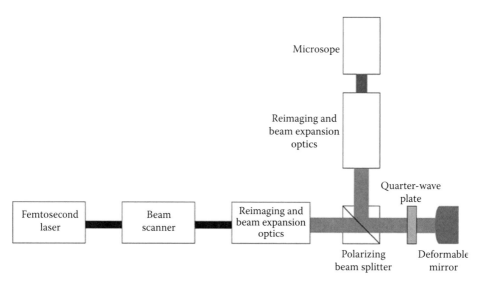

FIGURE 11.2 Basic configuration for an adaptive optics system placed after the scanning optics.

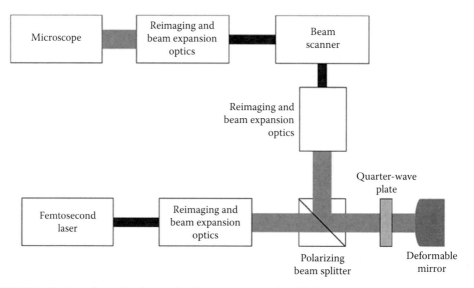

FIGURE 11.3 Basic configuration for an adaptive optics system placed before the scanning optics.

system before reexpansion and reimaging onto the back aperture of the objective. This would appear to be a more-complex configuration, but as will be discussed later, there are some practical advantages in some cases for such a setup. It should be noted here that a number of users have demonstrated that both methods work.

Before discussing which configuration is most suitable for a specific imaging system, an alternative to polarization coupling is considered. In the two systems described earlier, the beam hits the AO element at normal incidence but an alternative route is to mount the element at an angle such that the incoming and outgoing beams do not coincide (Chaigneau et al. 2011). Depending on the space available and the size of the mirror and associated optics, the mirror may be placed at 45° to the incoming light beam or at smaller angles. The disadvantage of using the AO at this angle is twofold. First, the travel available from the mirror is reduced by a factor of $\sqrt{2}$ (at 45°) and second, the actuators on the mirrors alter the light beam in different manners in the long and short axis. The clear advantage is that few optical components are needed and no complications associated with polarization are introduced. At smaller angles, clearly, these effects are reduced and thus consideration should be given to this alternative configuration and in a system using, for example, a diffractive liquid-crystal spatial light modulator (SLM) where the required modified beam naturally emerges at an angle via a blaze angle being applied to the device, there is a positive advantage.

11.2.3 Adaptive Optical Element

We now need to consider the optical device that is going to actually provide the aberration correction: the AO. As with any design parameter, the final choice of element rests on a large number of factors. The vast majority of AO systems have used reflective mirror devices and they will form the focus of this section but brief consideration is given to SLMs.

SLMs work by altering the relative phase of different parts of the incoming wave such that the correct wavefront is finally achieved. They have been used in microscopes for altering the wavefront to modify the point spread function at the focus of the objective (Neil et al. 2000a). Generally, these AO elements work as diffractive optical elements, which normally means that they are less efficient than reflective mirror–based devices. Conventionally, nematic SLMs are frequently slow with update rates rarely exceeding 100 Hz at present and are harder to control. Other liquid-crystal devices are available with higher update rates but these have other drawbacks including the need for additional polarizing optics. In the case of a deformable membrane mirror, voltages are simply applied to the individual actuators on the device and typically there are at most around 120 such actuators on a mirror. In the case of an SLM, they are driven as extension monitors with VGA or XVGA formats and thus a large number of pixel values need to be determined and then addressed—although the format makes the addressing easy, the more difficult aspect is to determine what values should be placed on the individual elements. As will be discussed later, if specific Zernike (or other orthogonal wavefront) terms are applied then the value determination is easier, for a more randomized search method the combination of possible terms is very large. Finally, when operating in a diffractive mode, the SLMs are clearly wavelength-dependent and thus with the broad spectrum of a femtosecond laser pulse, significant complications can occur and in a descanned configuration, the returned light will be altered in a totally different manner to the excitation light.

With these obvious disadvantages, the clear question is, would anyone consider using such a device? The answer is that SLMs have the potential of almost limitless correction of the wavefront. As will be discussed later, in the case of an AO mirror, there is a limitation to the movement of the mirror that is possible and hence on the level of aberration that can be corrected, although the number of actuators sets a limit on the order of aberration terms that can be corrected. In the case of the SLM, an almost infinite correction is possible by phase-wrapping, and high-order terms can also be easily corrected. With such devices either on their own (Neil et al. 2000b) or in combination with mirror AO elements (Wright et al. 2006) aberration correction is possible. Thus, such devices do have a potential role in nonlinear imaging systems, but will not be considered further here, as their use to date has been limited.

The main focus in this chapter is on the use of mirror devices and in the past 10 years a number of manufacturers of membrane mirror devices have become established. The exact method of actuation varies with some being electrostatically driven and some magnetically operated. Most devices appear as continuous mirror surfaces, whereas a few have a pixilated appearance because of their method of manufacture, although the actual mirror surfaces are continual. Some level of print-through has been seen with such devices, but it has not been reported that it actually affects the final performance, in particular in a nonlinear imaging system. The mirror price and the associated drive electronics also vary widely, and although clearly a final purchasing factor, the different drive methods and configurations will not be discussed further—neither will the detailed performance of the different devices, as these are changing continually and the reader is advised to search for the most-recent publications. The actual quantification of these devices will also not be considered, although again publications are readily available (Dalimier 2005). The consideration of the devices for a practical nonlinear system needs to cover the mirror size, actuator travel, actuator number, push and pull operation, speed of response, and update rate of the mirror and drive electronics along with the reflectivity.

The mirrors suitable for use in microscopy are now available from around 3 mm diameter (or the length of the diagonal for noncircular devices) up to 50 mm. Larger devices are available but they are generally not used for beam-scanned microscopy applications typically on the grounds of cost. Clearly, the larger the mirror, the greater the beam expansion required to cover the mirror so that all the actuators are used. It should also be remembered that the laser-beam size given is normally for the $1/e^2$ value—or full width half maximum—thus the beam expansion should take this into consideration. In a prescanned configuration, the smaller mirror size may be of interest, as it may be possible to direct the beam straight onto the mirror, reducing the number of optics needed. The smaller-size devices generally have fewer actuators and thus may be suitable only for correcting low-order aberrations.

The travel of each actuator sets the level of aberration that can be corrected, but this is linked to the mirror size. In the simplest case with a central actuator driven to its maximum movement, one will create a dimple in the mirror surface (for a pull mirror), which will appear as a concave surface thus providing a positive focus on the beam. The level of curvature will be determined by the distance of travel and the diameter of the mirror, where the smaller the mirror diameter, the steeper the curve (smaller radius of curvature) hence the greater level of focus, which is considered a low-order aberration. By simple geometry, it is thus possible to determine the maximum level of aberration that can be corrected for any given actuator travel and mirror diameter.

One then needs to consider the number of actuators that set the order of aberration that can be corrected. These currently range from 32 up to 1024, although for most applications at present there has been no demonstrated advantage in using more than 144 actuators and most work has been undertaken with between 32 and 52 actuators. As the number of actuators rises the complexity of mirror shapes and thus aberrations that can be corrected clearly increases but so does the control problem. If an optimized search routine is being used to determine the best mirror shape, there is clearly a time advantage in reducing the possible number of mirror patterns that can be applied.

Recently, a further parameter has been introduced into the decision-making process: that of having a mirror that can be made convex and concave via "push/pull" actuators. In the early work, the mirrors were initially biased with all actuators to a midpoint in their travel (which may not be at midvoltage setting because of the nonlinear nature of electrostatic attraction). From this position, the actuators could be moved either forward or backward to produce the required mirror shape—for example, to compensate for spherical aberration. The mirrors are now available that can be moved backward or forward from their unbiased position (position with zero drive volts present) and in recent, as yet unpublished work, in our laboratory these devices appear to work very well, although the optimization algorithms require some modification to account for a wider range of parameters as will be discussed later. The fact that a mirror does not need to be placed into an initial biased state also means that the effective travel, and hence level of correction possible, is increased as one should always consider the total travel available.

Most mirrors available have very similar update rates and impulse response times and are typically capable of operating at around 1–3 kHz. When selecting a mirror, the interface between the drive electronics and the computer should also be included in any speed calculations, and this includes the time that the computer may take to address the drive electronics. In this chapter, I am considering algorithm, or sensorless, AO, where the speed of mirror control does not set the overall speed performance of the system. For most biological imaging applications, the aberrations are not dynamic—in contrast, for example, to astronomical AO—and thus speed of response is not the first consideration in mirror selection. All manufacturers now supply their own drive electronics at a reasonable price and thus the use of homebuilt electronics is no longer a real consideration. The main questions to consider for the electronics, beyond speed, is the simple practical aspect of how easy it is to use the computer interface to the drive electronics and how the actuator pattern is sent to the mirror.

The final area of consideration is the mirror coating. Most devices are silver-, aluminum-, or gold-coated, all of which have acceptable reflectivity in the near-infrared wavelengths used for excitation in nonlinear microscopes. The gold mirrors are generally not as efficient in the visible portion of the spectrum, and thus, for systems in which the emission light also reflects off the AO element, this may be a consideration in the mirror selection. In particular, emission photons are precious as a great deal of effort has been taken to obtain them—and whereas losses on the excitation path can generally be compensated for by increasing the output from the laser, this is not possible for the emission without increasing the risk of sample damage and photobleaching in the case of fluorescence.

In summary, for a basic and easy-to-operate system, a deformable mirror is currently the method of choice for an AO beam-scanned microscope. The mirrors up to around 52 actuators are easy to control and perform well, and with the growth in push/pull mirrors, this option should be seriously considered. In terms of the coating, thought must be given to the range of wavelengths that the mirror is going to experience with the overriding consideration that emission photons are generally more precious than excitation photons. As one is generally in a nondynamic situation, the speed of response is likely to be the last factor in any design consideration.

11.2.4 Practical System

Figure 11.4 shows the full practical system for multiphoton fluorescence imaging that was used for the results presented later.

The output from a femtosecond Ti:Sapphire (Spectra Physics Mai Tai and Coherent Chameleon have both been used for the results shown here) was directed into a pair of scanning galvo mirrors that were imaged onto each other using 75 mm focal length concave mirrors. For the results presented, the laser was generally centered on 830 nm with a power on the sample of around 10 mW. The output from the scan head was then reimaged and the beam expanded via a polarizing beam splitter and quarter waveplate onto the deformable membrane mirror. For the work shown here, an Okotech 37 element mirror and 52 Imagine Optic mirror were used interchangeably as they both have a 15 mm diameter aperture, with the Imagine Optic mirror having a larger stroke and push/pull capability. The system has also been used with a small 3 mm^2 32 actuator mirror from Boston Micromachines with equal success. The reflection from the mirror then passes back through the waveplate and beam splitter to be reflected

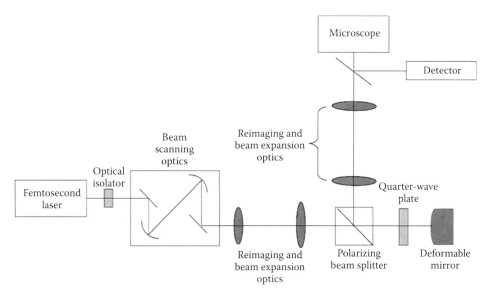

FIGURE 11.4 Basic nonlinear imaging system for multiphoton or harmonic imaging.

through the reimaging optics onto the back aperture of the microscope objective. For the results used later, the lens selected was a Nikon 0.75 NA air lens with a 20 times magnification. This lens is used frequently as it has a long working distance and the back aperture closely matches the 15 mm diameter of the AO mirrors, thus removing the need for further beam expansion. The fluorescent light is then collected from the dichroic mirror mounted close to the objective and the signal is digitized and recorded in the computer using home-written custom software. Z-axis (axial) movement is provided by mounting the objective on a piezo-driven objective scanner (PI Instruments) again controlled when required by the computer.

11.3 Determination of the Mirror Shape

Having discussed the design and build of a beam-scanned AO microscope, the next consideration is to determine what mirror shape will be used to correct for the aberration. In this book, there are several chapters that go into significant detail on the different options that are available and these should be read in conjunction with this specific section on nonlinear imaging. In the work described later, the approach adopted is to use an image metric-based optimization routine to determine the best possible mirror shape. This section will look at the different metrics that may be considered and which may be most suitable for different samples, a range of algorithms that can be used to optimize the system, and guidelines on how such methods may be implemented as well as some practical results. It is clear that computer programs will need to be written to control the mirror. Visual programming languages such as LabView provide an easy way to generate a simple and elegant user interface but generally are slow in undertaking mathematical calculations where languages such as Python, C, and C++ hold the upper hand. In our systems, we have used LabView, C++, and Python, and indeed combinations, and all have worked successfully. The final choice is clearly up to the individual user and their expertise and is not discussed further.

11.3.1 Image-Based Metrics and Optimization Algorithms

There are a number of metrics that can be used to assess the quality of an image but in the context of nonlinear beam-scanned microscopy, the considerations are clearly going to be different from a conventional fluorescence single-photon, wide-field microscope. One crucial point is that the time taken to image the full field of view is typically around 1/10th of a second for a beam-scanned microscope. Even with a guided search, this means that the optimization time will be very long if the system needs to undertake say 50 iterations. To overcome this problem, the user typically selects a small region of interest and then images only this area. The optimization metric then used has to be made robust against two significant parameters—noise and photobleaching—or, no matter how well the software is written or the system adjusted, the optimization will not converge. For this reason, absolute intensity, unless normalization is introduced, is not considered a viable option for conventional fluorescence but may be very suitable in harmonic or CARS imaging where there is no risk of photobleaching.

In this section, we compare the suitability of five metrics when used on nonlinear microscope images. These metrics are given in Table 11.1. They were selected based on providing a range of different types of metrics while also covering the most common metrics. The intensity squared metric, Equation 11.1, and the image variance metric, Equation 11.2, apply directly onto the image, without any preprocessing. $I(x,y)$ is the intensity of the (x,y)th pixel, $<I>$ is the average gray level of the image, and N is the number of pixels. The other metrics require some preprocessing of the image.

The metric in Equation 11.3 is the ratio of the filtered modulus of the Fourier amplitude over the unmasked modulus of the Fourier amplitude with $\mathcal{F}$ being the Fourier transform symbol. The mask used is typically a square mask of about 4 pixels (corresponding to around 240 mm^{-1} in the microscope object space) centered on the spatial frequency domain. The edge detection or Sobel filter metric,

TABLE 11. 1 Metrics Used to Assess Image Optimization

Metric Name	Analytical Formulation
Intensity squared Equation 11.1 (Langlois et al. 2002)	$\dfrac{\sum_{\text{pixels}} I^2(x,y)}{\left(\sum_{\text{pixels}} I(x,y)\right)^2}$
Image variance Equation 11.2 (Subbarao et al. 1993)	$\dfrac{\sqrt{\dfrac{1}{N}\sum_{\text{pixels}}(I(x,y)-<I>)^2}}{\sum_{\text{pixels}} I(x,y)}$
Fourier filter Equation 11.3 (Walker and Tyson 2009)	$\dfrac{\sum_{\text{pixels}}\left\|\mathcal{F}\{I(x,y)\}\right\|_{\text{masked}}}{\sum_{\text{pixels}}\left\|\mathcal{F}\{I(x,y)\}\right\|_{\text{unmasked}}}$
Sobel filter Equation 11.4	$\dfrac{\sum_{\text{pixels}}\sqrt{\left(\text{Sob}_x * I\right)^2(x,y)+\left(\text{Sob}_y * I\right)^2(x,y)}}{\sum_{\text{pixels}} I(x,y)}$
Wavelet filter Equation 11.5 (Kautsky et al. 2004)	$\sqrt{\dfrac{\left(\sum_{\text{pixels}} LH(x,y)^2+\sum_{\text{pixels}} HL(x,y)^2\right)}{\sum_{\text{pixels}} I(x,y)^2-\left(\sum_{\text{pixels}} LH(x,y)^2+\sum_{\text{pixels}} HL(x,y)^2\right)}}$

Equation 11.4, which we propose here, is calculated using the first derivative of the image whose approximation is obtained by the convolution of the image with two 3 by 3 kernels, S_x and S_y given by

$$S_x = \begin{bmatrix} -1 & -2 & -1 \\ 0 & 0 & 0 \\ +1 & +2 & +1 \end{bmatrix} \text{ and } S_y = \begin{bmatrix} -1 & 0 & +1 \\ -2 & 0 & +2 \\ -1 & 0 & +1 \end{bmatrix}$$

The metrics based on two-dimensional wavelet transforms have been suggested in Ferzli and Karam (2005) and Sherman et al. (2002). The wavelet functions are convenient to represent the local frequency content of an image. The first level of decomposition consists of applying a one-dimensional low- and high-pass filter—first vertically to each row of the image and second, horizontally to each column of the image. Thus, the result of a two-dimensional level-one wavelet decomposition is an array of four subimages: LL, LH, HL, and HH, where LL indicates a low-pass vertical filter followed by a low-pass horizontal filter, LH a low-pass vertical filter followed by a high-pass horizontal filter, and so on. LL is the approximation coefficient. HL, LH, and HH are, respectively, the horizontal detail, vertical detail, and diagonal detail of the original image. A second-level wavelet decomposition then consists of applying the same first-level decomposition twice, by reapplying the decomposition on LL. In Ferzli and Karam (2005), a third-level decomposition is performed and the edge widths are measured and summed up together on the absolute value of the vertical LH and horizontal HL coefficients image. The metric value is then the average of the edge widths summed on LH and HL. The readers who are interested in using this metric are strongly advised to read the references given and the further references within these articles. The full details of wavelet decomposition are well beyond the scope of this chapter on AO.

The other aspect of the optimization process to consider is the algorithm used to find the best solution. The original optimization-based nonlinear AO microscopy work used a modified hill-climbing algorithm (Marsh et al. 2003) or a genetic algorithm (Sherman et al. 2002), although subsequently results have been demonstrated using broader-range-optimization routines, including random search, adaptive random search, and simulated annealing (Poland et al. 2008). All of these methods work, although careful selection of the parameters within the algorithms is important to ensure that they converge efficiently. In the case of the genetic algorithm, for example, if too large a population is used, the initial search space is too large and convergence is slow, similarly if the mutation value is incorrect, the search is too random or in the case of a small mutation value too slow and only a small area of the optimization landscape is explored. Selecting the correct acceptable solution is also important. If too strict a limit is placed on the convergence, for example, the inherent noise in the system (laser fluctuations and detector noise) will mean that the system never reaches a true plateau. This is a particular challenge for nonlinear methods where small changes in excitation intensity clearly have a nonlinear effect in the signal. Thus, before running any optimization methods, users should develop an excellent understanding of the noise inherent in their system.

For the results presented later, we used a Simplex optimization routine (Nelder and Mead 1965) that had previously been used for astronomical AO. In brief, for a mirror with N actuators, we require $N + 1$ initial mirror shapes and these are generated through a lookup table after a series of experiments to determine the best basis set to start. A graphical interface is also used that enables the full image to be displayed and a region of interest selected for the optimization algorithm to analyze. Options are provided on the metric to be used and the metric value is displayed in a real-time graph. The mirror shape is also displayed with false color to express the wavefront distortion as a set of Zernike terms. The system is capable of recording all of the images and wavefronts produced for later analysis. One advantage of the Simplex routine in this context is that the original mirror shapes can be carefully selected to ensure that initial expected aberrations can be corrected quickly. Thus, the standard set of mirror shapes used initially specifically contains shapes to correct for spherical aberrations and astigmatism, which are known to be large in most samples. Specifically, when using the 0.75 NA air objective described earlier, we know that there will be spherical aberration caused when imaging samples mounted in water.

11.3.2 Results

The results presented below are taken from thick mouse backskin tissue stained with hematoxylin and eosin at a depth of around 100 μm. Although not normally used as a fluorescent label, hematoxylin and eosin provides a stale sample with minimal photobleaching thus ensuring that the metrics and not the experimental conditions are being tested. Figure 11.5 shows typical before and after images—in this case, using the Sobel metric. The gray line shows the position used for intensity cross sections.

Ten optimizations were run for each metric for the image shown in Figure 11.5. The average improvement of the metric value between the image before and after the optimization was calculated and a standard deviation was given to estimate how repeatable the optimization is.

The results are shown in Figure 11.6 demonstrating the relative improvement on an optimized image with all the different metric functions. To directly compare the metrics, once the image was optimized using each of the metrics, the improvement made was tested using every other metric. The horizontal axis in Figure 11.6 shows each of the metrics used to assess the final improvement. Clearly, the five sets of five vertical bars all have very different amplitudes. However, the bars within each block of five show

FIGURE 11.5 Sample of mouse back tissue imaged before and after optimization of the adaptive optics using a Sobel metric.

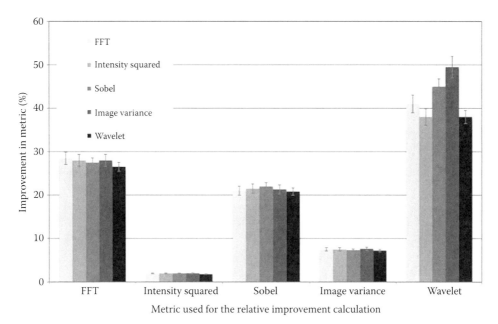

FIGURE 11.6 Relative improvement calculated with the metric given on the horizontal axis after an optimization performed by the metric given in the legend.

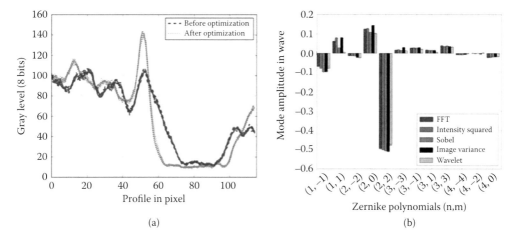

(a)

(b)

FIGURE 11.7 Typical output data of an optimization performed on the image in Figure 5. The profile (a) before (dark line) and after (gray line) optimization performed with the wavelet metric. The location of the profile is given on Figure 11.5 by the line. The histogram (b) shows the mirror Zernike mode amplitude by metrics.

the five different metrics used to actually perform the optimization. The height of each of the bars is very similar within each block (or using the same final metric). Although different algorithms produce different percentage improvements, the overall improvement in image quality was similar in all cases. In all cases, no matter which original metric was used, the final metric value was the same for each metric method. There is no difference in contrast or visual sharpness. The optimization is also repeatable and leads to the same mirror shape.

In Figure 11.7b, the mirror Zernike mode amplitudes are displayed and no large differences can be seen after optimizing with the different metrics. The error bars correspond to the standard deviation of the coefficient after 10 optimizations (they are there to show the variation of values, rather than the

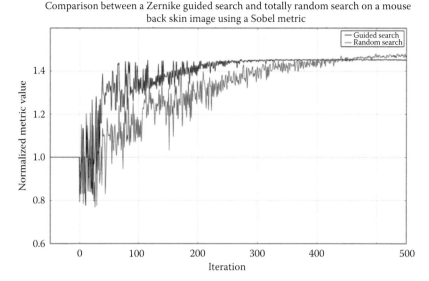

FIGURE 11.8 Normalized metric value shown against optimization iteration number using a standard mirror shape starting set and a set specific Zernike-mode shapes.

standard error). The profile comparison on Figure 11.7a shows a good improvement in the sharpness as the dark spot edges have become steeper. The 10 profiles before and after optimization have been superimposed and the overlap shows repeatability and stability of the optimization.

Typical optimizations require around 50–100 iterations to achieve acceptable results and up to 400 for the final optimal solution. To improve this using the Simplex algorithm, specific sets of Zernike terms were used in the initial set of mirror shapes. Using this approach, all metrics iterated to an optimal solution around four times faster and the plot of metric value against iteration number using the Sobel metric is shown in Figure 11.8. This demonstrates that if the original base of mirror shapes is carefully selected, then the optimization speed is significantly increased.

11.3.3 Three-Dimensional Optimization Methods

To the life-science researcher, the main advantage of nonlinear imaging is the ability to image deeply with inherent optical sectioning so that accurate and high-quality three-dimensional images can be obtained. When using AO, the question that has to be considered is, should the optics correct for every pixel in every image or can a correction for one area in one optical section then be applied to a set of optical sections before a new mirror shape is required? In addition, for a given type of sample—say, mouse brain—could a standard set of lookup tables be generated and used for all such samples?

Initially, let us consider the idea of correcting a single optical slice in one area and then applying this correction over the entire optical section (Poland et al. 2010). Using slices of mouse brain, labeled with fluorogold, images were taken and the image optimized at the center of the field of view. An image was then recorded using this mirror shape for the entire field of view and then optimized on a series of regions of interest across the sample to obtain the best possible solution for every small area rather than using the global correction. Using single isolated features in the sample as point sources, the resolution was then measured across the sample.

Figure 11.9 shows a plot of the resolution—initially with no correction, then with using one mirror correction across the entire sample, and finally with individual area corrections. It can be seen that the third option does provide the best resolution images but the improvement is small compared to the

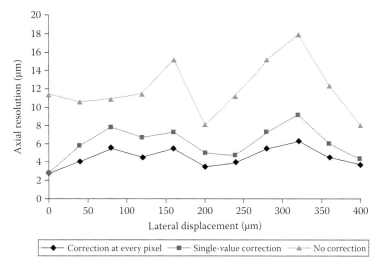

FIGURE 11.9 Spatial resolution measured across a sample for one optimal mirror shape across the sample and correction at specific points.

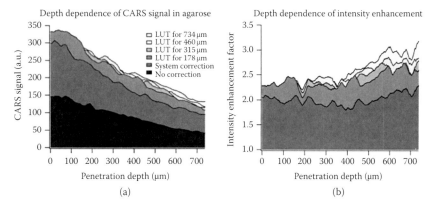

FIGURE 11.10 AO improvement of CARS intensity of 500-μm-thick agarose-bead sample. (a) Plot of CARS intensity as function of depth. (b) Plot of enhancement factors (intensity corrected/intensity uncorrected) as function of depth. Correcting for system aberrations only is suitable for small image depths, while sample- and system-induced aberration correction achieves factor three improvement at high depths.

single correction across the slice. Thus, using a single correction per slice appears to be a practical solution to selecting a mirror shape.

The other practical area to then consider is the use of lookup tables such that optimizations are undertaken at, say, every 50 μm and then these values used for the sections later recorded between the optimized planes (Wright et al. 2007). In an initial study using agarose and polystyrene beads, a CARS microscope image was optimized approximately every 125 μm and then that mirror shape was used for the next 125 μm.

Figure 11.10a shows the averaged intensity of 280 μm × 280 μm images (2.5X zoom) as a function of depth in the sample. The maximum sample depth was limited by the working distance of the objective to ~500 μm. It was observed that the heterogeneous distribution of the higher signal of the beads present in the sample did not change the average intensity substantially. In fact, the fluctuations in the curves arose from long-timescale intensity fluctuations of the laser used. Figure 11.10b shows the enhancement factors obtained from the same data by dividing the corrected by the uncorrected intensities. This again illustrates that one does not have to correct for every slice and that using one mirror shape either side of an optimized plane again provides a practical imaging solution. The work on AO within harmonic microscopes has also been undertaken and details on the configurations and methods used can be found in Jesacher et al. (2009) and Olivier et al. (2009).

11.4 Summary

In this chapter, we have described the basic requirements for a beam-scanned, nonlinear microscopy system incorporating AO. The considerations that should go into the optical system design have been explored specifically looking at the detailed requirements of where the AO element should be placed in relation to the scan optics and also emphasizing the importance of correctly reimaging the mirror, scanners, and back aperture of the objective so that they are all in conjugate planes. Some thoughts were then given as to what the users should look for in the actual AO element for specific tasks and the relative merits of different devices. We then examined five different metrics that can be used for optimization of nonlinear images and how these might be implemented in software. The results for the different metrics were then given clearly demonstrating that all the methods work and produce images that have both better image quality and resolution. Some thought was given to methods that can then be applied to use AO for improved imaging in full three-dimensional data sets and these methods were supported by experimental results.

It is clear that AO has a significant role to play in enabling life scientists to achieve the maximum possible from nonlinear imaging systems. Metric-based image optimization is a methodology that works in such microscopes and further research is needed to enable the best solution for any specific application to be determined but within the next few years it can be expected that AO nonlinear microscopes will help scientists to uncover new biological mechanisms through improved in-depth imaging.

Acknowledgments

The author acknowledges funding from the Engineering and Physical Science Research Council UK, a research excellence award from the British Heart Foundation in the United Kingdom, and European Union Framework 5 funding. In addition, much of the practical research and results were gathered by various team members over the past eight years: at Strathclyde University, Dr. Amanda Wright and Dr. Simon Poland; and at Durham University, Dr. Christopher Saunter and Cyril Bourgenot.

References

Chaigneau E., Wright A. J., Poland S. P., Girkin J. M., and Silver R. A. 2011. Impact of wavefront distortion and scattering on 2-photon microscopy in mammalian brain tissue. *Opt. Express*, 19(23), 6540–6552.

Dalimier E. and Dainty C. 2005. Comparative analysis of deformable mirrors for ocular adaptive optics. *Opt. Express*, 13, 4275–4285.

Denk W., Strickler J. H., and Webb W. W. 1990. Two-photon laser scanning fluorescence microscopy. *Science*, 248, 73–76.

Ferzli R. and Karam L. 2005. No-reference objective wavelet based noise immune image sharpness metric. In *Image Processing, ICIP, IEEE International Conference*, Genova, Italy 1, pp. 405–408.

Girkin J. M. 2008. Laser Sources for Non-linear microscopy. In B. R. Masters and P. So (Eds.), *Handbook of Biomedical Nonlinear Optical Microscopy, Oxford University Press*, pp. 191–216.

Girkin J. M. and McConnell G. 2005. Advances in laser sources for confocal and multiphoton microscopy. *Microsc. Res. Tech.*, 67, 8–14.

Girkin J. M., Poland S., and Wright, A. J. 2009. Adaptive optics for deeper imaging of biological samples. *Curr. Opin. Biotechnol.*, 20, 106–110.

Göppert-Mayer M. 1931. Ueber Elementar akte mit zwei Quantenspruengen [Elementary Events with two quantum jumps] *Ann. Phys.*, 9, 273–278.

Gualda E. J., Bueno J. M., and Artal P. 2010. Wavefront optimized non-linear microscopy of ex-vivo human retinas. *J. Biomed. Opt.*, 15, 026007.

Jesacher A., Thayil A., Grieve K., Débarre D., Watanabe T., Wilson T., Srinivas S., and Booth M. 2009. Adaptive harmonic generation microscopy of mammalian embryos. *Opt. Lett.*, 34, 3154–3156.

Kautsky J., Flusser J., Zitov B., and Simberov S. 2002. A new wavelet-based measure of image focus. *Pattern Recognit. Lett.*, 23, 1785–1794.

Langlois M., Saunter C., Dunlop C., and Love G. 2004. Multiconjugate adaptive optics: Laboratory experience. *Opt. Express*, 12, 1689–1699.

Marsh P., Burns D., and Girkin J. 2003. Practical implementation of adaptive optics in multiphoton microscopy. *Opt. Express*, 11(10), 1123–1130.

Neil M. A., Juskaitis R., Booth M. J., Wilson T., Tanaka T., and Kawata S. 2000b. Adaptive aberration correction in a two-photon microscope. *J. Microsc.*, 200, 105–108.

Neil M. A., Wilson T., and Juskaitis R. 2000a. A wavefront generator for complex pupil function synthesis and point spread function engineering. *J. Microsc.*, 197, 219–223.

Nelder J. A. and Mead R. 1965. A simplex method for function minimization. *Comp. J.*, 7, 308–313.

Olivier N., Débarre D., and Beaurepaire E. 2009. Dynamic aberration correction for multiharmonic microscopy. *Opt. Lett.*, 34, 3145–3147.

Poland S. P., Wright A. J., Cobb S., Vijverberg J. C., and Girkin J. M. 2010. A demonstration of the effectiveness of a single aberration correction per optical slice in beam scanned optically sectioning microscopes. *Micron*, 38, 200–206.

Poland S. P., and Wright A. J., and Girkin J. M. 2008. Evaluation of fitness parameters used in an iterative approach to aberration correction in optical sectioning microscopy. *Appl. Opt.*, 47, 731–736.

Sherman L., Ye J. E., Albert O., and Norris T. B. 2002. Adaptive correction of depth-induced aberrations in multiphoton scanning microscopy using a deformable mirror. *J. Microsc.*, 206, 65–71.

Subbarao M., Choi T., and Nikzad A. 1993. Focusing techniques. *J. Opt. Eng.*, 32, 2824–2836.

Walker K. N., and Tyson R. K. 2009. Wavefront correction using a Fourier-based image sharpness metric. *SPIE*, 7468, 74680O.

Wright A. J., Patterson B. A., Poland S. P., Girkin J. M., Gibson G. M., and Padgett M. J. 2006. Dynamic closed-loop system for focus tracking using a spatial light modulator and a deformable membrane mirror. *Opt. Express*, 14, 222–228.

Wright A. J., Poland S. P., Girkin J. M., Freudiger C. W., Evans C. L., and Xie X. S. 2007. Adaptive optics for enhanced signal in CARS microscopy. *Opt. Express*, 15(26), 18209–18219.

Zumbusch A., Holtom G. R., and Xie X. S. 1999. Three-dimensional vibrational imaging by coherent anti-Stokes Raman scattering. *Phys. Rev. Lett.*, 82, 4142–4145.

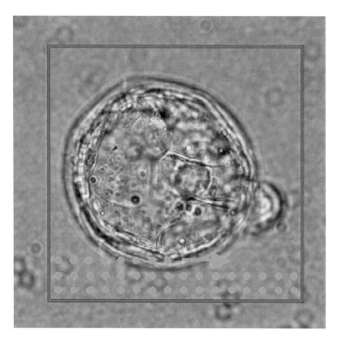

FIGURE 4.7 Transmitted light image of the mouse blastocyst sample. The recording of the wavefront at different raster positions is shown by the superimposed green dots. Two hundred and fifty-six wavefronts were recorded on a 16 × 16 grid. The size of the scan area indicated by the red frame is 130 × 130 μm.

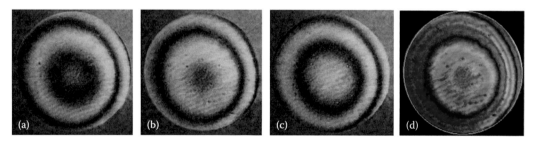

FIGURE 4.8 Phase stepping. (a through c) Raw interferograms where the relative phase is shifted by 0°, 120°, and 240°, respectively. (d) The wrapped phase (color coded) and amplitude (intensity) that was calculated from the raw images.

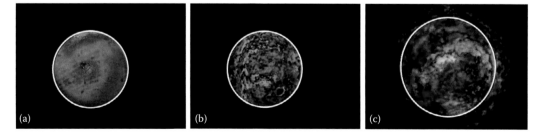

FIGURE 4.9 Examples for measured complex pupil plane wavefronts. The color represents the (wrapped) phase, while the brightness corresponds to the recorded intensity. (a) Wavefront recorded beside the cell (objective NA = 0.5, dry, condensor NA = 0.6). (b) Aberrating region of the sample (mouse blastocyst, objective NA = 0.5, dry, condensor NA = 0.6). (c) Aberrating region of the sample (mouse oocyte, objective NA = 0.8, oil, condensor NA = 0.6). The circles indicate the limiting aperture of the system.

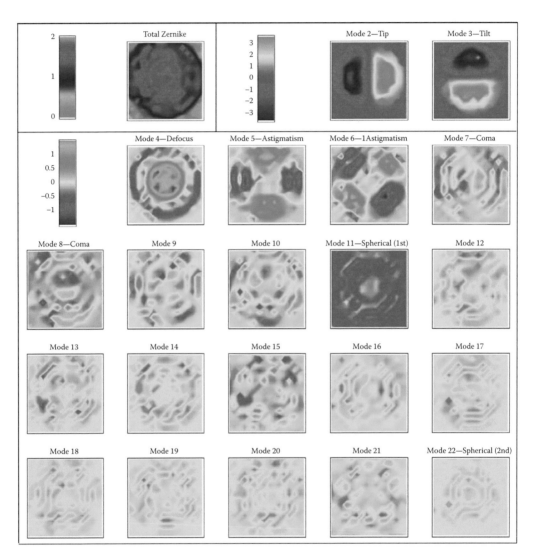

FIGURE 4.10 Zernike mode plots of the mouse oocyte sample, coefficients 2–22. For these images, the Zernike modal content was extracted from 256 wavefronts. Objective lens: Zeiss Plan-Neofluar 20×, 0.5 NA, dry lens. Condenser lens: Zeiss LD-Achroplan, 40×, 0.6 NA, correction ring. The scanned area was 100 × 100 μm.

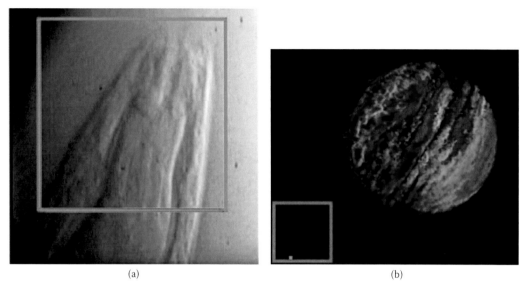

(a) (b)

FIGURE 4.14 (a) Transmitted light image of the specimen number 5 (*C. elegans*). The red box indicates the scanned region of 50 × 50 µm. (b) Video of the disturbance of the wavefront in the pupil plane of the lens as the focal spot scans across the specimen. Here the complex wavefront consisting of the amplitude $A(r,\theta)$ and the wrapped phase function $\phi(r,\theta)$ is displayed. The color encodes the phase, whereas the brightness corresponds to the amplitude of the wavefront. The green dot within the red frame in the lower left corner of the video indicates the relative position within the scanned area. (AVI video file online at http://www.opticsinfobase.org/oe/viewmedia.cfm?uri=oe-12-26-6540&seq=1.)

(1) (2) (3) (4) (5) (6)

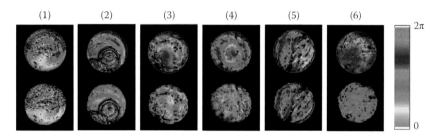

FIGURE 4.17 Specific interferogram examples from a particular position within the 16 × 16 grid that was recorded for each specimen. The color encodes the phase, and the brightness corresponds to the amplitude. The numbers (1)–(6) are the specimen numbers listed in Table 4.2. The upper part shows the measured initial wavefront; the lower part, a simulated correction of the Zernike modes up to $i = 22$.

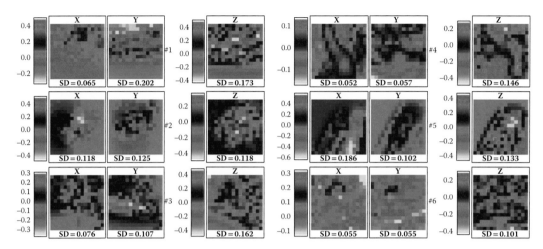

FIGURE 5.3 Specimen-induced distortion at a numerical aperture (NA) of 1.2. The specimen numbers 1–6 refer to the description in Table 4.2 of Chapter 4. The abbreviation SD denotes the standard deviation across the field of view (units are in micrometers).

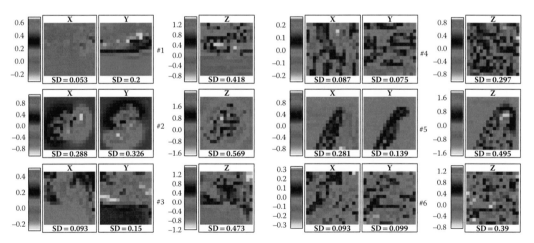

FIGURE 5.4 Specimen-induced distortion at an NA of 0.6. The specimen numbers 1 through 6 refer to the description in Table 4.2 of Chapter 4. The abbreviation SD denotes the standard deviation across the field of view (units are in micrometers).

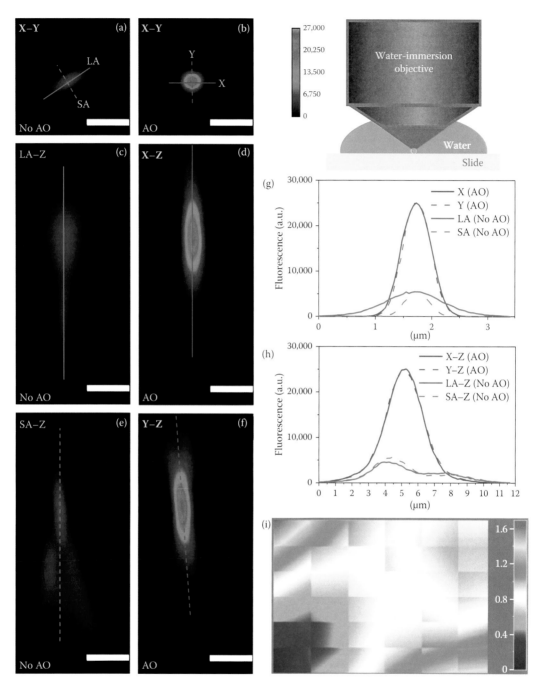

FIGURE 13.7 Correcting the system aberration improves images of a 500 nm fluorescent bead immersed in the design medium of water (inset, upper right). Lateral and axial images before (a, c, e) and after (b, d, f) AO correction. Z denotes the axial direction, and LA and SA denote the long and short axes, respectively, of the bead image before correction. (g) Intensity profiles along lines drawn in the lateral plane (red and green in a and b). (h) Similar profiles along lines drawn in the axial planes (red and green lines in c–f). (i) The final corrective wavefront for system aberration when using a Zeiss 20x, 1.0NA objective, in units of wavelength ($\lambda = 850$ nm). Scale bar: 2 μm. (Ji, N., et al., *Nat. Meth.*, 7, 2, 141–147, 2010.)

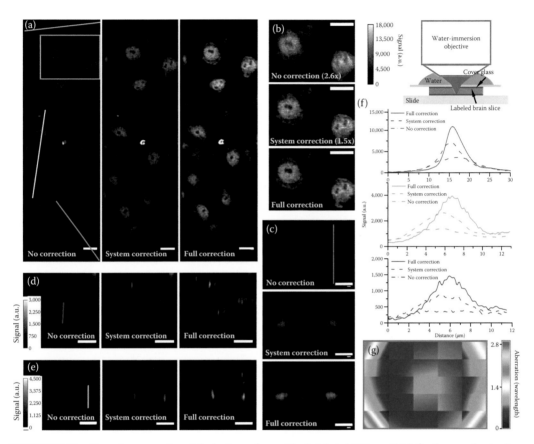

FIGURE 13.11 Aberration correction at the bottom of an antibody-labeled 300-μm-thick fixed mouse-brain slice. (a and b) Lateral images of a field of neurons acquired with and without correction as indicated (a), and magnified images from one subfield marked by the rectangle in a, with all images normalized to the same peak intensity (b). (c–e) Images in the axial planes defined by the yellow (c), green (d) and blue (e) lines in a. (f) Intensity profiles along the gray, purple, and orange lines in c–e. (g) The corrective wavefront in units of excitation light wavelength (850 nm), after subtraction of system aberrations, obtained with 36 subregions and direct phase measurement. Scale bars: 10 μm. (Ji, N., et al., *Nat. Meth.*, 7, 2, 141–147, 2010.)

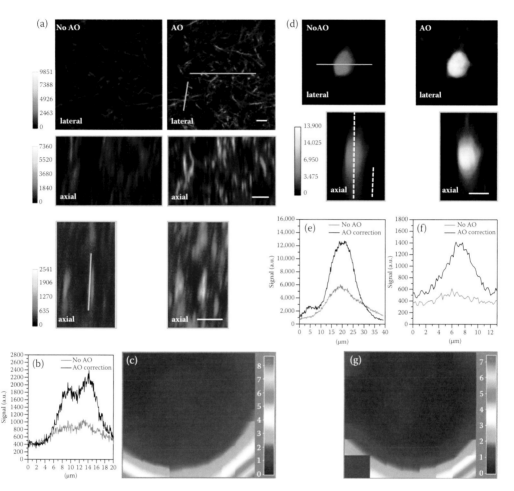

FIGURE 13.12 AO improves imaging quality for in vivo mouse-brain imaging: (a) Lateral and axial images of GFP-expressing dendritic processes 170 μm below the surface of the brain before and after AO correction; (b) Axial signal profiles along the white line in (a); (c) Measured aberrated wavefront in units of excitation wavelength; (d) Lateral and axial images of GFP-expressing neurons 110 μm below the surface of the brain with and without AO correction; (e, f) Axial signal profiles along the white dashed lines in (d) (left dashed line goes through a soma, and the right dashed line goes through a dendrite); (g) Aberrated wavefront measured in units of excitation wavelength. Scale bars: 10 μm. (Ji, N., et al., *Proc. Natl. Acad. Sci. USA*, 109, 22–27, 2012.)

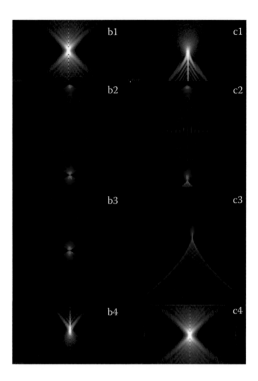

FIGURE 15.9 (b1) Calculated PSF of an ideal microscope. (b2) Plotted ray paths emerging from a point under the microscope (bottom fan) and exiting the microscope to focus at the image plane (top fan of rays). (b3) Enlarged view of rays converging onto the focal point. (c1) The PSF computed with a layer of water in the imaging path, showing the typical spherical aberration. (c2) The geometrical optics pattern due to refraction in a layer with refractive index different from the immersion oil (depicted in the top ray fan). (c3) Enlarged view of (c2) near the focus, with the geometrical optics pattern corresponding to the aberrated PSF of (c1). (b4) The PSF computed for the ideal microscope optics in which the adaptive element introduces the phase corrections, according to Equation 15.1, for the layer of water. The PSF shows a spherical aberration "inverted" to that in (c1). (c4) The PSF for imaging into water with phases corrected by the adaptive element, showing the recovery of the nonaberrated PSF (Kam et al. 2007).

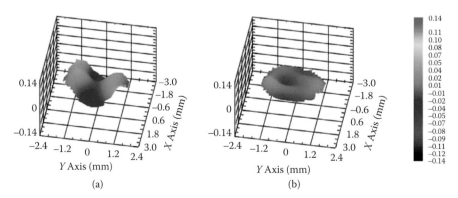

FIGURE 16.2 Wavefront change after AO compensation. The sample is a mouse heart, as described in Section 16.3.4 and imaged at 20 μm depth. (a) Distorted wavefront without the compensation. (b) Wavefront after AO compensation.

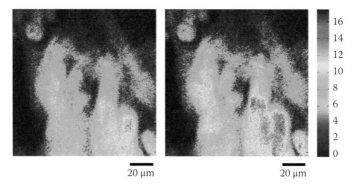

FIGURE 16.8 Mouse-tongue images at 80 μm depth without and with AO compensation, respectively.

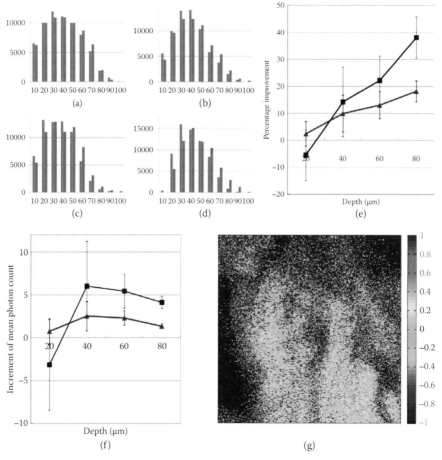

FIGURE 16.9 Signal improvement after AO compensation. (a–d) Histograms for the number of pixels according to their intensity; the x-axis represents intensity of the pixels. For example, 10 means that the pixels have 0–10% intensity of the maximum in the whole image, and 100 means 90–100% intensity pixels. The y-axis represents the number of pixels in the intensity range. The blue bars show the number of pixels before the compensation, and red bars show the result after compensation. Each histogram was normalized to itself. The distribution at (a) 20 μm, (b) 40 μm, (c) 60 μm, and (d) 80 μm imaging depth. (e) Percentage improvement according to imaging depth. The red line shows the improvement with background rejection (only the fluorescent area was calculated), and the blue line shows the improvement.

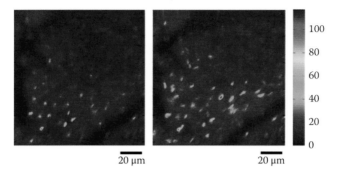

FIGURE 16.10 Mouse-heart images at 80 μm depth without and with AO compensation, respectively.

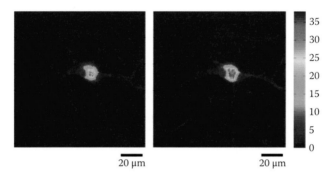

FIGURE 16.12 Mouse-brain images at 50 μm depth without and with AO compensation, respectively.

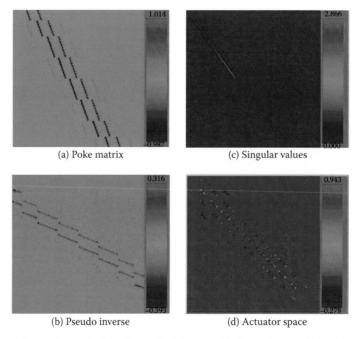

FIGURE 17.4 (a) Poke matrix obtained by the method described in the main text. (b) SVD inverse of the matrix in (a). (c) Singular-value modes with modes lower than 15% of maximum mode have been removed. (d) Actuator space obtained by multiplying (a) and (b).

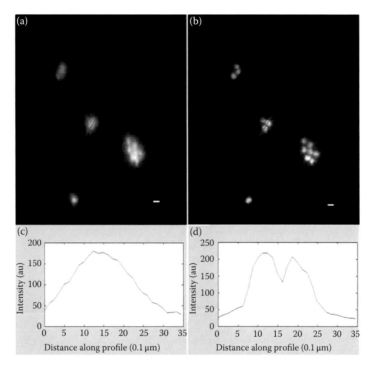

FIGURE 17.14 Real-time AO correction of 1 μm green fluorescent microspheres 20 μm beneath the surface of a fruit-fly embryo, using a 1 μm crimson fluorescent microsphere guide-star located at the center of the image. The size of the scale bar in the figures is 1 μm.

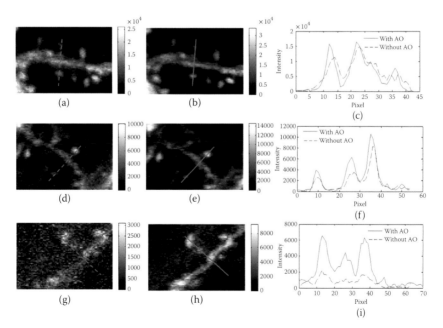

FIGURE 17.20 Confocal images with and without wavefront error correction, respectively, and the intensity profiles along the line indicated in the confocal images. (a) and (b) The images before and after correction for brain tissue with 15 μm thickness. (c) The intensity profile along the lines indicated in (a) and (b). (d)–(f) The images before and after correction and intensity profile for brain tissues with 50 μm thickness. (g)–(i) The results for brain tissue with 100 μm thickness.

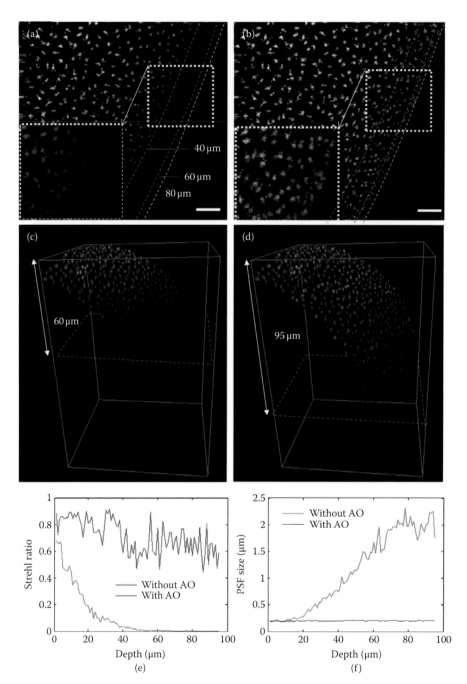

FIGURE 17.27 Comparison of penetration depths between without and with correction for imaging of cycle 13 fly embryos with EGFP-Cnn label. (a)–(b) The maximum intensity projection of the scan series from the top surface to 100 μm with and without adaptive optics. (c)–(d) The three-dimensional reconstructions with and without AO. (e)–(f) The Strehl ratio and PSF size change for different depth. The red and blue lines indicate without and with AO, respectively. The scale bar is 10 μm.

12

AO Two-Photon Fluorescence Microscopy Using Stochastic Parallel Descent Algorithm with Zernike Polynomial Basis

Yaopeng Zhou
Abbott Laboratories

12.1 Introduction

The two-photon fluorescence microscope (TPFM) provides a powerful tool for deep-tissue imaging. Its unique ability of seeing through the tissue surface allows researchers to acquire high-resolution images of cell-level details up to a few millimeters below the tissue surface. Various biological tissues such as brain [1], heart [2], kidney [3], and skin [4] have been studied by imaging with TPFMs. The TPFM with adaptive optics (AO) described in this chapter applies to the study of stem cells and microenvironment in the bone marrow of mouse skull bone. The constructed AO TPFM is a collaboration effort between Boston University (BU) and Wellman Center for Photomedicine.

Hematopoietic stem cells (HSCs) reside in bone marrow and are responsible for the maintenance of all cells found in blood (white blood cells, red blood cells, and platelets) and the immune system (B cells, T cells, natural killer cells, and dendritic cells) [5]. HSCs and their clinical applications have been widely studied [6, 7].

Much remains to be studied about HSCs and the microenvironment that HSCs reside in. Ex vivo experiments on HSCs are limited by the incomplete understanding of the biological functionality of the HSCs and the microenvironment. In vivo imaging of the HSCs has been explored by utilizing single- and two-photon scanning laser microscopes [5].

Because of optical aberrations, the TPFM is limited by both imaging depth and imaging resolution. The presented results show the imaging depth to a few tens of micrometers. However, it is desirable to image through the entire flat bone, which can be several hundreds of micrometers.

12.1.1 Advantages of Two-Photon Fluorescence Microscopy in Tissue Imaging

Compared to single-photon fluorescence microscopy, the two-photon fluorescence imaging technique has three major advantages to allow imaging at the cellular level in deep tissue. First, two-photon fluorescence imaging provides intrinsic sectioning ability when applied to thick-tissue imaging. The two-photon fluorescence excitation is directly proportional to the square value of the power density of the excitation light at the focal volume. Out-of-focus fluorescence excitation drops off significantly faster than that of single-photon excitation; in turn, less background is produced in the imaging.

Second, two-photon fluorescence imaging uses near-infrared light for fluorescence excitation. The near-infrared light suffers significantly less light scattering when traveling through tissues, compared to visible light, ranging from 400 to 600 nm. Near-infrared light ranges from 800 to 1500 nm.

Both scattering and absorption contribute to light attenuation in tissue traveling. The main absorbers of light in blood-perfused tissues are oxyhemoglobin and deoxyhemoglobin [8]. Their absorption spectrum shows a minimum between 700 and 1000 nm.

Third, the two-photon fluorescence imaging technique has less photobleaching than single-photon imaging. Both single- and two-photon fluorescence imaging suffer photobleaching, a photochemical process that destroys the fluorescence agents. For two-photon fluorescence excitation, photobleaching is directly proportional to the time-averaged square value of power density at the light illumination path. Photobleaching is confined to the vicinity of the focal volume, since photobleaching drops quadratically below or above the focal plane [9]. However, for single-photon fluorescence imaging, photobleaching occurs at all regions of the light illumination path.

12.1.2 Two-Photon Fluorescence Excitation and Emission

A fluorescent molecule in a simple form can be described as having two different energy states: a ground state and an excited state. A ground state electron jumps to the excited state by absorbing external energy such as that carried by photons from an illumination light. When the excited electrons shift back to the ground state, they can release energy by emitting fluorescence photons. For the single-photon fluorescence excitation, a lone photon carries enough energy to excite a fluorescent molecule, resulting in the emission of a fluorescent photon. Two-photon excitation is based on the idea that two photons of lower energy can excite a fluorescent molecule, resulting in the emission of a fluorescent photon, typically at a higher energy than either of the two absorbed photons. In general, the emitted photon in two-photon excitation has twice the frequency and half the wavelength as that of the absorbed photon.

Two-photon fluorescence excitation is a nonlinear fluorescence imaging technique, which also includes other techniques such as three- or four-photon fluorescence excitation, second- or third-harmonic generation, and coherent anti-Stokes Raman spectroscopy (CARS).

The most commonly used fluorescent agents such as 4',6-diamidino-2-phenylindole, fluorescein iso-thiocyanate, green fluorescent protein (GFP), Cy3, Cy5, Texas red, and quantum dots have excitation spectra in the 400–700 nm range, and the laser used for two-photon fluorescence excitation lies in the 700–1400 nm range. The most commonly used laser source is the mode-locked, femtosecond pulsed laser. The high-rate pulsed laser provides both temporal and spatial concentrations of light photons optimized for two-photon fluorescence excitation.

12.1.3 Optical Resolution and Tissue Optical Aberrations

The resolution for the two-photon excitation is determined by the size of the illuminated focal volume. The excited fluorescence intensity is proportional to the square of the illumination light intensity as shown in Figure 12.1. An empirical approximation can be used to calculate the optical resolution of the TPFM [10].

Tissue-induced optical aberrations reduce both imaging depth and resolution. To conduct deep bone-marrow imaging, the illumination laser beam needs to pass through several layers inside the mouse skull bone. The skull structure has a "sandwich" characteristic; that is, the mouse skull consists of two thin layers of bone, sandwiching cavities. Inside the cavity, there are bone marrow and blood vessels.

To reach the bone marrow, the illumination laser beam needs to pass through the surface bone layer first. From the optical imaging standpoint, these multiple layers of different materials affect the image quality. The index of refraction for each layer in the bone structure varies. The refractive index of decalcified, air-dried bone, at room temperature, is 1.530. Air has a refractive index of about 1.0003, and water has a refractive index of about 1.33. When the converging illumination laser light passes through

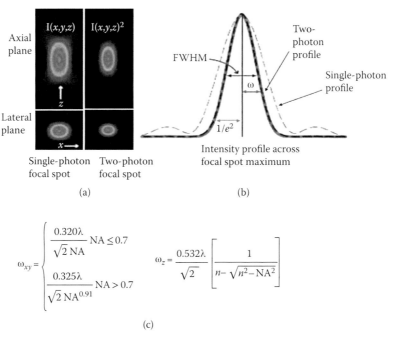

$$\omega_{xy} = \begin{cases} \dfrac{0.320\lambda}{\sqrt{2}\,\text{NA}} & \text{NA} \leq 0.7 \\[2em] \dfrac{0.325\lambda}{\sqrt{2}\,\text{NA}^{0.91}} & \text{NA} > 0.7 \end{cases} \qquad \omega_z = \dfrac{0.532\lambda}{\sqrt{2}} \left[\dfrac{1}{n - \sqrt{n^2 - \text{NA}^2}} \right]$$

(c)

FIGURE 12.1 (a) Normalized intensity maps at the axial and lateral planes of the focal spots from single-photon and two-photon fluorescence imaging. (b) Intensity profile across focal-spot maximum in axial plane and full width at half maximum (FWHM) of the fitted Gaussian function (dashed black line). (c) The PSF of a two-photon excitation process can be fitted to the Gaussian function, and the FWHM can be calculated. $2\sqrt{\ln 2}\,\omega_{xy}$ is the FWHM of the fitted Gaussian function in the lateral direction. $2\sqrt{\ln 2}\,\omega_z$ is the FWHM of the fitted Gaussian in the axial direction. NA is the numerical aperture of the objective, n is the refractive index of the imaging media, and λ is the wavelength of the illumination laser. (From Zipfel, W.R., et al., *Nat. Biotechnol.*, 21, 11, 1369–1377, 2003.)

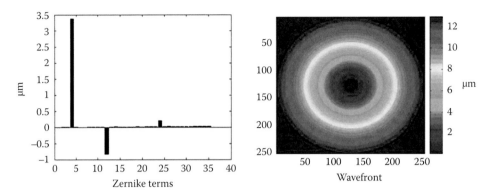

FIGURE 12.2   MATLAB simulation of optical aberrations from the index mismatch.

these materials, its path will be locally affected by their refractive indices. The local structure of the bone cavities can add extra complexity to the beam path. All these contribute to optical aberrations.

While the complex nature of heterogeneous materials in mouse-skull tissue can lead to complex aberrations, it might be expected that one type in particular, spherical aberration, will be prominent. Spherical aberrations are caused by index mismatches that are oriented perpendicularly to the optical axis. Booth et al. characterized such an aberration in a simplified geometric approach [11]. In attempting to show the effect of the index mismatch in a dry mouse skull, we quantified the optical aberrations at 50 μm below the bone surface in MATLAB® (MATLAB, MA) according to Booth's geometric model, and the final results are shown in Zernike modes [12] to identify the aberration.

The three terms standing out on the Zernike polynomials are *defocus* (Zernike term no. 4), *first-order spherical* (Zernike term no. 12), and *second-order spherical* (Zernike term no. 24). One should be aware that the bone structure is a much more complicated imaging environment, and various other optical aberrations are presented besides spherical aberrations (Figure 12.2).

Despite its capacity to image deeply into tissue, two-photon fluorescence excitation cannot avoid signal degradation of the image with deep subsurface penetration, due to tissue scattering, absorption, and optical aberrations. Both scattering and absorption ultimately limit the depth that can be imaged successfully. However, aberrations can, in principle, be measured and compensated through an AO control system.

12.2 Background

Much work has been done to analyze the effects of imaging with a high NA objective on a specimen. Hell et al. illustrated the index mismatch–caused aberration in a confocal microscope [13]. Several methods have been suggested to overcome this decrease in signal and resolution. Sheppard et al. illustrated inserting weak aberration correction lenses and also altering the tube length of the objective [14]. The latter employed a static correction that worked to restore much of the resolution at a certain depth, but it cannot be easily adjusted for varying depths. Hellmuth et al. showed that if one overcorrects for the cover-glass thickness or uses a less-than-nominal cover-glass thickness, this would compensate for spherical aberration at a certain depth in a specimen with index mismatch [15]. Wan et al. changed the immersion medium to do the same [16]. Booth and Wilson used both of these methods to compensate for spherical aberration in skin as well as employing an iris to reduce the pupil area [17]. All of the methods discussed use static means to correct for specimen-induced spherical aberration at a specific depth, not dynamic means of compensation as a beam is scanned deeper into a sample.

With recent advances in AO, it is now possible to employ dynamic compensation of aberrations, which can be applied during the scanning process for a continuous volume of aberration-free imaging. AO has been applied to correct optical aberration in a multiphoton microscope setup by several groups with different approaches. Neil et al. have demonstrated adaptive correction on fluorescent beads by first measuring the aberrations of the beam with a modal wavefront sensor and then applying the phase conjugate of the measured aberrations with a liquid-crystal spatial light modulator [18]. Rueckel et al. applied adaptive wavefront correction based on sensing the wavefront of coherence-gated backscattered light [19]. The system combined a two-photon scanning laser microscope with an interferometer for wavefront measurement.

Several groups also tried to take the approach of correcting optical aberrations without measuring the wavefront error. Sherman et al. demonstrated AO correction without a wavefront sensor by using a genetic learning algorithm [20]. Wright et al. used AO in a CARS microscope with a random search algorithm [21]. The setup employs a deformable membrane mirror and a random search optimization algorithm to improve signal intensity and image quality at large sample depths. Marsh et al. showed the technique of "hill climbing" to drive the AO. Ji et al. used image deviation between a reference image and an aberrated image to optimize a subregion of a spatial light modulator [22]. The optimization process is repeated to allow the correction of optical aberrations across the entire pupil. Débarre et al. used a sequential mode–based optimization algorithm on a deformable mirror–based AO system to optimize images in two-photon fluorescence excitation imaging [23].

12.3 Method

Typical AO systems use closed-loop feedback control to correct the optical aberrations of the illumination and imaging optics. The closed-loop feedback system includes a deformable mirror, a wavefront sensor, and a computer-controlled feedback loop. It is challenging to apply the wavefront sensor–enabled AO system to the TPFM. The main reason is that the encoded wavefront aberration is wavelength-dependent. The collected fluorescence light contains undesired wavefront information for aberration correction applied to the illumination light beam. Nevertheless, without the measured wavefront aberration, the AO correction can also be achieved on a trial-and-error type of approach in the two-photon fluorescence imaging setup.

In the sensorless AO system, the control system is set to maximize the fluorescence intensity by arbitrarily perturbing multiple degrees of freedom of the deformable mirror (DM). The AO controller uses a scalar measure, the metric, of the fluorescence signal (e.g., intensity, contrast, and resolution) as the feedback signal. The deformable mirror (DM) shape in the sensorless AO can gradually converge to an optimum steady state by reading the feedback fluorescence metric. The sensorless AO is inefficient in comparison to conventional AO, but has proven to be effective for complex AO applications in which wavefront sensors are impractical [24, 25]. At the same time, the sensorless AO system can significantly reduce the complexity of the optical design and construction.

Due to the nonlinear fluorescence excitation nature of two-photon microscopy, the sensorless AO can be very effective for optical aberration correction. A small improvement in the focal spot radius (r) can result in a large improvement in signal intensity, due to the nonlinear excitation mechanism. The illumination intensity is inversely proportional to r^2, and the fluorescent intensity is proportional to the value of illumination intensity squared. Thus, the fluorescence intensity goes up quadratically with decreasing r.

Various sensorless AO algorithms have been developed in recent years and they can be divided into two categories: sequential and parallel methods. Hill climbing, genetic algorithms, and random searches are a few examples of algorithms that achieve the optimal state by sequentially perturbing the DM degrees of freedom. On the other hand, stochastic parallel gradient descent (SPGD) is a more suitable approach for AO in vivo imaging. The SPGD, proposed by Vorontsov, perturbs the DM degrees of freedom simultaneously, yielding faster convergence than the sequential method [24].

The multi-dithering method also allows gradient estimation in parallel. Small perturbations are applied simultaneously to all channels in the form of harmonic signals with different dithering frequencies for each channel. To determine gradient components, dithered carriers are demodulated by synchronous detectors and passed through low-pass filters [26]. The time required for metric gradient estimation is independent of the number of control channels, which is a significant advantage in the control efficiency over the sequential gradient method. However, there is a significant drawback in the multi-dithering method. The signal-to-noise ratio in the measurement channel decreases proportionally with the number of control channels [27]. A high signal-to-noise system is required in this application. Unfortunately, two-photon fluorescence imaging is inherently a low-light imaging modality, particularly at depths for which aberration correction is required. Thus, signal-to-noise ratios are relatively low.

12.3.1 Stochastic Parallel Gradient Descent Algorithm with Zernike Polynomial Basis

The stochastic parallel gradient descent (SPGD) algorithm maximizes or minimizes a metric signal corresponding to system performance in an iterative control loop based on randomized perturbations of the system's controllable inputs. The control loop for AO includes temporarily changing the mirror shape by applying perturbations on its independent control inputs (e.g., DM actuators), assessing the effect of these perturbations on the metric (e.g., fluorescence intensity), estimating a metric gradient with respect to control input perturbations, and finally updating the mirror shape to a state that should incrementally increase the metric. A block diagram is shown in Figure 12.3. First, the current deflection state of each input channel of the DM is perturbed by a unit amount, in a direction that is randomly signed for each input channel. Note that the obvious basis for input to the DM is the DM actuators, with each actuator's input serving as an independent control channel. Other basis sets can be established as well. For example, one could control the DM based on basis sets coordinated over all actuators, with independent inputs corresponding to Zernike modes. The fluorescence signal is measured and stored as the metric value. Then the control channels are perturbed in the opposite direction, and the metric is recorded again. The difference in recorded metrics for positive and negative perturbations provides a gradient in the multidimensional input basis space, and a new state for the DM is determined by multiplying each input channel's signed perturbation by the global gradient and adding this value to the previous input channel state. When the measured metric reaches a steady state, the loop is stopped.

Figure 12.4 shows the concept of the application of AO in a fluorescence microscope. The AO loop utilizes the measured fluorescence intensity as the metric and builds the DM iteratively based on the gradient information from each loop cycle.

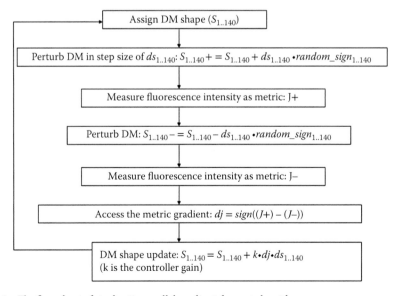

FIGURE 12.3 The flow chart of stochastic parallel gradient descent algorithm.

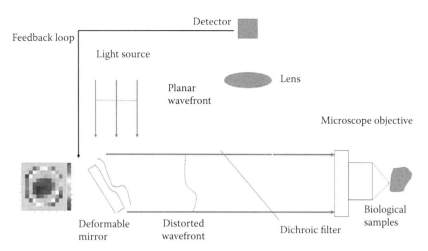

FIGURE 12.4 The deformable mirror is used to manipulate the incoming light beam wavefront to compensate the aberrations in the microscope objective and biological sample. A planar wavefront, originated from the light source, is bouncing off the DM, and forms a focal plane on the biological sample, where the fluorescent light is emitted. The fluorescent signal, generated at the sample, is reflected by the dichroic filter and collected by a detector through a lens. An AO feedback loop builds the compensating DM shape based on maximizing the detected fluorescent signal.

The SPGD controller has one output—the fluorescence signal intensity—and multiple–inputs: the mirror actuators. Such a structure for the controller, while simple, may not be efficient. By coordinating the action of the actuators, one can establish an output controller basis that is potentially more effective in this application.

In this section, three choices for the input control basis are introduced. The three choices are compared based on the closed-loop performance of the control loop, as measured by the final value of the fluorescence signal (i.e., the metric).

The most straightforward control basis for the SPGD controller is to treat individual actuators as independent, local control channels. In the BU-Wellman Center for Photomedicine two-photon fluorescent microscope setup, the DM has about 100 actuators in the pupil plane, and each actuator essentially defines a sub-aperture of the pupil. The advantage for this actuator basis is that it can produce the finest spatial resolution achievable with the DM. On the other hand, this actuator basis has the most control channels and it makes no use of the information about relative positions of the actuators in the array. The SPGD algorithm efficiency decreases with the number of control channels, so this basis will have lower efficiency, characterized by slower optimization of the metric.

Since the main aberrations requiring corrections are expected to have relatively low spatial frequencies [28], actuator binning can be used to decrease the number of control channels. Groups of adjacent actuators in two-by-two arrays, controlled in common, were examined as an alternative basis. In this case, the control basis is reduced to ~25 channels in the pupil plane. Higher convergence speed is expected using this basis.

Considering that the main aberrations known to degrade deep-tissue imaging performance are spherical aberrations, a third approach was taken as well—namely, to use a Zernike polynomial control basis. Zernike polynomials are often used to represent optical aberrations. They comprise an infinite set of orthonormal equations defined over a circular aperture. An arbitrary wavefront can be represented by a truncated set of Zernike polynomials, and this representation will closely approximate aberrations composed of low spatial frequencies. Zernike polynomials have been widely used in optics, primarily because low-order Zernike terms define well-known optical defects (e.g., defocus, astigmatism, and coma). These, along with radially symmetric terms (e.g., spherical aberrations), are the primary aberrations expected in deep-tissue-imaging microscopy. A truncated set of Zernike polynomials can be

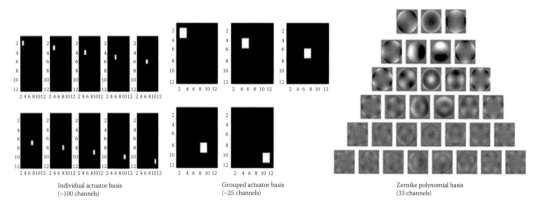

FIGURE 12.5 Three control bases for SPGD. Left: diagonal control channels are highlighted in an individual–actuator control basis; Center: diagonal control channels are highlighted in the grouped actuator control basis; Right: the 2nd–7th of Zernike polynomial shapes in the Zernike basis.

chosen for SPGD control parameters. In this approach, the actuators are perturbed in a coordinated fashion, corresponding to superposition of Zernike polynomial shapes.

Booth and Wilson demonstrated in theory that only a few Zernike terms need to be corrected with AO to improve optical quality significantly in two-photon microscopy [28]. In practice, the first 35 Zernike terms, without the lowest two, which affect only image shift, are used as the control basis. Figure 12.5 shows the three choices of the control basis.

With the appropriate perturbation step size selected for each candidate SPGD control basis, each was used to optimize the fluorescent signal from the microsphere bead embedded in a mouse skull at 200 μm below the surface, under control of the AO loop. For each control test, the SPGD algorithm was run for over 100 iterations. Figure 12.6 shows the metric response during control for each of the three control basis. The metric value, fluorescence intensity (vertical scale), is normalized by the metric value at the initial state (mirror in neutral, flat state). The best control basis is the one that converges to the highest metric value. Clearly, the Zernike-based controller produced the best results out of the three control bases. The controller produced 40% improvement in the metric. By contrast, the grouped actuator–based controller produced ~4% improvement, and the individual actuator–based controller produced essentially no improvement.

The SPGD algorithm for AO in an actual fluorescence microscopy experiment in deep-tissue imaging performed best when the control basis was Zernike polynomials with perturbations applied as changes in the Zernike polynomial coefficients. Key factors in the effectiveness of the Zernike control basis in this SPGD application were that the Zernike polynomials closely approximate the expected aberrations in the microscope, and the Zernike-based controller required fewer control degrees of freedom. To represent Zernike polynomial shapes on the DM accurately, it is also necessary to drive each actuator precisely for the desired deflection profile. This is nontrivial, since the fundamental electrostatic actuation mechanism is nonlinear. An empirical calibration procedure is used to derive a voltage map to drive the DM into Zernike shapes [29].

12.3.2 Metric Selection for Stochastic Parallel Gradient Descent Algorithm

We chose to use the average fluorescence intensity over the whole scanning field as the feedback metric in the SPGD closed-loop AO. The selection of this metric is based on the assumption that stronger two-photon fluorescence intensity can be generated in the excitation process from an aberration-free optical setup than that from aberrated optical setup.

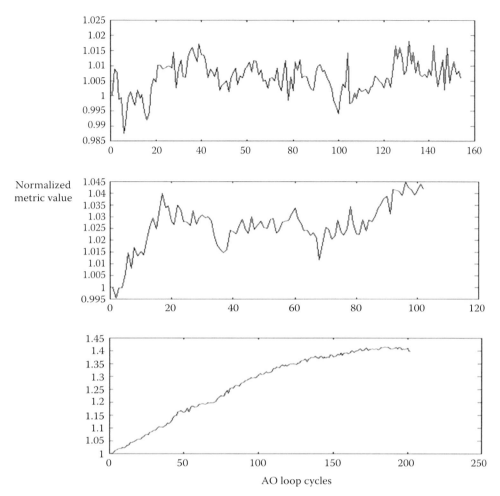

FIGURE 12.6 The measured metric as a function of AO loops based on three control bases. Top, the measured metric as a function of AO controller based on individual actuator SPGD; middle, the measured metric as a function of AO controller based on grouped actuator SPGD; and bottom, the measured metric as a function of AO controller based on Zernike polynomial SPGD.

In the focal plane of the illumination laser in the two-photon scanning laser fluorescence micro-scope (TPSL-FM), the fluorescent intensity is highly dependent on the average illumination intensity in the lateral plane of the focal volume. The average illumination intensity decreases with an increase in the width of the point spread function (PSF). This is especially significant for the two-photon fluores-cence excitation, due to the nonlinear nature of the excitation.

Stronger fluorescence intensity can be generated from the focal plane in two-photon fluorescence excitation in the absence of aberrations [30]. The generated fluorescence intensity can be simplified as

$$F(z) = W^2(z) \int \mathrm{PSF}^2\left(\vec{\rho}, z\right) d^2\vec{\rho} \tag{12.1}$$

where $W(z)$ is the total illumination power in the focal plane; PSF is the point spread function in the focal plane [30]; $\vec{\rho}$ is the vector corresponding to the radial coordinates in a plane perpendicular to the beam axis; and z is the coordinate aligned with the beam axis, which is defined as 0 at the sample surface and is positive for the planes extending into the sample. This expression has dropped pre-factors such as fluorophore concentration and cross section, which Leary and Mertz assumed to be constant.

Alternatively, the two-photon fluorescence intensity at depth z can be expressed as a function of the illumination laser optical transfer function, defined by Parseval's theorem:

$$\text{OTF}(\vec{k}_\perp, z) = \int \text{PSF}(\vec{\rho}, z) e^{i\vec{\rho}\cdot\vec{k}_\perp} d^2\vec{\rho} \tag{12.2}$$

where $\vec{k}_\perp$ represents the Fourier form wave vector coordinates. Following Leary and Mertz, this leads to

$$F(z) = \frac{W^2(z)}{(2\pi)^2} \int \left| \text{OTF}(\vec{k}_\perp, z) \right|^2 d^2\vec{k}_\perp \tag{12.3}$$

From Schwarz's inequality [31], we obtain:

$$\left| \text{OTF}_\phi \vec{k}_\perp; z) \right| \leq \left| \text{OTF}_0(\vec{k}_\perp; z) \right| \tag{12.4}$$

where subscripts ϕ and 0 indicate the conditions where there are optical aberrations and no optical aberrations, respectively. Thus, we can conclude as follows:

$$F_\phi(z) \leq F_0(z) \tag{12.5}$$

Based on this simplified view of the effect of aberrations on the PSF, it is reasonable to expect that optical aberrations reduce the two-photon fluorescence signal generated at the focal plane. Conversely, compensation of those aberrations should increase the two-photon fluorescence signal in the focal plane. These assumptions ignore the possibility that aberration-induced enlargement of the PSF might generate additional two-photon fluorescence signals from structures above and below the focal plane. Nevertheless, the analysis of Leary and Mertz offers some expectation that two-photon fluorescence signal intensity would serve as a reasonable metric for control of aberrations.

Other measurement metrics, such as sharpness, have also been used in wavefront correction systems to improve image quality. Muller used image sharpness as the measurement metric to correct image distortion from atmosphere turbulence in the telescope application [32]. The sharpness has the following definition:

$$S = \int I^2(x, y) dx dy \tag{12.6}$$

where $I(x, y)$ is the intensity value at each pixel on the imaging plane. Variables x and y are the coordinates of the pixel. We believe the image sharpness is a valid measurement metric for the AO SPGD application and could be used as an alternative or complimentary metric besides the intensity metric.

12.3.3 Application of Stochastic Parallel Gradient Decent Algorithm with Zernike Polynomial Basis

In the approach using the SPGD algorithm, the time required for metric gradient estimation is independent of the number of control channels, and the imaging aperture is fixed in the control loops. The SPGD algorithm presents an attractive solution for wavefront aberration correction with a much higher efficiency and faster convergence speed. Formal analysis of the convergence rate shows that the parallel stochastic method takes a factor $\sqrt{N}$ (N is the number of control channels), fewer iterations than the sequential perturbation method to reach the same level of performance—owing to the mutually uncorrelated statistics of the parallel perturbations [33].

In SPGD, the true metric gradient can be described as the projection of measured metric on the wavefront error; for example, dJ/ds, where J is the measured metric value and s is the wavefront error. In the case of small perturbation from DM, dJ/ds can be approximated as $\delta J/\delta s$ [34]. From Ref. [35], the metric gradient can be related to the time-dependent evolution of wavefront error in the following expression:

$$\tau \frac{ds}{dt} = \gamma \frac{\delta J}{\delta s} \tag{12.7}$$

where τ is a time constant and γ is the update coefficient. The wavefront error can be updated as follows:

$$s^{n+1} = s^n + \gamma \frac{\delta J}{\delta s} \tag{12.8}$$

If γ is positive, the measured metric maximizes or is negative otherwise. From Equation 12.7, we have

$$\frac{dJ}{dt} = \frac{\delta J}{\delta s} \frac{ds}{dt} = \gamma \left(\frac{\delta J}{\delta s} \right)^2 \tau^{-1} \geq 0 \tag{12.9}$$

The above inequality indicates that the measured metric should increase with continued operation of the control loop; that is, the controller should always converge to a steady-state value. Like all gradient-based controllers, it is possible that this steady-state value will be a local and not a global maximum.

In practice, the metric gradient can be approximated as the product of the perturbation step size ($\delta(\vec{s})$) and the sign of the gradient difference (J). In other words, the metric gradient slope in Equation 12.8 is being treated as unity. The main reason to use unity as the gradient metric is to reduce the effect of the measurement noise. The relatively low signal-to-noise ratio of the two-photon fluorescence signal in deep-tissue fluorescence imaging made estimation of the gradient particularly challenging and occasionally resulted in severe convergence difficulties. This effect was mitigated by the zero-order approximation used for gradient estimation, at a cost of slower convergence.

For actual imaging experiments, we used Zernike polynomials as our control basis. The perturbation step size for each Zernike term was chosen from experimental results. In practice, the perturbation step size for all 35 terms was chosen as 23 nm, or $(1/40)\lambda$, where λ is 920 nm. The perturbation step sizes can be adjusted to larger values for certain Zernike terms to improve the correction efficiency, should these Zernike terms represent significant aberrations in the optical setup.

The SPGD algorithm would require perturbation of all the Zernike terms simultaneously in the closed loop. However, in practice, we found out that performance could be improved by partitioning the controller into sets of Zernike terms of successively higher spatial frequency.

In the practical application of Zernike-based SPGD algorithm for wavefront correction, we divided the 35 Zernike terms into three different categories based on their order in the American National Standards Institute (ANSI) standard: Zernike terms no. 3 to no. 9 are in the first category, which are the second and third orders of Zernike polynomials. Piston, tip, and tilt terms are ignored in the closed-loop control system. The second category includes Zernike terms no. 10 to no. 20, which are the fourth and fifth orders of Zernike polynomials. The third category includes Zernike terms no. 21 to no. 35, which are the sixth and seventh orders of Zernike polynomials.

The SPGD control algorithm was applied to each category separately. Each full control sequence started with compensation of the Zernike terms in categories one, then two, and then three. This process was repeated until the closed loop reached a steady state.

These partitions are not entirely arbitrary. In the first group, there are the low-order aberrations that dominate many optical microscopy systems, including astigmatism and coma. In the second group, there is the first-order spherical aberration that is expected to be particularly relevant to subsurface imaging. In the third group, there are higher-order aberrations, including the second-order spherical aberration, that define the highest relevant spatial frequency that is controllable with the DM.

12.4 Optical System Design

The first TPSL-FM was pioneered by Denk and Webb [36]. The noninvasive imaging technique has been used in an exponentially increasing number of applications in recent years for biomedical imaging [37–46].

Figure 12.7 shows the actual optical layout of the BU-Wellman Center for Photomedicine two-photon fluorescence imaging system with AO and the optical system magnification. The constructed optical system occupies 1.5X1 square meters optical bench. The apparatus includes four main components: the light source, raster scanning, light detection, and wavefront correction.

Due to the nonlinear nature of the fluorescence excitation, a high-power illumination laser density is essential for the practical two-photon microscope setup. In practice, a pulsed laser can provide a factor of 10^4 or 10^5 times more fluorescence excitation than a continuous wave laser with the same amount of average power, due to the temporal focusing.

Light is delivered from a solid-state mode-locked Ti-sapphire laser and an optical parametric oscillator with a combined tuning range from 700 to 1300 nm (Newport Corp., CA). Laser power varies, depending on the output wavelength. For example, at 920 nm, the laser is capable of providing up to 1.3 W output power. The pulse width is 100 fs, with a repetition rate of 80 MHz. A pair of linear rotatable polarizers was installed to perform beam intensity control at the output of the laser. The constructed system has a throughput of 7%. The maximum power that can be reached at the surface of biological sample is close to 100 mW.

The entrance pupil is defined by the MEMS DM (Boston Micromachines, MA). The MEMS DM has clear aperture of 4.8 × 4.8 mm occupied by the 12 × 12 actuator array. To create precise Zernike term

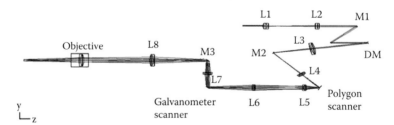

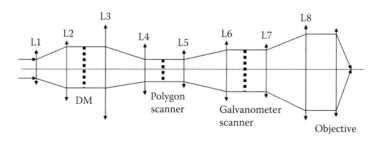

FIGURE 12.7 The optical layout of two-photon scanning laser fluorescence microscope with adaptive optics. Top: the design model in ZEMAX. Bottom: the system magnification of the lens system, shown unfolded. Lens: L1–L8. Mirror: M1–M3.

shapes, the 10×10 actuator array is chosen to be the active area that defines the pupil. The entrance pupil is 4×4 mm. The entrance pupil is conjugated to the pivot points of the raster scan, which are the horizontal scanner, the vertical scanner, and the back aperture of the objective. The 4×4 mm entrance pupil is imaged onto the back aperture of objective, and the illumination light overfills the back aperture of the objective.

L1 and L2 are f40 (mm) and f100 (mm) lenses, respectively. They function as telescope lens pair for beam expansion. L3 and L4 are f150 (mm) and f75 (mm), respectively. They function as relay lenses to image the DM pupil onto the horizontal scanner (polygon scanner). Since the polygon scanner has a fixed scanning angle of 20° (optical), this scanning angle needs to be demagnified; L5 and L6 were chosen to be f40 (mm) and f100 (mm). L5 and L6 create an image plane of the polygon scanner on the vertical scanner (galvanometer scanner) and have a magnification factor of 2.5. Consequently, they also demagnify the scanning angle of the polygon scanner to 2.5 times smaller, which is 8°. The scanning angle of the galvanometer is also adjusted to 8° to create a square scanning field. L7 and L8 are f60 (mm) and f150 (mm), respectively. The magnification of 2.5 between L7 and L8 is created to match the large pupil size at the back aperture of the objective (XlUMPLanFI, 20×, 0.95 NA, Olympus, Japan), which is roughly about 12.5 mm (diameter). The scanning angle coming out of the objective is $3.2 \times 3.2°$.

The XY scanner is constructed using a polygon scanner (Lincoln Laser, AZ) and a galvanometer scanner (Cambridge Technology, MA). The scanning system scans the focused beam across the specimen in a raster pattern. The polygon, capable of delivering constant scanning speed over a 20° optical angle, is in charge of the horizontal scan (fast scan). On the other hand, the galvanometer has an adjustable scanning angle, which provides the vertical scan (slow scan). The slow scan angle is changed to match the fast scan to provide a square field of illumination on the imaging samples. Both the horizontal scanner and the vertical scanner are optically conjugated to the entrance pupil. A frame grabber board is synchronized to the scanning mirrors and is capable of grabbing up to 30 frames per second.

A photomultiplier tube (PMT, Hamamatsu, Japan) is used for signal detection. The high gain and the low read-out noise make the PMT a universal choice in photon-limited applications. The nonlinear signal is collected in a "non-descanned" fashion to maximize the detection of the scattered signal from the focal volume. The detector has a 3×12 mm active area, which requires focusing optics in the detection channel.

Figure 12.8 shows the direct signal collection without de-scanning. The dichroic filter (Semrock, NY) allows the illumination beam pass through, usually in the range of 800–1100 nm (infrared) and reflects the fluorescence signal in the range of 400–600 nm. The infrared light is focused on the biological sample through the XlUMPLanFI microscopic objective. The fluorescence light reflects off the dichroic filter and is detected by a PMT.

The AO software suite includes two main components: one for the PMT data acquisition and the other for the AO loop control. Both components were written with MATLAB version 7.0 (MATHWORKS, MA).

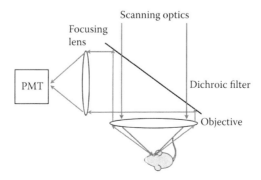

FIGURE 12.8 The fluorescence signal is collected directly through a focusing lens. The illumination light, originated from the scanning optics, passes through the dichroic filter. The fluorescence signal, generated at the sample, is reflected by the dichroic filter, and detected by the PMT.

A data acquisition board (National Instruments, TX) is synchronized with raster scanning mirrors to feed the PMT signal to the AO loop. The typical AO speed is 1 Hz. A Pentium 4 IBM compatible computer is used to control the DM.

The optical filters are selected to accommodate the absorption and emission spectrums of the GFP. GFP has a relatively wide absorption spectrum. The absorption spectrum peaks at 488 nm, and the fluorescence emission peaks at 509 nm. In the two-photon process, the illumination wavelength will be 920 nm. To match the GFP model, the chosen dichroic filter is a long-pass laser transmitting filter (Semrock, NY). The transmitting window is from 750 to 1100 nm, and the reflecting window is from 350 to 720 nm. An optical band-pass filter (Semrock, NY) was also installed in front of the PMT to eliminate the photon noise. The transmission spectrum for the band-pass filter is from 400 to 700 nm.

12.5 Experiments and Results

12.5.1 Point Spread Function Quantification

The intrinsic three-dimensional imaging ability from the TPFM provides a significant advantage in tissue imaging. The effects of AO on a TPFM can be demonstrated by the change of the PSF of constructed TPFM with and without AO correction. AO is used to correct both the inherent optical system and the sample aberrations.

The PSF defines the optical resolution of a microscope, and it can be measured based on the light intensity distribution at the focal volume. The PSF from a TPFM can be approximated by a Gaussian function both laterally and axially. The full width at half maximum (FWHM) of the fitted Gaussian function is then a good measure of the optical quality of a TPSL-FM. The theoretical FWHM under diffraction-limited conditions for the planned TPFM are 0.37 μm laterally and 1.4 μm axially. The illumination light wavelength is 0.92 μm.

To measure the excitation light intensity distribution of the focal volume from a TPFM, it was necessary to design a scanning system that had sub-micrometer imaging resolution. A "simplified" version of the TPFM with AO was developed for this purpose, as shown in Figure 12.9. A single tip-tilt scanning

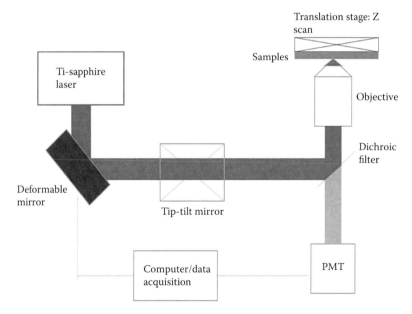

FIGURE 12.9 The two-photon scanning laser microscope setup for PSF measurement. The tip-tilt mirror is used for the x–y scan, and translation stage is used for the z scan. The line in dark shade marks the illumination beam. The line in pale shade indicates the fluorescence signal.

mirror (Newport Corp., CA) capable of two-dimensional scans is used to replace the polygon and the galvanometer. The main reason for this change is that the adjustable scanning angle from the tip-tilt mirror can create a relatively small scanning field with sufficient pixel resolution. For example, in the setup of Figure 12.9, the constructed scanning mechanics is able to provide a 5×5 µm scanning field, and a data acquisition board (National Instruments, TX) is able to grab a single frame at 50×50 pixels per second. The resulting pixel resolution is 0.1 µm/pixel. To achieve a three-dimensional measurement, the axial direction movement is controlled by a linear translation stage (Sutter Instrument, CA). The stage is able to achieve 0.2 µm resolution.

The fluorescence agents used are green fluorescent microspheres (Duke Scientific Corp., CA). The microsphere has a diameter of 0.1 µm and can be excited at 920 nm with the two-photon excitation process. The microspheres are diluted in water with agarose powder in a heated environment (85°C). The cooled final product, agarose gel, is used as the imaging target. The agarose gel has a refractive index of 1.33, which is well matched to the water immersion objective.

Photobleaching is a physical process that will damage the fluorescence agent in the two-photon excitation process. Consequently, the fluorescence signal will decrease due to photobleaching. To improve the signal-to-noise ratio in the PSF detection, the data acquisition (DAQ) is running at a speed of acquiring 1,000,000 samples every second. Each pixel in a single acquired frame will be the average value of 400 samples.

Figure 12.10 shows the measured three-dimensional intensity distribution. To identify the FWHM the in x–z plane, a MATLAB program was used to detect the maximum pixel value of a single frame in the x–y intensity distribution for each measured axial position. This maximum intensity cross section in the axial direction was then fitted to a Gaussian function. The peak value from the Gaussian function also indicates the position of the focal plane of the measured focal volume. The x–y intensity measurement of the focal plane presents the information for the lateral resolution. A centroiding program is used to find the center of the mass of the x–y intensity measurement at the focal plane, and Gaussian functions were fit to the one-dimensional intensity measurements.

The measured system aberrations are dominated by low-order aberrations: astigmatism, trefoil, and coma. A set of five PSF measurements are shown in Figure 12.11. The PSF FWHM averages are 0.52 µm (lateral) and 3.09 µm (axial) without AO, and the improved PSF FWHM averages are 0.43 µm (lateral) and 2.12 µm (axial). The lateral and axial PSF FWHMs are improved 17% and 31%, respectively.

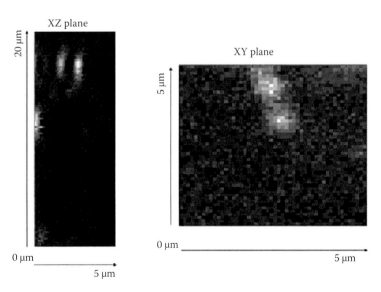

FIGURE 12.10 Images of fluorescent microspheres. Left, the fluorescent microspheres in the *xz* plane. Right, The fluorescent microspheres imaged in the *xy* plane.

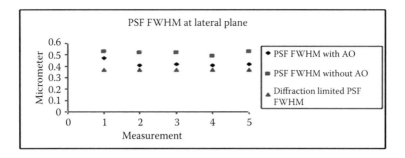

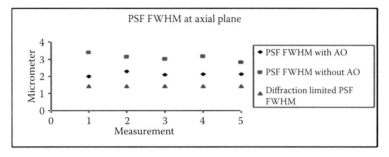

FIGURE 12.11   Point spread function full width at half maximum. Top: the five measurements of the PSF FWHM with AO on/off at lateral plane. Bottom: the five measurements of the PSF FWHM with AO on/off at axial plane.

12.5.2 Mouse Bone-Marrow Imaging with Adaptive Optics

To be able to use AO to correct the tissue aberration in mouse skull bone imaging, light needs to travel to the bone structure below the skull surface and generate enough florescence signal for AO feedback and imaging purposes. Generally, this is a light-starving environment, and the signal gets weaker with deeper imaging depth. The biggest challenge is to find a strong signal source in the mouse model for the feedback metric. The chosen samples are genetically modified mice whose osteoblasts express GFP, developed by Dr. David Rowe from the University of Connecticut and maintained by Dr. Cristina Lo Celso from Massachusetts General Hospital.

The test mouse is anesthetized by an injection of 80 mg/kg ketamine + 12 mg/kg xylazine. The hair on the dorsal scalp is removed by applying a small quantity of hair remover cream, letting it work for 10–15 minutes, followed by gently wiping off the area with a cotton swab. The area is then cleansed with sterile phosphate-buffered saline, disinfected by Betadine solution, followed by an excision of some of the skin covering the area. The imaging gel (Methocel 2%) is applied to the exposed area and covered by a class-1 cover glass (Fisher Scientific, PA). All protocols were approved by the Subcommittee on Research Animal Care Application from Massachusetts General Hospital and performed by the researcher from the Wellman Center for Photomedicine.

Figures 12.12 and 12.13 show the AO correction results at 145 and 184 μm below the surface. To avoid the focus shift caused by the defocus term created on the DM, the focus term (Zernike term no. 4) is not adjusted during the closed-loop SPGD correction. The images with AO on at both locations clearly show brighter features and sharper contrast.

Figure 12.14 shows a plot of the average intensity over the whole image frame measured every micrometer over a 200 μm imaging depth in the skull bone. The three data sets correspond to conditions when the DM was held in its initial flat shape, when the DM was corrected using AO at an imaging depth of 145 μm, and when the DM was corrected using AO at an imaging depth of 184 μm. The intensity plot shows the improvements of intensity at the optimized locations as well as the neighboring depth. For example, the intensity for the data set corrected at 145 μm depth is larger at 145 μm depth than the other

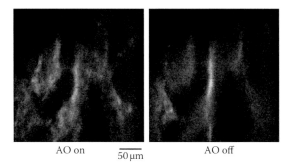

AO on $\overline{\quad}$ 50 μm AO off

FIGURE 12.12 The bone cavities imaged with AO on/off at 145 μm below the surface.

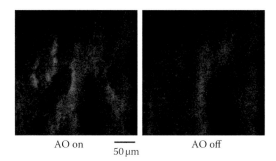

AO on $\overline{\quad}$ 50 μm AO off

FIGURE 12.13 The bone cavities imaged with AO on/off at 184 μm below the surface.

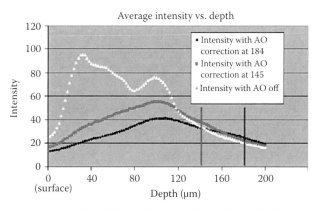

FIGURE 12.14 The average intensity plot of the nonlinear signal at different depths with AO on/off.

two data sets. It is also notable that the two intensity plots with AO on showed weaker intensity at shallow depths (e.g., 0–120 μm). This is because the DM shapes corrected at depths of 145 and 184 μm create optical aberrations in the shallower depths.

If the images taken from 145 μm below the surface are used as the basis for quantified analysis, the average pixel value increases to 36.74 with AO on from 32.80 with AO off. However, if only the pixels that are larger than 100 (gray scale) are counted in the image with AO on, which roughly represents the two-photon excitation signal from the GFP, the average pixel value jumps to 130.50 with AO on from 72.12 with AO off. This represents an 81% increment in terms of intensity.

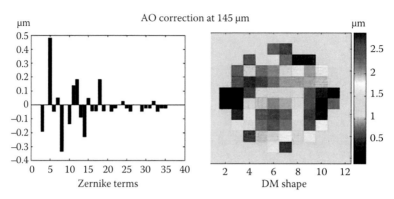

FIGURE 12.15 The Zernike terms and resulting DM shape for the optical aberration correction for the mouse bone cavity imaging at 145 µm below the surface.

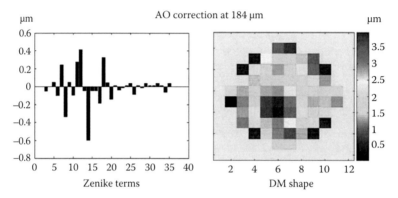

FIGURE 12.16 The Zernike terms and resulting DM shape for the optical aberration correction for the mouse bone cavity imaging at 184 µm below the surface.

Figures 12.15 and 12.16 show the resulting DM shapes and Zernike terms for the aberration correction at the two different depths. The correction shape is largely affected strongly by the bone structure and depth. As predicted, the considerable spherical aberration is corrected in the beam path at both locations. At 184 µm, the first spherical aberration amplitude is 0.41 µm; however, Zernike term 14 is the most significant aberration at −0.6 µm RMS. At 145 µm, astigmatism stands out at 0.48 µm RMS.

12.6 Critical Discussion

The imaging depth is limited at about 200 µm for the GFP mouse, which is about half the desired imaging depth for the entire mouse skull bone. Deeper detection is not achievable because of the diminishing fluorescence signal. Higher illumination intensity can be used to increase the fluorescence intensity signal. However, such an option is limited, to avoid thermal damage of the sample that is being imaged.

The collected fluorescence signal can also be increased by enlarging the numerical aperture of the collection optics. Combs et al. use parabolic mirrors, combined with the imaging objective, to collect the solid angle around the excitation spot when imaging the sample [47]. The enlarged collection angle allows twofold improvement of the fluorescence signal in their prototype instrument.

In an attempt to demonstrate AO in deeper bone-marrow imaging, we used an unconventional approach to create a strong fluorescence signal in a deep layer (200 µm or deeper below the surface) in the mouse skull bone structure for AO correction.

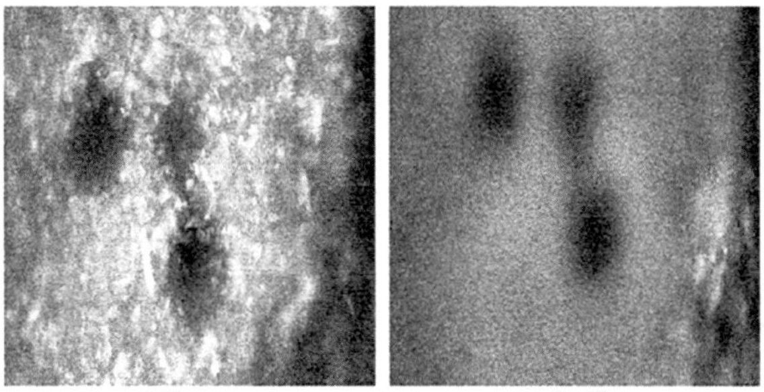

AO correction on microsphere fluorescence signal at 280 μm

AO on —— AO off
50 μm

FIGURE 12.17   The microsphere fluorescence signal images from 280 μm below the surface in the mouse skull bone with AO on/off.

To create the fluorescence signal, a single piece of mouse skull bone was bathed in the green fluorescent microspheres solution (0.1 μm, Duke Scientific Corp, CA) for 30 minutes. This process allows a large number of microsphere beads to penetrate into the bone cavity and act as the fluorescence agents. It was discovered that these microspheres located in the deep layer are capable of generating enough fluorescence signal for imaging as well as for AO feedback. The AO correction shows tremendous improvement in the quality of the images. The clusters of microspheres can be resolved with the AO correction, as shown in Figure 12.17.

With increasing imaging depth, the high-frequency aberrations—for example, the second spherical aberrations—become significant. To further improve from the current AO system, a woofer-tweeter AO approach, used in retinal imaging aberration correction [48], can potentially be used to improve the AO correction in the two-photon fluorescence imaging. The optical aberrations are compensated by two deformable mirrors. The compensations are split based on their frequency content. A dedicated DM is used to correct the high-frequency aberrations, while a separate DM is used to correct the low-frequency aberrations.

The newly developed open-loop DM control provides another potential improvement for the future development of the sensorless AO [49]. The DM is able to create Zernike shapes in diffraction-limited precision in open loop, which will in turn improve the overall residual errors of the SPGD-based closed-loop AO.

12.7 Summary

The BU-Wellman Center for Photomedicine AO TPFM is used to correct optical aberrations to extend the imaging depth in mouse bone-marrow imaging. The instrument that has been developed includes a Ti-sapphire pulsed laser, a MEMS DM, a raster scanning system, and a PMT as core components. The SPGD algorithm–based AO uses the fluorescence intensity as a feedback signal to drive the DM to compensate for the optical aberrations. To improve the AO loop efficiency for the single-input–multi-output system, a Zernike polynomial basis is used to control the DM. The Zernike polynomial–based AO can compensate and enhance the microscope's performance. The PSF of the microscope was measured by using a simplified optical configuration with a single scanning mirror. A single microsphere fluorescent bead was used to measure the PSF. Both the lateral and the axial resolutions were found to be improved with the AO correction of system aberrations. The lateral and the axial PSF FWHMs were improved 17% and 31%, respectively. AO was applied to deep imaging of mouse skull tissue in a genetically modified GFP mouse. The fluorescence signal was improved 81% at a depth of 145 μm below the surface from AO correction. The bone cavity structure was clearly resolved with AO at a depth of 184 μm. Deeper correction was explored using microspheres as fluorescence agents in the mouse skull bone.

Acknowledgments

The research described in this chapter was done as part of the PhD dissertation research by Yaopeng Zhou, advised by Professor Thomas Bifano from Boston University and Dr. Charles Lin from Wellman Center for Photomedicine. Parts of this work have appeared in *Adaptive Optics Two-Photon Scanning Laser Fluorescence Microscopy: Design, Construction, Implementation*, Yaopeng Zhou, VDM Verlag, 2009.

References

1. Denk, W. and Svoboda, K. "Photon upmanship: why multiphoton imaging is more than a gimmick." *Neuron*, 18, 351–357 (1997).
2. Rubart, M. "Two-photon microscopy of cells and tissue." *Circulation Research*, 95, 1154–1166 (2004).
3. Molitoris, B.A. and Sandoval, R.M. "Intravital multiphoton microscopy of dynamic renal processes." *American Journal of Physiology-Renal Physiology*, 288, F1084–F1089 (2005).
4. Laiho, L.H., Pelet, S., Hancewicz, T.M., Kaplan, P.D. and So, P.T. "Two-photon 3-D mapping of ex vivo human skin endogenous fluorescence species based on fluorescence emission spectra." *Journal of Biomedical Optics*, 10, 024016 (2005).
5. Lo Celso, C., Wu, J.W., and Lin, C.P. "In vivo imaging of hematopoietic stem cells and their micro-environment." *Journal of Biophotonics*, 2(11), 619–631 (2009).
6. Thomas, E.D., Lochte Jr, H.L., Lu, W.C., and Ferrebee, J.W. "Intravenous infusion of bone marrow in patients receiving radiation and chemotherapy." *New England Journal of Medicine*, 257, 491–496 (1957).
7. Scothorne, R.J. and Tough, J.S. "Histochemical studies of human skin atuografts and homografts." *British Journal of Plastic Surgery*, 5, 161–170 (1952).
8. Kokhanovsky, A.A. *Light Scattering Review 4* Springer. 295 (2009).
9. Denk, W., Strickler, J.H., and Webb, W.W. "Two-photon laser scanning fluorescence microscopy." *Science*, 248(4951), 73–76 (1990).
10. Zipfel, W.R., Williams, R.M., and Webb, W.W. "Nonlinear magic: multiphoton microscopy in the bioscience." *Nature Biotechnology*, 21(11), 1369–1377 (2003).
11. Booth, M.J., Neil, M.A.A., and Wilson, T. "Aberration correction for confocal imaging in refractive-index-mismatched media." *Journal of Microscopy*, 192(Pt 2), 90–98 (1998).
12. American National Standards Institute (ANSI), American National Standard for Ophthalmics—Methods for reporting optical aberrations of the eye (ANSI Z80.28-2004), Washington, DC, American National Standards Institute (2004).
13. Hell, S.W., Rainer, G., Cremer, C., and Stelzer, E.H.K. "Aberrations in confocal fluorescence microscopy induced by mismatches in refractive index." *Journal of Microscopy*, 169, 91–405 (1993).
14. Sheppard, C.J.R. and Gu, M. "Aberration compensation in confocal microscopy." *Applied Optics*, 30, 3563–3568 (1991).
15. Hellmuth, T., Seidel, R., and Seigal, A. "Spherical aberration in confocal microscopy." *Proceedings of SPIE*, 1028, 28–32 (1988).
16. Wan, D.-S., Rajadhyaksha, M. and Webb, R.H. "Analysis of spherical aberration of a water immersion objective: application to specimens with refractive indices." *Journal of Microscopy*, 197, 274–284 (2000).
17. Booth, M.J. and Wilson, T. "Strategies for the compensation of specimen induced spherical aberration in confocal microscopy of skin." *Journal of Microscopy*, 200, 68–74 (2000).
18. Neil, M.A.A., Jusi Kaitis, R., Booth, M.J., Wilson, T., Tanaka, T., and Kawata, S. "Adaptive aberration correction in a two-photon microscope." *Journal of Microscopy*, 200, 105–108 (2000).
19. Rueckel, M., Mack-Bucher, J.A., and Denk, W. "Adaptive wavefront correction in two-photon microscopy using coherence-gated wavefront sensing." *Proceedings of the National Academy of Sciences of the United States of America*, 103(46), 17137–17142 (2006).

20. Sherman, L., Ye, J.Y., Albert, O., and Norris, T.B. "Adaptive correction of depth-induced aberrations in multiphoton scanning microscopy using a deformable mirror." *Journal of Microscopy*, 206, 65–71 (2002).

21. Wright, A.J., Poland, S.P., Girkin, J.M., Freudiger, C.W., Evans, C.L., and Xie, X.S. "Adaptive optics for enhanced signal in CARS microscopy." *Optics Express*, 15(26), 18209–18219 (2007).

22. Ji, N., Milkie, D.E., and Betzig, E. "Adaptive optics via pupil segmentation for high-resolution imaging in biological tissues." *Nature Methods*, 7, 141–147 (2009).

23. Débarre, D., Botcherby, E.J., Watanabe, T., Srinivas, S., Booth, M.J., and Wilson, T. "Image-based adaptive optics for two-photon microscopy." *Optics Letters*, 34, 2495–2497 (2009).

24. Vorontsov, M.A. and Sivokon, V.P. "Stochastic parallel-gradient-descent technique for high-resolution wave-front phase-distortion correction." *Journal of the Optical Society of America, A*, 15(10), 2745–2758 (1998).

25. Wright, A.J., Burns, D., Patterson, B.A., Poland, S.P., Valentine, G.J. and Girkin, J.M. "Exploration of the optimisation algorithms used in the implementation of adaptive optics in confocal and multiphoton microscopy." *Microscopy Research and Technique*, 67, 36–44 (2005).

26. O'Meara, T.R. "The multi-dither principle in adaptive optics." *Journal of the Optical Society of America*, 67, 306–315 (1977).

27. Vorontsov, M.A., Carhart, G.W., Pruidze, D.V., Ricklin, J.C., and Voelz, D.G. "Image quality criteria for an adaptive imaging system based on statistical analysis of the speckle field." *Journal of the Optical Society of America A*, 13, 1456–1466 (1996).

28. Booth, M.J. and Wilson, T. "Refractive-index-mismatch induced aberrations in single-photon and two-photon microscopy and the use of aberration correction." *Journal of Biomedical Optics*, 6(3), 266–272 (2001).

29. Zhou, Y. and Bifano, T.G. "Characterization of contour shapes achievable with a MEMS deformable mirror." *Proceedings of SPIE*, 6113, 123–130 (2006).

30. Leray, A. and Mertz, J. "Rejection of two-photon fluorescence background in thick tissue by differential aberration imaging." *Optics Express*, 14(22), p. 10565 (2006).

31. Goodman, J.W. *Introduction to Fourier Optics*. Roberts & Company Publishers, Greenwood Village, CO (2005).

32. Muller, R.A. and Buffington, A. "Real-time correction of atmospherically degraded telescope images through image sharpening." *Journal of the Optical Society of America*, 64(9), 1200–1210 (1974).

33. Kokorowski, S.A., Pedinoff, M.E., and Pearson, J.E. "Analytical, experimental and computer simulation results on the interactive effects of speckle with multi-dither adaptive optics systems." *Journal of the Optical Society of America*, 67, 333–345 (1977).

34. Vorontsov, M.A., Carhart, G.W., Cohen, M., and Cauwenberghs, G. "Adaptive optics based on analog parallel stochastic optimization: analysis and experimental demonstration." *Journal of the Optical Society of America*, 17, p. 8 (2000).

35. Vorontsov, M.A., Carhart, G.W., and Ricklin, J.C. "Adaptive phase-distortion correction based on parallel gradient-descent optimization." *Optics Letters*, 22(12), 907–909 (1997).

36. Denk, W., Strickler, J., and Webb, W. "Two-photon laser scanning fluorescence microscopy." *Science*, 248(4951), 73–6 (1990).

37. Rose, C.R., Kovalchuk, Y., Eilers, J. and Konnerth, A. "Two-photon Na^+ imaging in spines and fine dendrites of central neurons." *Pflugers Archive*, 439, 201–207 (1999).

38. Stosiek, C., Garaschuk, O., Holthoff, K., and Konnerth, A. "In vivo two-photon calcium imaging of neuronal networks." *Proceedings of the National Academy of Sciences of the United States of America*, 100, 7319–7324 (2003).

39. Helmchen, F. and Waters, J. "Ca(2+) imaging in the mammalian brain in vivo." *European Journal of Pharmacology*, 447, 119–129 (2002).

40. D'Amore, J.D., Kajdasz, S.T., Mclellan, M.E., Bacskai, B. J., Stern, E.A., and Hyman, B.T. "In vivo multiphoton imaging of a transgenic mouse model of Alzheimer disease reveals marked thioflavine-S-associated alterations in neurite trajectories." *Journal of Neuropathology & Experiment Neurology*, 62, 137–145 (2003).

41. Brown, E.B., Campbell, R.B., Tsuzuki, Y., Xu, L. Carmellet, P., Dukumura, D., and Jain, R. "In vivo measurement of gene expression, angiogenesis and physiological function in tumors using multi-photon laser scanning microscopy." *Nature Medicine*, 7, 864–868 (2001).

42. Cahalan, M.D., Parker, I., Wei, S.H., and Miller, M.J. "Two-photon tissue imaging: seeing the immune system in a fresh light." *Nature Reviews Immunology*, 2, 872–880 (2002).

43. Miller, M.J., Wei, S.H., Parker, I., and Cahalan, M.D. "Two-photon imaging of lymphocyte motility and antigen response in intact lymph node." *Science*, 296, 1869–1873 (2002).

44. Miller, M.J., Wei, S.H., Cahalan, M.D., and Parker, I. "Autonomous T cell trafficking examined in vivo with intravital two-photon microscopy." *Proceedings of the National Academy of Sciences of the United States of America*, 100, 2604–2609 (2003).

45. Squirrell, J.M., Wokosin, D.L., White, J.G., and Bavister, B.D. "Long-term two-photon fluorescence imaging of mammalian embryos without compromising viability." *Nature Biotechnology*, 17, 763–767 (1999).

46. Zipfel, W.R., Williams, R.M., Christie, R., Nikitin, A.Y., Hyman, B.T., and Webb, W.W. "Live tissue intrinsic emission microscopy using multiphoton-excited native fluorescence and second harmonic generation." *Proceedings of the National Academy of Sciences of the United States of America*, 100, 7075–7080 (2003).

47. Combs, C.A., Smirnov, A., Chess, D., Mcgavern, D.B., Schroeder, J.L., Riley, J., Kang, S.S., Lugar-Hammer, M., Gandjbakhche, A., Knutson, J.R., and Balaban, R.S. "Optimizing multiphoton fluorescence microscopy light collection from living tissue by noncontact total emission detection (epiTED)." *Journal of Microscopy*, 241(Pt 2), 153–161 (2011).

48. Zou, W., Qi, X., and Burns, S.A. "Wavefront-aberration sorting and correction for a dual-deformable mirror adaptive-optics system." *Optics letters*, 33(22), 2602–2604 (2008).

49. Diouf, A., Legendre, A.P., Stewart, J.B., Bifano, T.G., and Lu, Y. "Open-loop shape control for continuous microelectromechanical system deformable mirror." *Applied Optics*, 49(31), 148–154 (2010).

13

Pupil-Segmentation-Based Adaptive Optics for Microscopy

Na Ji
*Howard Hughes
Medical Institute*

Eric Betzig
*Howard Hughes
Medical Institute*

13.1 Introduction

Optical microscopy has long been used to study objects too minuscule for the naked human eye [1]. These minute bodies almost always have spatially heterogeneous optical properties. For example, their compositions differ in reflectivity, absorption, scattering, or fluorescence. These inhomogeneities provide contrast in micrographs and are often themselves the center of investigation. Because of the wave nature of light, however, the resolving power of conventional optical microscopy is limited by diffraction. Even in a microscope devoid of defects, a point source appears as a blob on its image plane and the illumination light can be focused only into a volume rather than a single point, with the extent of its positional uncertainty dictated by the point spread function (PSF) of the imaging system. For a high numerical aperture (NA) microscope operating in the visible range, the diffraction-limited resolution is on the order of hundreds of nanometers.

In practice, however, diffraction-limited resolution is often not achieved. This is because microscopes are designed to operate at the diffraction limit only in samples with specific optical properties. For example, an oil-immersion objective forms perfect images only for objects on the distant surface of a coverglass of a certain thickness and with its other surface immersed in oil; a water-dipping objective can focus light only to a diffraction-limited volume in water. Deviations from such rigorous requirements are common—coverglasses have variations in their thickness, and the immersion media may be at a temperature that causes its refractive index to differ from the specification. More inevitably, the structures we would like to visualize degrade image quality by themselves—the

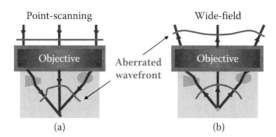

FIGURE 13.1 Schematics depicting the effects of optical inhomogeneities on the excitation light in a point-scanning microscope (*left*) and the emission light in a wide-field microscope (*right*). Arrowed lines describe light ray propagation direction and the curves intersecting the arrowed lines denote wavefronts.

sample heterogeneities that give rise to image contrast also cause spatial variations in their optical properties. For example, biological samples are comprised of structures (i.e., proteins, nuclear acids, lipids) with different refractive indices [2]. It is all but impossible to design a microscope with an immersion medium matching these varying refractive indices simultaneously. Therefore, these samples induce optical aberrations on both the incoming and outgoing waves. For point-scanning microscopes, such as a two-photon fluorescence microscope, the aberrations of the excitation light result in an enlarged focal spot within the sample and a concomitant deterioration of signal and resolution. For a wide-field microscope, the wavefront from the fluorescence emission at each point in the sample is similarly distorted, preventing a diffraction-limited image from being formed at the image detector (Figure 13.1).

In either case, diffraction-limited performance may be restored using active optical components, such as a spatial light modulator (SLM) or a deformable mirror, to modify the excitation or emission wavefront in such a way as to cancel out the sample-induced aberrations [3]. Such approaches, collectively named adaptive optics (AO), were initially developed in astronomy to combat the aberrations caused by the earth's atmosphere on the wavefronts arriving from extraterrestrial sources [4–6]. A wavefront sensor, usually made of a 2D array of lenses focused on the different subregions of a camera and incorporated into a feedback loop with a deformable or segmented mirror, measures the wavefront directly and dictates the surface form of the mirror.

In microscopy, such direct wavefront sensing is not generally applicable. For in vivo imaging with a point-scanning microscope, it is usually not possible to use a wavefront sensor to measure the excitation light wavefront directly after the aberrating sample. For samples with high optical transparency, such as the human eyes, direct wavefront sensing schemes using reflected light have been successful [7, 8, Section IV of this book]. However, the strong scattering of most biological samples prevents direct wavefront sensing from being widely applicable. Indirect wavefront-sensing schemes are needed as a result. In this chapter, we describe two pupil-segmentation-based AO methods, where we measure the wavefront aberration segment by segment. This zonal approaches they differ from the zonal wavefront sensing method in astronomy in that they are indirect, image/signal-based methods, which makes them applicable to strongly scattering samples, such as mouse-brain tissues.

13.2 Adaptive Optical Two-Photon Microscopy Using Pupil Segmentation

Our approach is based on a simple physical picture of focal formation—a focus arises from a spherical wavefront, wherein all light rays are bent to converge at a common point with a common phase, resulting in maximally constructive interference, and thus a diffraction-limited focus [9] (Figure 13.2a). Within an optically inhomogeneous sample, however, rays traversing regions with different refractive

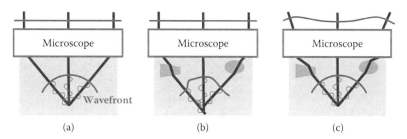

(a)   (b)   (c)

FIGURE 13.2 Illustrations depicting (a) an ideal imaging system, (b) an aberrated imaging system, and (c) an imaging system with pre-compensation for sample-induced aberrations. The thin, sinusoidal curves denote the phase relationship among the light rays. (Ji, N., et al., *Nat. Meth.*, 7, 2, 141–147, 2010.)

indices have both their directions and phases modified such that they do not meet at a common spot and, where they do intersect, they interfere less constructively. Expressed alternatively, the wavefront near the focal region is no longer spherical, but rather aberrated (Figure 13.2b).

For two-photon fluorescence microscopy, the only aberrations affecting the image quality are those experienced by the focused excitation light, since the fluorescence emission is detected by a non-image-forming detector and does not convey spatial information. To recover the diffraction-limited resolution and the optimal signal, we need to modify the wavefront of the excitation light before it enters the objective, so that each ray, even after deflection and phase retardation by local refractive index inhomogeneities, still reaches the focal point at the optimal phase (Figure 13.2c). To accomplish this, a SLM or a deformable mirror is placed along the excitation light path at a plane conjugate to the back pupil plane of the objective [9]. By doing so, any change of phase at a given point on the SLM is transferred to a unique corresponding point at the back pupil plane, and thus on the excitation wavefront.

There still remains, however, the issue of how to measure the deflection and phase offset of each ray without a wavefront sensor. In one approach, we illuminated only one segment of the rear pupil, denoted as the "on" subregion, by applying a flat phase to the corresponding section of the SLM and a binary grating of phase 0 and π everywhere else (causing the light in these regions to be diffracted outside the rear pupil). An image is then obtained in the usual manner by scanning only the light from the "on" subregion. If the light ray from this subregion is deflected by refractive index inhomogeneities along its path, the fluorescence image formed thereby is shifted laterally from that obtained under full-pupil illumination. From the magnitude and direction of this shift, the two-dimensional phase gradient needed to shift the single-ray image back to the position of the full-pupil one is calculated and then added to the "on" subregion of the SLM. As a result, the original ray of interest is tilted such that, even after traversing the sample inhomogeneities, it intersects the original desired focal point. The same procedure is then repeated for other subregions of the SLM till the complete pupil is sampled, and all rays intersect at a common point (Figure 13.3a–e). Either the intrinsic fluorescence in the sample can be used to measure the image shift, or small (0.1–2.0 μm) fluorescent beads can be used as bright and photostable "guide stars" to measure the shift.

Two methods can then be used to determine the relative phases of these rays that maximize constructive interference. In the direct-measurement approach, one reference subregion is kept on at all times, and each of the other subregions is turned on in sequence. The phase of each of these other rays is varied when each is on, until the interference with the reference ray produces maximum intensity at the focal point (Figure 13.3f–h). In the phase-reconstruction approach, the phase gradients determined above are used along with the constraint that the wavefront is continuous, to calculate the phase algorithmically [10]. The direct approach can be more accurate, since an error in one pupil subregion does not propagate to others, whereas the phase-reconstruction method is insensitive to photobleaching and requires relatively few fluorescence photons for accurate measurement as well as substantially fewer

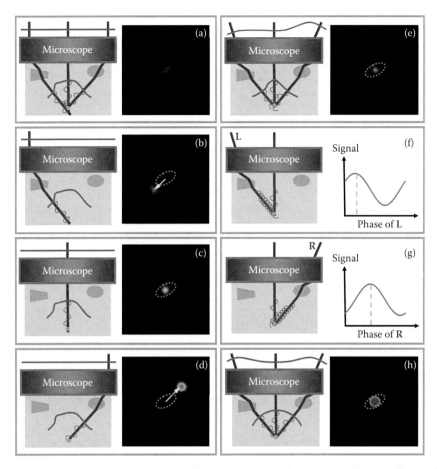

FIGURE 13.3 Schematics illustrating the pupil-segmentation-based AO approach using three subregions, represented by three beamlets. (a) Aberrated wavefront leads to an aberrated image. (b–d) Images acquired with the left, center, and right subregions illuminated, respectively, allow the tilt of each beamlet to be measured from the displacement of the image. (e) Beamlets intersect at a common point, after proper phase gradients are applied at the SLM to steer each beam. (f) Left beamlet interferes with central reference beamlet, determining the optimal phase offset (dashed vertical line). (g) The same procedure is used for the right beamlet. (h) Final corrected wavefront and recovered diffraction-limited focus provide an aberration-free image. If phase reconstruction is used, the phase-measurement steps (f and g) are skipped. (Ji, N., et al., *Nat. Meth.*, 7, 2, 141–147, 2010.).

images (the direct approach requires 6N images for N pupil subregions—N images for beam deflection measurement and 5N images for the fitting of interference curves—whereas the phase-reconstruction method needs N images for N pupil subregions).

The above pupil-segmentation approach is zonal by nature, reminiscent of a Shack-Hartmann sensor in that it divides the wavefront into zones and measures the local wavefront tilt of each zone by image shift, but without the requirement for a transparent sample. Similar to the Shack-Hartmann sensor, it works well with point sources or sparsely labeled structures. However, by illuminating one pupil segment at a time, each image is taken under a much lower NA. The resulting increase in the focal volume of the excitation light, in particular the increase in its axial extent, causes a problem in densely labeled samples—the fluorescent structures that are originally outside the excitation volume under full-pupil illumination now contribute to the images taken at different pupil segments, and thus may make image shift measurement difficult.

We have developed a different approach, one that is implemented with the entire pupil illuminated at all times and thus can be used for the AO correction of densely labeled samples [11]. It is based on the same simple physical picture of focal formation—that is, an ideal focus is formed with all light rays converging at a common point with a common phase. Different from the single-segment-illumination approach, it does not use image shift to determine the local wavefront tilt. Instead, it relies on the fact that a light ray will interfere only at the focus, and thus modulate the focal intensity and, consequently, the signal strength—most effectively when it intersects the focus formed by all the other light rays. Thus, scanning one ray across a range of angles while keeping all the others fixed would vary the signal strength (obtained either at one sample position by parking the beam or as the averaged signal of images by scanning the beam). Visualizing the signal modulation by plotting the signal relative to the scanning field position of the ray as an image (not a true image of the sample, but rather a map reflecting the variation of the focal intensity), we can determine the local wavefront tilt. If the scanned ray faces no perturbing inhomogeneities, this image has a maximum at its center, corresponding to zero applied phase ramp (Figure 13.4a and c). Any wavefront tilt causes this image to take on one of three forms—an image with a shifted maximum, if the scanned ray constructively interferes with the fixed rays (Figure 13.4b and d); an image with a shifted minimum, if this interference is destructive; or an image with relatively flat intensity, if the phase of the scanned ray relative to the fixed ones is near $\pm\pi/2$. In the final case, an additional offset of $\pi/2$ can be applied to the scanned ray, and the measurement repeated, yielding an image with a shifted extremum. For any of the three scenarios, ray deflection—that is, local wavefront tilt—can be measured from the displacement of the extremum, just as in a Shack-Hartmann sensor or the above single-segment illumination algorithm. The same procedure is then sequentially applied to all rays/pupil-segments across the entire rear pupil. After all the wavefront tilts are measured, we can then determine the phase of each ray by either direct measurement or phase reconstruction, as described above. Because this approach fully illuminates the pupil and thus utilizes the whole NA of the microscope objective, it can be applied to samples of arbitrary 3D complexity.

It is worth noting here that even though both approaches we outlined here are zonal AO approaches and both rely on measuring shifts, they are fundamentally different. The single-segment illumination

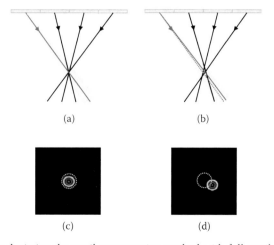

(a)   (b)

(c)   (d)

FIGURE 13.4 Schematics depicting the pupil-segmentation method with full-pupil illumination. (a) An ideal focus has all rays (arrowed lines) intersect at the same point. (b) For an aberrated focus, rays do not intersect at the same point. Scanning one of the rays (e.g., leftmost ray) through a range of angles (shaded cone) varies the intensity of the signal excited at the focus. (c) Plotted as an image, this data exhibits an intensity extremum centered (dashed circle) over the scan region, if the ray already intersects the ideal focus. (d) Shift of the extremum from the center indicates local wavefront tilt. The dashed line in b depicts the original ray direction, while the solid line in b depicts the direction at which signal extremum is reached. (Milkie, D.E., et al., *Opt. Lett.*, 36, 21, 4206–4208, 2011.)

approach relies on measuring the shift of sample images, conceptually similar to the principles underlying a Shack-Hartmann sensor, while the full-pupil illumination approach relies on the intensity modulation of the focus, independent of image contrast mechanism and image structure. In the 13.4, we will show how both these approaches can recover diffraction-limited resolution through highly aberrating samples.

13.3 Experimental Setup

Even though our approach can be easily adapted to other point-scanning modalities, such as those relying on second-harmonic generation, third-harmonic generation, and Raman scattering, we implemented the pupil-segmentation-based AO approach in a two-photon fluorescence microscope because of its widespread usage in bioimaging (Figure 13.5). Near-infrared femtosecond pulses generated by a titanium-sapphire laser (Chameleon Ultra II; Coherent Inc.) are raster-scanned in 2D by a pair of galvanometers (X and Y) (6215H; Cambridge Technology Inc.). The galvanometers are optically conjugate to each other with two custom-made 30 mm focal-length telecentric f-θ lenses (F1) (Special Optics). A third F1 lens and a custom-made 150 mm focal-length telecentric f-θ (F5) lens (Special Optics) conjugate the Y galvanometer to a liquid-crystal phase-only SLM (1,920 × 1,080 pixels; PLUTO-NIR; Holoeye Photonics AG). The conjugation of the galvanometers to the SLM ensures that the intensity at each subregion of the SLM remains constant during beam scanning. The SLM is itself either conjugated by a pair of F5 lenses to the rear pupil of a 20 ×, NA 1.0 water-dipping objective (W Plan-Apochromat; Carl Zeiss Inc.), or by a custom-made 120 mm focal-length telecentric f-θ (F4) lens and a custom-made 240 mm focal-length telecentric f-θ (F8) lens (Special Optics) to the rear pupil of a 16 ×, NA 0.8 water-dipping objective (LWD 16 × W; Nikon Corp.). For the underfilled Zeiss objective, the SLM area used in AO correction is rectangular, whereas for the overfilled Nikon objective, the SLM area is square. Conjugation of the SLM to the objective rear pupil ensures that the corrective phase pattern applied at the SLM would be transferred to the rear pupil and remain stationary during beam scanning. The objective is mounted to a piezo-flexure stage (P-733.ZCL; Physik Instrumente, GmbH) for 2D and 3D imaging in the axial direction. The fluorescence signal is split into red and green components and detected by two

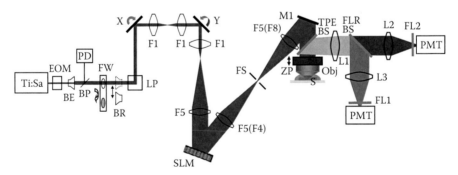

FIGURE 13.5 Schematic of an AO two-photon fluorescence microscope. From left to right: Ti: Sapphire laser (Ti:Sa); electro-optical modulator (EOM); 2× beam expander (BE), beam pickoff (BP) that reflects ~3% of the light into a photodiode (PD) and a neutral-density filter wheel (FW) for excitation power control; motorized beam reducer (BR); two-axis laser positioner (LP); X galvanometer (X); two 30 mm focal-length telecentric *f*–θ lenses (F1); Y galvanometer (Y); 30 mm (F1) and 150 mm (F5) focal-length telecentric *f*–θ lenses; spatial light modulator (SLM); two 150 mm focal-length lenses (F5) for Zeiss objective, one 120 mm (F4) and one 240 mm (F8) lens for Nikon objective; field stop (FS); mirror (M1); dichroic beamsplitter (TPE BS); Zeiss or Nikon water-dipping objective (Obj) mounted on a Z-piezo stage (ZP); sample (S); fluorescence collimating lens (L1); dichroic beamsplitter (FLR BS) to separate green and red fluorescence; two lenses (L2, L3) to focus fluorescence through two filter sets (FL1, FL2) onto two photomultiplier tubes (PMT). (Ji, N., et al., *Nat. Meth.*, 7, 2, 141–147, 2010.)

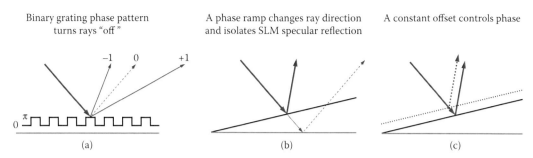

Binary grating phase pattern turns rays "off"

A phase ramp changes ray direction and isolates SLM specular reflection

A constant offset controls phase

(a)   (b)   (c)

FIGURE 13.6 Schematics illustrating the different operational modes of SLM. (a) A binary grating phase pattern on the SLM allows energy to be diverted away from the corresponding region on the back pupil; (b) A phase ramp changes the direction of the reflected ray; (c) A constant offset on the SLM graylevel changes the phase of the reflected ray.

photomultiplier tubes (PMT). Other important components of our setup include a motorized beam reducer (BR) (56C-30-1X-4X, Special Optics) and a 2D beam positioning system (LP) made of two beam-steering mirrors mounted on two fast translation stages (M-663; Physik Instrumente, GmbH). Together, they allow the light to be concentrated at the "on" subregions during the beam deflection measurement, the advantage of which is discussed in the following section. The entire optical path was designed and optimized with ray-tracing software: OSLO (Sinclair Optics, Inc.) and Zemax (Zemax Development Corp.).

The phase-only SLM is used to both measure and then correct any aberrations. We chose a SLM instead of a deformable mirror because it offers several advantages. With 1,920 × 1,080 pixels, it can be readily divided into hundreds of mechanically uncoupled, independent subregions, each with a smoothly varying linear phase ramp. With phase wrapping, it can produce >100 wave amplitude of phase change and >60 wavelengths·mm^{-1} phase gradients, much greater than with popular large-stroke deformable mirrors [12]. The SLM alone allows us to execute our AO procedure. Specific subregions are turned "off" by applying a phase grating consisting of alternate rows of 0 and π phase shift (Figure 13.6a), which diffracts most of the light to a field stop at an intermediate image plane ("FS" in Figure 13.5), where it is blocked. For the "on" subregions, a gentler, global phase ramp is applied to separate the light specularly reflected from the front surface of SLM, which cannot be controlled, from the large fraction of light modulated by the SLM. Ray tilting is achieved by superimposing a phase ramp unique to each subregion on the global phase ramp so that the rays intersect at a common point (Figure 13.6b). The phases of all rays are controlled by superimposing a constant offset for each subregion, according to the measured or reconstructed phases determined during the execution of the AO algorithm (Figure 13.6c).

13.4 Results

13.4.1 System Aberration

Aberration affecting the performance of a microscope can come from anywhere along the optical path. It is therefore necessary to characterize the intrinsic aberration of the optical system, so that we can later isolate the sample-induced aberration. This is especially important because many active optical elements, such as SLMs or deformable mirrors, are not optically flat at zero control voltage. In our system, system aberration comes mostly from the SLM, whose silicon substrate has a potato-chip-shaped surface profile causing a peak-to-valley variation of ~1.7λ at 850 nm. For a 500 nm-diameter fluorescent bead under the water immersion for which the Zeiss objective is designed, this aberration caused a four-fold signal reduction (Figure 13.7). After applying our AO algorithm by segmenting the pupil into 36

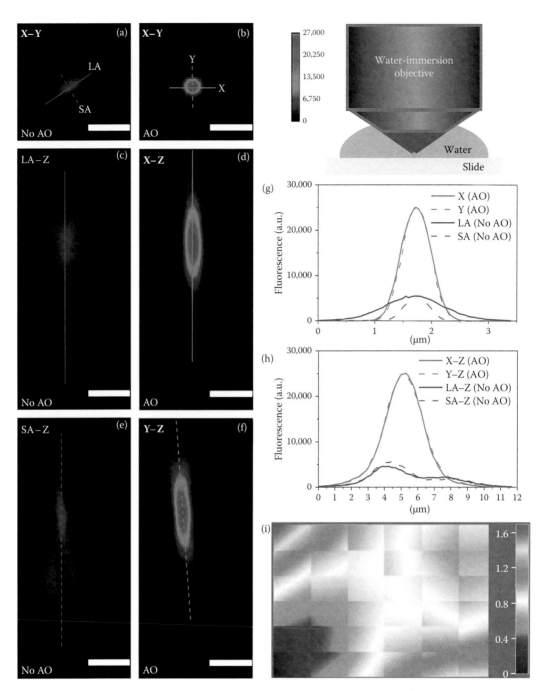

FIGURE 13.7 **(See color insert.)** Correcting the system aberration improves images of a 500 nm fluorescent bead immersed in the design medium of water (inset, upper right). Lateral and axial images before (a, c, e) and after (b, d, f) AO correction. Z denotes the axial direction, and LA and SA denote the long and short axes, respectively, of the bead image before correction. (g) Intensity profiles along lines drawn in the lateral plane (red and green in a and b). (h) Similar profiles along lines drawn in the axial planes (red and green lines in c–f). (i) The final corrective wavefront for system aberration when using a Zeiss 20x, 1.0NA objective, in units of wavelength ($\lambda = 850$ nm). Scale bar: 2 μm. (Ji, N., et al., *Nat. Meth.*, 7, 2, 141–147, 2010.)

subregions and direct phase measurement, the full width at half maximum of the bead images in both the lateral and axial directions approached their diffraction-limited values.

13.4.2 Nonbiological Samples

Our pupil-segmentation-based AO approach is very effective when used to correct optical aberrations caused by refractive index mismatch in nonbiological samples. In the first example, we used it to correct for the aberrating effects when the water-dipping 1.0 NA Zeiss objective was used to image a 500 nm-diameter fluorescent bead in air (Figure 13.8a–d). The pupil was segmented into 49 subregions, and the phase correction was measured after the deflection measurement was carried out on each subregion. Compared to the images taken only with system aberration correction, the peak signal after full AO correction increased eightfold, and both the lateral and the axial full width at half maximum values approached their diffraction-limited values. The sample-induced aberration was clearly dominated by spherical aberration, with a peak-to-valley value of 20 wavelengths. Furthermore, the correction remained valid over a 98 × 98 μm field of view [9]. In other words, we transformed this water-dipping objective into an air objective. With AO correction, one can now use the same objective in immersion media of different refractive indices by simply applying the corresponding corrective wavefront to the SLM. In another example, we imaged a 500 nm diameter bead placed on the inside surface of a glass capillary tube that was immersed in water. The bead positioned well away from the centerline of the capillary tube yielded a highly asymmetric wavefront, which could nevertheless be corrected by pupil segmentation. After such correction, the signal increased about 3.5-fold, and near-diffraction–limited

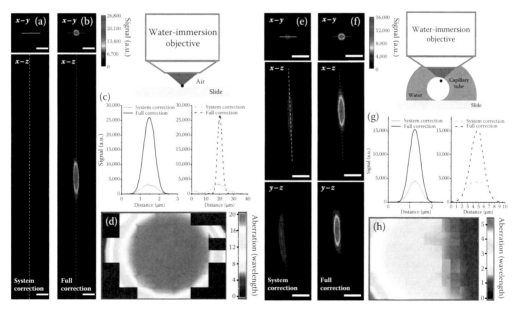

FIGURE 13.8 Correction of aberrations in nonbiological samples. (a and b) Lateral and axial images of a 500 nm fluorescent bead viewed in air using a water-dipping objective, with correction for only system aberrations (a) or all aberrations (b). (c) Signal profiles in the lateral (*x–y*) and axial (*x–z*) planes along the solid and dashed lines in (a) and (b). (d) The final corrective wavefront in units of wavelength (850 nm), after subtraction of system aberrations, obtained with 49 independent subregions and direct phase measurement. (e and f) Images of a bead in an air-filled capillary tube in water, with system aberration correction only (e) and full correction (f). (g) Signal profiles in the lateral (*x–y*) and axial (*x–z*) planes along the solid and dashed lines in (e) and (f). (h) The final corrective wavefront in units of excitation wavelength (850 nm), after subtraction of system aberrations, obtained with 36 subregions and direct phase measurement. Scale bars: 2 μm. (Ji, N., et al., *Nat. Meth.*, 7, 2, 141–147, 2010.)

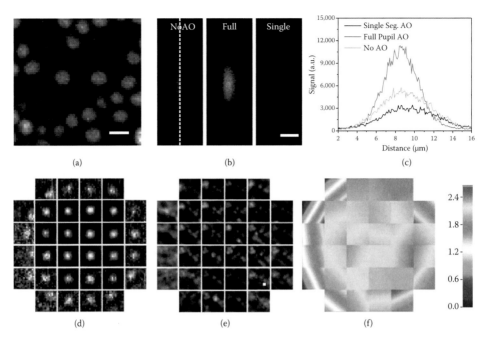

FIGURE 13.9 **(See color insert.)** (a) Maximal intensity projection across a depth of 30 μm in a dense sample of fluorescent beads. (b) Axial images of a 2 μm diameter bead without AO, with full-pupil illumination, and with single-segment illumination AO. (c) Signal profile along the dotted line in b. (d) Signal modulation during full-pupil illumination AO for different pupil segments. (e) Images measured with single-pupil illumination AO. (f) Aberration in units of wavelength. Scale bar: 2 μm. (Milkie, D.E., et al., *Opt. Lett.*, 36, 21, 4206–4208, 2011.)

performance was again attained (Figure 13.8e–h). Here, the pupil was segmented into 6 × 6 subregions and the phase was obtained by direct measurement. (The discontinuities incommensurate with the 6 × 6 segmentation were caused by subtracting a 7 × 7 segmented system aberration.)

The above examples utilized the single-segment illumination approach. The full-pupil illumination approach has also been applied to a nonbiological sample [11]. In particular, to demonstrate its ability to achieve accurate AO correction in densely labeled samples, we prepared a dense aggregate of 2 μm diameter fluorescent beads in agarose (Figure 13.9a) and applied an aberration pattern on the SLM (Figure 13.9f). While the full-pupil illumination AO improved both the signal and the image resolution (Figure 13.9b and c), correction using the single-segment illumination actually degraded the image quality beyond that in the noncorrected case, due to the confusion caused by structures originally out of focus, which led to erroneous displacement measurements (Figure 13.9d and e).

13.4.3 Biological Sample Examples

Interferometric measurements have shown that aberrations caused by biological samples are much more complex than what is typical in simple nonbiological samples, due to their extreme optical inhomogeneity [13, 14]. However, as the example in Figure 13.10 shows, even segmenting the pupil into 12 subregions provided a significant improvement in the image quality of a 1 μm fluorescent bead imaged through a 250 μm-thick fixed mouse-brain cortical slice. Indeed, even though the corrective wavefront, with a peak-to-valley value of four wavelengths, appeared quite disjointed, our AO correction removed the ghost images and improved the maximal signal by fourfold. For most

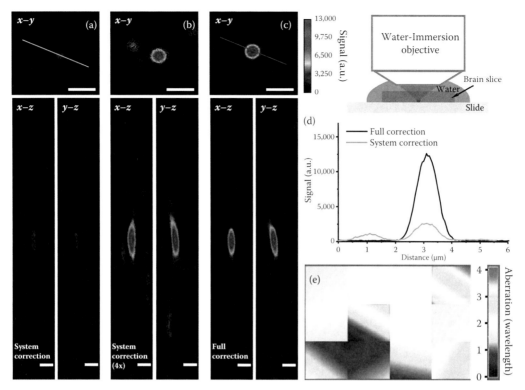

FIGURE 13.10 Correction of aberrations induced by 250 μm thick fixed mouse brain slices. (a–c) Lateral (*x–y*) and axial (*x–z* and *y–z*) images of a single 1 μm diameter bead under a brain slice, with only system correction (a), with a 4 × display gain (b) and after full AO correction (c). (d) Lateral intensity profiles along the light gray and dark gray lines in (a) and (c). (e) The corrective wavefront, after subtraction of system aberrations, in units of excitation light wavelength (850 nm), using 12 independent subregions and direct phase measurement. Scale bars: 2 μm. (Ji, N., et al., *Nat. Meth.*, 7, 2, 141–147, 2010.)

biological applications, it would be desirable if the intrinsic fluorescence from the sample, rather than a guide star like a fluorescent bead, can be used for AO correction. Instead of using the centroid shift of a point object, such as the fluorescent bead, to measure the beam deflection, one can use image correlation to measure the image shift when different pupil segments are turned on. In Figure 13.11, fluorescent images of several neurons at the bottom of a fixed brain slice were used to measure the aberration caused by a cover slip and the 300-μm-thick brain tissue. Three sets of image data are shown—without correcting the system aberration, with system aberration correction, and with correction for both the system and sample-induced aberration—and the image quality improves progressively. (It is not always true that correcting the system aberration improves image quality. Depending on the nature of the aberrations, system aberration may, in some cases, partially cancel the sample-induced aberration. As a result, correcting the system aberration may actually reduce the image quality.) The improvement is especially striking in the axial plane, because axial resolution is usually more susceptible to aberration than lateral resolution—a long-standing observation in microscopy [15] that can be simply explained with our focus model. For most of the light rays within the excitation beam, as the rays start to deviate from and miss the focus, the volume they form has a longer dimension axially than laterally.

In addition to fixed in vitro samples, AO can be used to improve imaging quality in vivo. Using the pupil-segmentation-based AO, we systematically characterized and corrected the optical aberrations

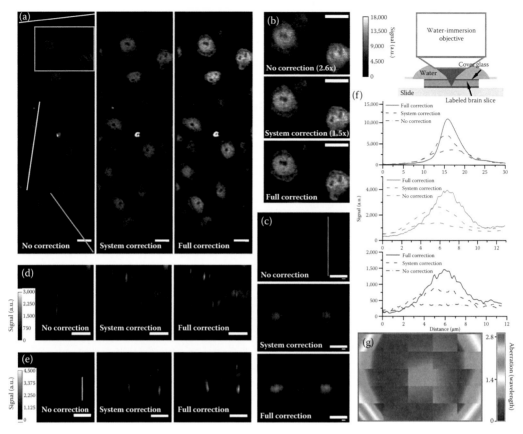

FIGURE 13.11   **(See color insert.)** Aberration correction at the bottom of an antibody-labeled 300 μm thick fixed mouse brain slice. (a and b) Lateral images of a field of neurons acquired with and without correction as indicated (a), and magnified images from one subfield marked by the rectangle in a, with all images normalized to the same peak intensity (b). (c–e) Images in the axial planes defined by the yellow (c), green (d) and blue (e) lines in a. (f) Intensity profiles along the gray, purple, and orange lines in c–e. (g) The corrective wavefront in units of excitation light wavelength (850 nm), after subtraction of system aberrations, obtained with 36 subregions and direct phase measurement. Scale bars: 10 μm. (Ji, N., et al., *Nat. Meth.*, 7, 2, 141–147, 2010.)

encountered during the in vivo two-photon imaging in the mouse cortex [16]. We found that the brain-induced aberrations are temporally stable over a few hours, that diffraction-limited resolution can be recovered at a depth of 450 μm in the cortex of the living mouse, and that, if the aberrating structure is devoid of large blood vessels, the AO correction obtained at one point inside the sample can improve the image quality over a surrounding volume of hundreds of microns in dimension. The improvement in both the signal and the resolution allowed more dendritic structures to be detected (Figure 13.12a), with fine dendritic processes showing up to fivefold gain in signal strength (Figure 13.12f). As a general rule, smaller structures show a larger signal gain upon AO correction than larger structures (Figure 13.12d–f). This is because the enlarged focal volume caused by aberration allows more fluorophores to be excited in a large fluorescent structure, such as a soma, thus partly compensating for the decreased focal intensity; whereas for small fluorescent features such as dendrites, the full impact of the reduced focal intensity is felt more acutely. Indeed, for an aberration-free imaging system, the total amount of two-photon-excited fluorescence in a fluorescent sea is independent of NA—the decline of peak intensity is exactly compensated by an increase in the focal volume, and thus the number of fluorophores in the excitation region [17].

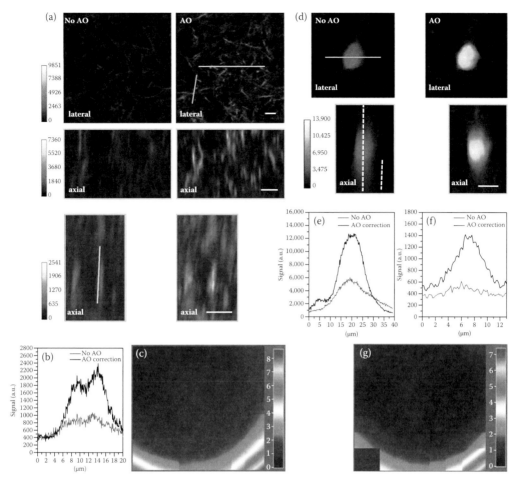

FIGURE 13.12 **(See color insert.)** AO improves imaging quality for in vivo mouse-brain imaging: (a) Lateral and axial images of GFP-expressing dendritic processes 170 μm below the surface of the brain before and after AO correction; (b) Axial signal profiles along the white line in (a); (c) Measured aberrated wavefront in units of excitation wavelength; (d) Lateral and axial images of GFP-expressing neurons 110 μm below the surface of the brain with and without AO correction; (e, f) Axial signal profiles along the white dashed lines in (d) (left dashed line goes through a soma and the right dashed line goes through a dendrite); (g) Aberrated wavefront measured in units of excitation wavelength. Scale bars: 10 μm. (Ji, N., et al., *Proc. Natl. Acad. Sci. USA*, 109, 22–27, 2012.)

13.5 Discussion

13.5.1 Pupil Segmentation Strategy

Since we approximate the local wavefront of each pupil segment as a plane, the greater the number of subregions into which we segment the pupil, the more accurate the wavefront can be described. Figure 13.13 demonstrates, for a 1 μm fluorescent bead below a 250-μm-thick fixed brain slice, how the lateral resolution, axial resolution, and signal vary as a function of the number of pupil subregions. For this fluorescent bead, slight improvement in lateral resolution is seen for all subregion numbers, consistent with the above-mentioned fact that lateral resolution is more resistant to aberration. However, even by segmenting the pupil into 12 subregions, we improved both the axial resolution and the signal significantly. With increasing number of subregions thereafter, the improvement in the axial resolution rapidly saturated, indicating that even with modest numbers of subregions, most rays already intersect near

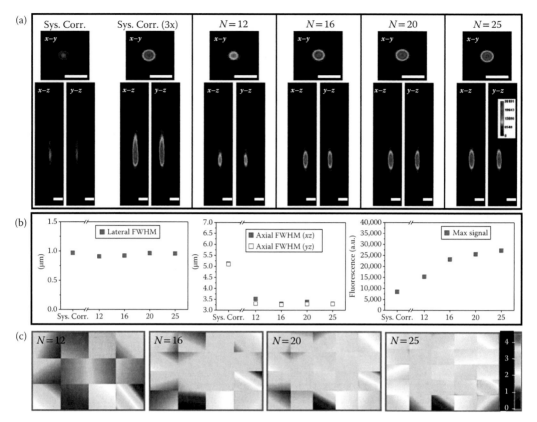

FIGURE 13.13 (a) Lateral (*top*) and axial (*bottom*) images of a 1 μm bead as viewed through a 250-μm-thick fixed mouse brain slice under different correction conditions. (b) Lateral and axial FWHM (full width half maximum) and peak signal for each of the correction conditions in (a). (c) The corrective wavefront in units of wavelength ($\lambda = 850$ nm), after subtraction of system aberrations, for the cases $N = 12$, 16, 20, and 25, obtained using the independent mask algorithm with direct phase measurement. Scale bar: 2 μm. (Ji, N., et al., *Nat. Meth.*, 7, 2, 141–147, 2010.)

a common point and formed a near-diffraction limited volume within which two-photon fluorescence generation is largely confined. The signal, however, continued to increase, indicating greater sensitivity of the two-photon signal to even small wavefront errors. This is hardly surprising—for a point object, a phase error φ in the electric field over a small fraction a of the rear pupil will yield a signal that is $\left|1 - a(1 - \exp(i\varphi))\right|^4$ of the optimal value. For example, the signal is more than halved (41%), even if only 10% of the wavefront is 180° out of correct phase.

The number of subregions required depends on several factors—the specifics of the sample under investigation, the parameter being optimized, and the extent of optimization desired. Because we correct the pupil segment by segment, our zonal approach allows an initial low-resolution map of the aberrated wavefront to be made at a modest number of pupil subregions, and only then do the areas suggestive of fine structure need to be sampled with smaller subregions. For the single-segment illumination method, the reduced NA of excitation ultimately limits the number of subregions into which the pupil can be segmented (which is not a problem for the full-pupil illumination method). Under single-segment illumination, one can densely sample the wavefront slope at the pupil while still using relatively large illuminated subregions, by making these subregions overlap. Figure 13.14 shows how overlapping subregions lead to more distinct (but not independent) pupil areas. We find the method illustrated in Figure 13.14d to be particularly useful, for it produces a 2D array of local wavefront tilt measurement,

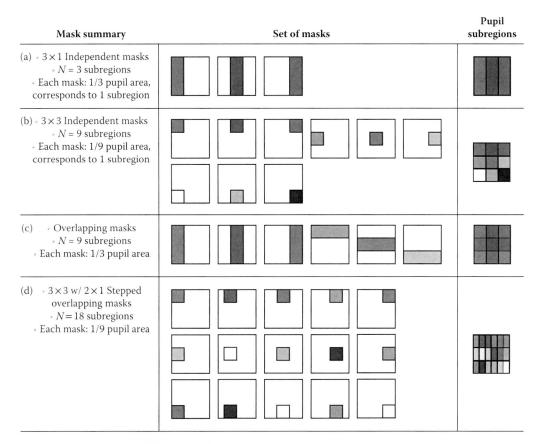

Mask summary	Set of masks	Pupil subregions

(a) • 3×1 Independent masks • N = 3 subregions • Each mask: 1/3 pupil area, corresponds to 1 subregion

(b) • 3×3 Independent masks • N = 9 subregions • Each mask: 1/9 pupil area, corresponds to 1 subregion

(c) • Overlapping masks • N = 9 subregions • Each mask: 1/3 pupil area

(d) • 3×3 w/ 2×1 Stepped overlapping masks • N = 18 subregions • Each mask: 1/9 pupil area

FIGURE 13.14 Examples of the independent, overlapping, and stepped overlapping mask approaches to AO correction. The objective rear pupil is represented by the large square, and the shaded rectangles represent the fraction of the pupil turned on at specific points during beam deflection measurement. (a) Independent mask approach: three nonoverlapping masks, each covering 1/3 of the total pupil area. (b) Independent mask approach: nine nonoverlapping masks, each covering 1/9 of the pupil area. (c) Overlapping mask approach: six overlapping masks, each covering 1/3 of the pupil area, lead to unique wavefront estimations in nine pupil subregions. (d) Stepped overlapping mask approach: a mask covering 1/9 of the pupil area is translated in horizontal steps equal to half the width of the mask, which, in vertical steps equal to its height, lead to 18 unique subregions. (Ji, N., et al., *Nat. Meth.*, 7, 2, 141–147, 2010.)

which can then be used directly for wavefront reconstruction, using algorithms long established for Shack-Hartman wavefront sensors.

13.5.2 Phase Measurement vs Phase Reconstruction

Figure 13.15 shows an example of AO correction through a 250 um brain slice with the phase determined either by direct measurement or phase reconstruction. For the same number of distinct pupil areas, phase measurement provides a better correction than phase reconstruction, because the direct interferometric measurement affords a higher accuracy. Keeping the "on" subregion size the same, the performance of phase reconstruction approach can be improved to the same level by overlapping the subregions to increase the sampling density of the wavefront. A critical advantage of the phase reconstruction approach, however, is that it is insensitive to sample photobleaching and more tolerant of low signal-to-noise conditions.

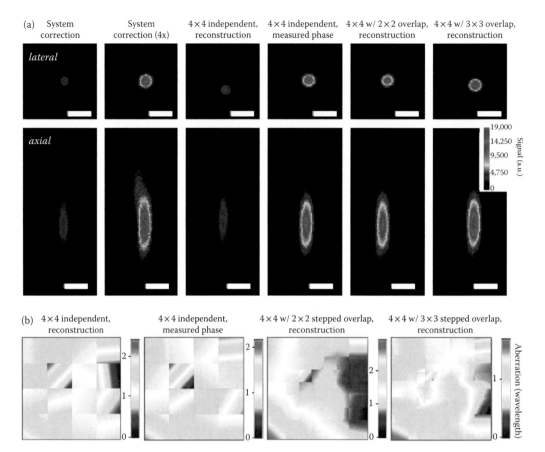

FIGURE 13.15 AO correction on a 1-μm-diameter bead under a 250 μm fixed brain slice, using different variations of our pupil segmentation algorithm. (a) Lateral and axial images of the bead obtained with (left to right): system aberration correction only; system aberration correction with 4 × display gain; 4 × 4 independent subregions with reconstructed phase; 4 × 4 independent subregions with measured phase; 4 × 4 stepped overlap with 2 × 2 steps and phase reconstruction; and 4 × 4 stepped overlap with 3 × 3 steps and phase reconstruction. Scale bars: 2 μm. (b) Final corrective wavefronts in units of excitation light wavelength (850 nm) after subtraction of system aberrations. (Ji, N., et al., *Nat. Meth.*, 7, 2, 141–147, 2010.)

13.5.3 Field Dependence and Averaged Correction

A two-photon fluorescence image is acquired by scanning the laser focus across the desired field of view (Figure 13.16). During the scan, the excitation light travels through different tissue environments and thus experiences different aberrations. If the fluorescent feature used to measure image shift during AO is spatially extended (e.g., Figure 13.16a), the resulting correction will reflect an averaged aberration across the image field, rather than a more local correction, as would be the case if the corrective feature were point-like (e.g., Figure 13.16b). Certain applications may require a locally optimized correction, while for others, an averaged correction may be more desirable, especially when a large field of view is covered during imaging (e.g., calcium imaging of a population of neurons). For both single-segment illumination and full-pupil illumination methods, one can obtain AO corrections applicable either locally or over a broad area by choosing different-sized sample features during measurement.

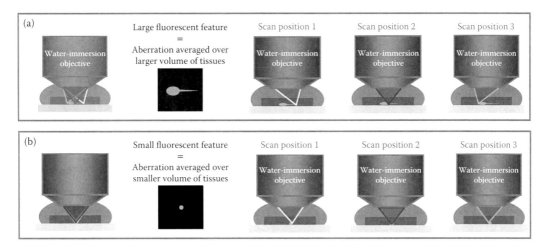

FIGURE 13.16 (a) For fluorescent features occupying a large field of view, the excitation beam probes a large volume of biological tissue during correction. The measured aberration will represent an average over this large volume. (b) When AO correction is performed over a small fluorescent object, the excitation light probes similar tissue volume, leading to a correction that is more accurate locally, but may be less optimal elsewhere than the wider, field-averaged correction.

13.5.4 A Single-Image Approach

Our pupil-segmentation-based approach with single-segment illumination has many similarities with the zonal wavefront sensing approach used in astronomy. They both segment the pupil and measure the local wavefront tilt by image shift and can derive the phase using phase reconstruction. In all the previous examples, we measure the local wavefront tilt one at a time by turning on pupil subregions sequentially. As a result, we take multiple images, while the zonal approach in astronomy needs only one image, since in a Shack-Hartman sensor, the different areas of the wavefront are measured simultaneously. It is possible, however, to achieve one-image correction using our method as well. Because the SLM can generate a large wavefront tilt with minimal coupling between neighboring regions, we can apply distinct, large phase ramps to each subregion. The excitation light reflecting off the SLM then forms an array of foci inside the sample. Scanning the excitation light then gives rise to an image composed of multiple images of the fluorescent feature. If there is no aberration, the multiple images will center at image coordinates specified by the applied phase ramps. The optical aberrations will manifest themselves as image shift away from those coordinates. Figure 13.17 shows an example of such one-image AO correction for system aberration.

13.5.5 Application to Superresolution Point-Scanning: STED

Recent years have seen the rapid development of a suite of superresolution techniques, both point-scanning and wide-field, which can provide imaging resolutions of tens of nanometers [18]. Compared to diffraction-limited techniques with resolution of hundreds of nanometers, superresolution techniques are even more sensitive to aberration. For the point-scanning superresolution method based on stimulated emission depletion (STED), the focus of a fluorescence excitation beam is overlapped with a donut-shaped-beam that depletes fluorescence except at its very center, thus generating an excitation volume below the diffraction limit. In our system, if we do not correct for system aberration; the focus of such a depletion beam is severely distorted, so that STED microscopy would be severely compromised (Figure 13.18). It is reasonable to say that, in general, the same aberrations would impact superresolution

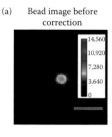

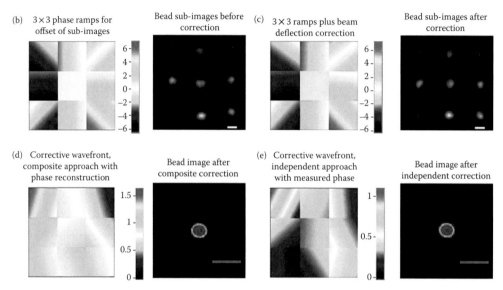

FIGURE 13.17 AO correction for system aberration using only a single image. (a) Image of a 1 μm bead without system aberration correction. (b) After simultaneously applying different phase gradients to each of the 3 × 3 subregions of the SLM, nine spatially offset images of the same bead were seen. The deviations from the ideal 3 × 3 array indicate the aberration-induced tilt error in each subregion. (c) Correcting these tilt errors causes the bead subimages to become aligned with the ideal 3 × 3 array. (d) The corrective wavefront obtained using the tilt error data and phase reconstruction, and the resulting corrected bead image, both of which were similar to those obtained by 3 × 3 independent masks and direct phase measurement (e). Scale bar: 2 μm.

imaging methods more than diffraction-limited imaging methods. For example, a two-photon fluorescence image could still be obtained under the same system aberration shown in Figure 13.18, while a STED microscope would operate poorly.

13.5.6 Zonal-Based Approaches for Scattering Control

In this chapter, we described how segmenting the pupil into different zones allows us to achieve AO correction for images degraded by refractive index variations in the sample. In other words, in strongly scattering samples such as brains, AO corrects the light weakly deflected from refractive bodies of multiwavelength scale, but not the light multiply scattered by sub-wavelength particles, for which much finer correction patterns are required. In recent years, zonal wavefront shaping has also been used to control multiply-scattered light to achieve focusing through turbid media [19] and has been demonstrated recently to form diffraction-limited fluorescence images through samples that are 16 mean-free scattering lengths in thickness [20]. The experimental approaches in these methods are zonal, like ours—the wavefront is segmented into hundreds or thousands of zones, and the phase in each zone is varied while monitoring the resulting signal change. Unlike ours, they are iterative—many passes are required to

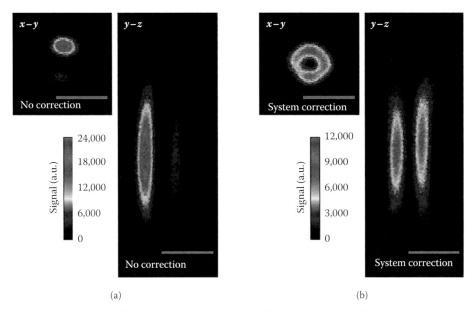

(a)   (b)

FIGURE 13.18   Generation of an annular PSF (a) without and (b) with correction for system aberration. Lateral (*x–y*) and axial (*y–z*) images of a single 500-μm-diameter bead were taken with a helical phase ramp from 0 to 2π on the SLM. Scale bar: 2 μm. (Ji, N., et al., *Nat. Meth.*, 7, 2, 141–147, 2010.)

converge to the corrective wavefront. However, several phase-control algorithms have been developed [21], with a most recent addition holding the promise of rapid convergence [22]. It is conceivable that these methods may be adapted for AO correction of sample aberrations, although it remains to be seen whether they work well for aberrations with large wavefront slopes under the stringent photon budgets of typical biological imaging experiments.

13.6 Summary

In this chapter, we described a zonal AO approach to correct sample-induced aberration in two-photon fluorescence microscopy. Because a perfect focus is nothing but a spot of optimal constructive interference, we first direct rays to intersect at the same point by measuring the image shifts when the corresponding subregions of back pupil are illuminated, and then find the optimal phase for those rays by either interference measurement or phase reconstruction. Alternatively, to avoid the NA reduction associated with single-segment illumination AO, we illuminate the full pupil and recover a perfect focus by scanning one ray around the focus formed by all other rays while monitoring the focal intensity variation. Our approaches work in both nonbiological and biological samples, and have allowed us to recover diffraction-limited resolution hundreds of microns deep inside living specimens. The same principle may be applied to other imaging modalities, including the superresolution methods.

References

1. Hooke, R. *Micrographia or some physiological descriptions of minute bodies made by magnifying glasses with observations and inquiries thereupon.* 1665, London: Royal Society.
2. Choi, W., et al. Tomographic phase microscopy. *Nat Meth*, 2007. **4**(9): 717–719.
3. Booth, M.J. Adaptive optics in microscopy. *Phil. Trans. R. Soc. A*, 2007. **365**(1861): 2829–2843.
4. Hardy, J.W. *Adaptive optics for astronomical telescopes.* 1998, Oxford: Oxford University Press.
5. Tyson, R.K. *Principles of adaptive optics.* 1991, San Diego, CA: Academic Press.

6. Babcock, H.W. Adaptive optics revisited. *Science*, 1990. **249**(4966): 253–257.
7. Liang, J., D.R. Williams, and D.T. Miller. Supernormal vision and high-resolution retinal imaging through adaptive optics. *J. Opt. Soc. Am. A*, 1997. **14**(11): 2884–2892.
8. Rueckel, M., J.A. Mack-Bucher, and W. Denk. Adaptive wavefront correction in two-photon microscopy using coherence-gated wavefront sensing. *Proc. Natl. Acad. Sci. USA*, 2006. **103**(46): 17137–17142.
9. Ji, N., D.E. Milkie, and E. Betzig. Adaptive optics via pupil segmentation for high-resolution imaging in biological tissues. *Nat. Meth.*, 2010. **7**(2): 141–147.
10. Panagopoulou, S.I. and D.R. Neal. Zonal matrix iterative method for wavefront reconstruction from gradient measurements. *J. Refract. Surg.*, 2005. **21**(5): S563–S569.
11. Milkie, D.E., E. Betzig, and N. Ji. Pupil-segmentation-based adaptive optical microscopy with full-pupil illumination. *Opt. Lett.*, 2011. **36**(21): 4206–4208.
12. Devaney, N., et al. Characterisation of MEMs mirrors for use in atmospheric and ocular wavefront correction. *SPIE*, 2008. **6888**: 02.
13. Schwertner, M., et al. Measurement of specimen-induced aberrations of biological samples using phase stepping interferometry. *J. Microsc.*, 2004. **213**: 11–19.
14. Schwertner, M., M.J. Booth, and T. Wilson. Characterizing specimen induced aberrations for high NA adaptive optical microscopy. *Opt. Express*, 2004. **12**(26): 6540–6552.
15. Sheppard, C.J.R. and M. Gu. Aberration compensation in confocal microscopy. *Appl. Opt.*, 1991. **30**(25): 3563–3568.
16. Ji, N., T.R. Sato, and E. Betzig. Characterization and adaptive optical correction of aberrations during in vivo imaging in the mouse cortex. *Proc. Natl. Acad. Sci. USA*, 2012. **109**: 22–27.
17. Pawley, J., ed. *Handbook of biological confocal microscopy*. 2nd ed. 2006, New York: Springer.
18. Ji, N., et al. Advances in the speed and resolution of light microscopy. *Curr. Opin. Neurobiol.*, 2008. **18**(6): 605–616.
19. Vellekoop, I.M. and A.P. Mosk. Focusing coherent light through opaque strongly scattering media. *Opt. Lett.*, 2007. **32**(16): 2309–2311.
20. Vellekoop, I.M. and C.M. Aegerter. Scattered light fluorescence microscopy: imaging through turbid layers. *Opt. Lett.*, 2010. **35**(8): 1245–1247.
21. Vellekoop, I.M. and A.P. Mosk. Phase control algorithms for focusing light through turbid media. *Opt. Commun.*, 2008. **281**(11): 3071–3080.
22. Cui, M. Parallel wavefront optimization method for focusing light through random scattering media. *Opt. Lett.*, 2011. **36**(6): 870–872.

Applications

Part 2: Direct Wavefront Sensing

14

Coherence-Gated Wavefront Sensing

Jonas Binding
*Max Planck Institute
for Medical Research*

Markus Rückel
BASF SE

14.1 Introduction

In optical microscopy, signal size and resolution can be impaired, particularly in biological specimens, by absorption and scattering loss and by specimen-induced aberrations. Although absorption loss and high-angle scattering can be reduced to some extent by choosing the excitation wavelength appropriately, they are also frequently compensated by increased excitation intensities. Optical aberrations, caused by refractive index inhomogeneities within the sample, can be compensated by wavefront shaping with active elements, which has intrigued researchers for several decades (Sheppard and Cogswell 1991). Nowadays, an increasing supply of deformable mirrors (DMs) and liquid crystals allows the correction of optical aberrations with varying temporal and spatial resolution.

In astronomy, where atmospheric turbulences produce optical aberrations, limiting the performance of large terrestrial telescopes, methods for fast wavefront measurement have been developed that allow for the determination of the optimal correction settings from the analysis of a strong (natural or artificial) guide star in the field of view. Such a guide star serves as a point-like emitter, whose light passes through the unknown atmospheric distortions. This light is then analyzed by a wavefront sensor—for example, a Shack-Hartmann wavefront sensor. The such-determined aberrations can serve to increase the optical resolution within a certain patch of the sky around the location of the guide star, making much weaker and closely spaced objects discernible. In particular for solar astronomy, variants of these methods have been developed to allow the use of laterally extended objects as a reference (von der Lühe 1983; Soltau et al. 1997; Rimmele and Radick 1998; Poyneer 2003) based on image cross-correlation.

In most biomedical applications of adaptive optics, one of the key questions is how to measure the specimen-induced aberrations that need to be corrected. Generally, the samples do not contain sufficiently strong and well-isolated point-like light sources as they are used in astronomy. However, in the case of retinal imaging, point-like light sources can be created by the diffusive backscattering of the retina (Liang et al. 1994). Typically, cornea and eye lens distort the wavefront of the light in such a way that high-resolution imaging of the retina is not possible. Using adaptive optics based on the measurement of the wavefront of the diffusively backscattered light spot projected onto the retina, aberrations can be compensated and the cones of the retina can be resolved.

However, many other biological samples scatter throughout the depth profile; in other words, they lack a predominant layered structure that could be exploited to create a "guide star" in the desired depth. Aberrations are not necessarily dominated by certain layers but can originate anywhere in the sample, and the backscattered light comes from many different depths. Since even the methods used in solar astronomy can deal only with laterally extended objects, but not with axially extended ones, the light emanating from the sample cannot be easily analyzed to directly measure aberrations corresponding to imaging in any given depth inside the sample.

Because of these difficulties to perform direct wavefront sensing in biomedical microscopy, considerable work has been done to implement different image-based wavefront optimization schemes that avoid direct wavefront sensing. They rely on the ability to modify the unknown wavefront in a known manner with the active element and to analyze the microscope images taken with different settings of the active element to reconstruct, more or less efficiently, the wavefront. This category includes genetic learning algorithms (Sherman et al. 2002), hill-climbing algorithms (Marsh et al. 2003), adaptive random-search algorithms (Wright et al. 2005), and modal wavefront sensing (Neil et al. 2000; Booth et al. 2002).

The downside of all of these image-based methods is that they require a number (which can be rather larger) of microscope images to be taken to optimize the optical resolution. This is often an issue, since extensive imaging typically leads to deterioration of the contrast agent used and even of the sample itself. In particular, fluorescence microscopy frequently uses chemical or genetically encoded fluorophores as contrast agents, which are subject to bleaching, so that any gain in signal and/or resolution by the use of adaptive optics is no longer usable as the fluorophores have been bleached during the optimization routine. Even if bleaching is not limiting, some image-based algorithms need so many images for the estimation of the wavefront that the imaging frame rate is severely decreased.

Given the disadvantages of image-based optimization schemes, it seems worthwhile to think about direct wavefront sensing and how it could be made to work. As mentioned earlier, in many samples, a sufficient amount of light is scattered back from all depths. If the incoming light is focused at a certain depth and the backscattered light from this particular depth is selected interferometrically in a low-coherence interferometer, this effectively creates a point source inside the sample. This idea is at the heart of coherence-gated wavefront sensing (CGWS) (Feierabend 2004; Feierabend et al. 2004; Rueckel and Denk 2005, 2006, 2007; Rueckel 2006; Rueckel et al. 2006), which is the topic of this chapter.

14.2 Background/Prior Work: Coherence Gating

The idea to probe the structure of biological tissue using coherence gating of the backscattered light is not new, even though the previous application had nothing to do with adaptive optics. Coherence gating was introduced (Huang et al. 1991) as a new biological imaging modality under the name of optical coherence tomography (OCT). In short, a light source with low temporal coherence is sent into a sample and the backscattered light is analyzed interferometrically. This essentially provides time-of-flight information about the backscattered light, which can be translated into depth information using a known or assumed refractive index depth profile of the sample. In addition to the axial resolution, which depends on coherence gating, a certain lateral resolution is usually achieved by focusing of the

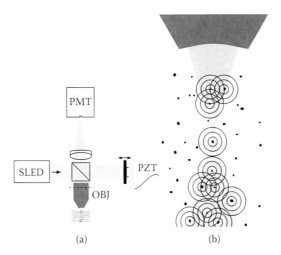

(a) (b)

FIGURE 14.1 (a) OCM setup: Michelson interferometer with piezo-mounted (PZT) reference mirror in one arm and objective plus sample in the other arm; the PMT is conjugated to the sample focus; (b) each scatterer in the illumination double cone produces a secondary wave.

light, allowing OCT to provide three-dimensional imaging with decoupled axial and lateral resolution. A multitude of different variations of OCT has since been developed, using different light sources, detectors, and interferometer layouts, based on time-of-flight scanning (time-domain OCT) or wavelength scanning (frequency-domain OCT); for a review, see Schmitt (1999).

Let us analyze a variant of OCT further, which is usually referred to as optical coherence microscopy (OCM); see Figure 14.1a. A spatially coherent but spectrally broad light source, such as a superluminescent diode (SLED) or femtosecond laser, is used to illuminate a Michelson interferometer. In one arm of the interferometer—that is, the sample arm—the light is focused by an objective lens into the sample. The other arm, called the reference arm, consists of a mirror mounted on a piezo element. The reference arm length is chosen to match the optical path to the geometric focus of the objective lens in the sample arm. Such a system can be used to image a single point; the sample needs to be scanned in x-, y-, and z-axis to acquire a three-dimensional image stack. Using this simplest possible OCM setup as an example, we will develop an intuitive picture of low-coherence interferometry.

The light is scattered by point-like scatterers everywhere in the illuminated double cone of the sample, producing secondary spherical wavelets at the position of each scatterer (Figure 14.1b). Some of this scattered light is collected again by the objective lens, recombined with the reference arm light by the beam splitter, and focused by the tube lens onto the detector.

For a single scatterer precisely at the focus (Figure 14.2a), the optical path length to the detector matches exactly the optical path length in the reference arm so that all wavelengths in the spectrum interfere constructively with the light from the reference arm.

For single scatterers in positions that cause the optical path to be longer or shorter by $\lambda/2$ (Figure 14.2b), we have the opposite situation: nearly all the light is reflected back to the source; hardly any reaches the detector.* By changing the reference-arm length by $\lambda/2$, the situation for each of the mentioned scatterers can be reversed.

In contrast, for scatterers where the path difference between the arms is long with respect to the coherence length of the source, constructive and destructive interference alternate rapidly as a function of wavelength. As a result, the total light intensity from this scatterer (integrated over all wavelengths)

---

* Because of the finite width of the wavelength spectrum, the condition $\delta = \lambda/2$ can only be approximately fulfilled for all wavelengths at the same time. As a consequence, wavelengths at the edge of the spectrum still cause a small constructive interference in the direction of the detector.

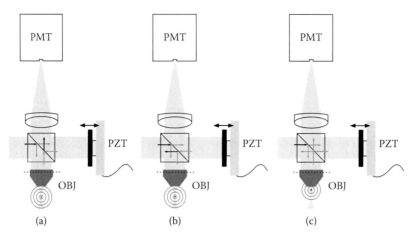

FIGURE 14.2   The precise position of a single scatterer determines whether its scattered light reaches the detector. (a) A scatterer exactly in the focus. (b) A scatterer axially displaced by $\lambda/4$ from the focus. (c) A scatterer far from focus.

leaving the interferometer is almost identical in both directions and hardly changes when the reference arm length is modulated by $\lambda/2$.

In addition, for a scatterer outside the depth of field of the objective (Figure 14.2c), the backscattered wave is noticeably curved, causing constructive and destructive interference to depend on the radial position in the light beam. This decreases the modulation of the signal on the detector to decrease for scatterers outside the depth of field, which is often referred to as the confocal effect of OCT.

Since the signal from a scatterer near the optimal path length is modulated when the reference-arm length is modulated, lock-in detection can be used to separate signal from the scatterers within a coherence length of the focus from signal, from scatterers further away, and from excess reference light or any stray light the detector might receive.

When we later discuss the coherence volume (CV), we refer to the volume where scatterers can contribute to the lock-in signal, keeping in mind that there is no sharp edge to this volume, but that detection efficiency drops off gradually for typical spectral shapes. The CV is limited laterally by the focusing of the incident light and axially by the coherence length.

Two important effects need to be taken into account when thinking about image formation in OCM. First, the density of scatterers inside biological samples is generally high enough such that many of them are found inside the CV. Since their backscattered signals superimpose coherently, they can either add up to form a strong signal or interfere destructively and thus effectively cancel backscattering in certain directions, depending on the relative phases of their individual contributions to the backscattered electromagnetic field. In OCT images, this effect is visible as speckle whenever the distribution of scatterers is random, causing zones of constructive and destructive interference to alternate rapidly.

Second, when high-numerical-aperture (NA) objective lenses are used, the difference between paths with different angles to the optical axis needs to be taken into account even for small distances from the focus. For a scatterer at the focus, the optical path length to the back focal plane (BFP) of the objective lens is by definition independent of the ray angle, but as scatterers are further and further above or below the focus, the path length will increasingly depend on the direction. In other words, the spherical wave emitted by a scatterer in focus is transformed into a plane wave by the objective, whereas all spherical waves with axially displaced origins are transformed into more or less curved spherical waves (Figure 14.2c). On the detector, these waves are not all in focus, so, due to lack of spatial overlap, the interference term diminishes, decreasing the signal modulation and thereby detectability for scatterers far from focus. This effect of z-selectivity is completely independent of the coherence length and depends solely on the NA. It has been termed confocal effect,* since the sample and the reference light need to have the same focus to gain maximum signal. The total axial response function of OCT is determined by the combined effects of coherence gating and confocal filtering. For short coherence lengths and low-NA systems, the axial resolution of a confocal microscope will essentially be determined by the coherence length, while for long coherence lengths and high-NA systems it is determined by the NA. The highest axial resolution can be achieved for large spectral widths in combination with high NAs.

OCT is usually concerned with determining the density of scatterers and thus measures the amplitude of the coherence-gated backscattered light, which is the coherent sum of contributions from all scatterers in the CV, as described earlier. The amplitude is taken as a measure of the local backscattering coefficient of the sample, which can be linked to structures of interest such as cell boundaries or intracellular organelles such as nuclei or mitochondria.

However, due to the interferometric nature of the detection, not only the amplitude but also the phase of the backscattered light can be measured. It is the phase resulting from integrating the electromagnetic field over the full acceptance angle of the objective lens and summing over all scatterers in the CV. This phase will change if the optical path length to the CV changes or if the contributing scatterers move inside the CV. However, when scanning through a sample, the phase measurements from different spatial positions will have no relation with each other, unless the optical path lengths of the scatterers in the two volumes are highly correlated. As such, a phase measurement in OCT usually does not contain much spatial information about the sample and is only interesting when dynamic changes in optical path are to be observed, due to either mechanical movement or refractive index changes (Akkin et al. 2009).

14.3 Background/Prior Work: Wavefront Sensing Using a Shack-Hartmann Sensor

The Shack-Hartmann sensor (Figure 14.3) is based on the principle that an aberrated wavefront which contains components only up to a certain spatial frequency can be decomposed into many tiny wavefront patches, each of which is well approximated by a flat wave with a certain wavefront tilt. If the local tilt of all these patches is measured individually, the global wavefront can be reconstructed by fitting the data to a suitable functional form. For example, a least-squares fit to the local slopes of a certain number of Zernike modes can provide a Zernike decomposition of the wavefront. To measure the local tilts simultaneously, the wavefront is passed through an array of tiny lenslets, each corresponding to one of the aforementioned patches of the wavefront. Each lenslet focuses its incoming wavefront on a position that depends linearly on wavefront tilt. Without tilt, the light is focused on the optical axis defined by this lenslet, whereas strongly tilted wavefronts will focus at a certain lateral distance from the optical

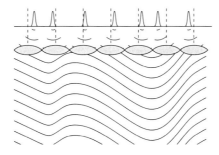

FIGURE 14.3 Principle of a Shack-Hartmann sensor.

---

* This effect is comparable to the confocal pinhole in a confocal laser scanning microscope, with the difference that the size of the "pinhole" in OCT is defined by size of the reference beam focused on the detector—that is, diffraction-limited at best.

axis. By placing a charge-coupled device (CCD) camera in the focal plane of the lenslets, the diffraction pattern from each lenslet can be recorded and its shift with respect to the optical axis determined. This provides the local wavefront tilt measurements that allow reconstruction of the full wavefront.

14.4 Methods

14.4.1 Basic Coherence-Gated Wavefront Sensing Setup

We now want to perform wavefront sensing on the backscattered light, in a manner similar to a Shack-Hartmann sensor. Coherence gating will be used for z-selectivity of the analyzed light, so the new setup is based on the OCM setup (Figure 14.1a). Although OCM used confocal detection, wavefront sensing is best performed in a plane conjugated to the BFP of the objective. Therefore, the CGWS setup (Figure 14.4) contains a telescope that is used to image the BFP of the objective lens onto the detector. At the same time, the point detector is replaced by a camera, allowing simultaneous measurement of the spatial intensity distribution in the BFP, after interference with the reference light.

The backscattered light from the sample entering the objective under a certain angle to the optical axis will reach the BFP at a position corresponding directly to the ray angle, so that the camera effectively provides angular resolution of the backscattered light. Phase-shifting interferometry (Malacara 1992) works in the same way as in OCM: although the sample arm remains fixed, the length of the reference arm is varied, and several camera images are acquired. For simplicity, the sample arm is changed in fixed steps of $\lambda/4$; with four such images I_1, I_2, I_3, and I_4, amplitude and phase of the coherence-gated sample light, E_{smp}, can be determined for each pixel of the camera individually:

$$E_{smp} = (I_1 - I_3) + i(I_4 - I_2) \tag{14.1}$$

where i is the imaginary unit.

To understand in detail the information content of this amplitude and phase measurement in the pupil,[*] we will analyze sequentially different samples, as depicted in Figure 14.5. For an aberration-free sample containing one single scatterer positioned precisely at the focus, the backscattered spherical

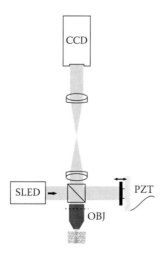

FIGURE 14.4 The CGWS setup.

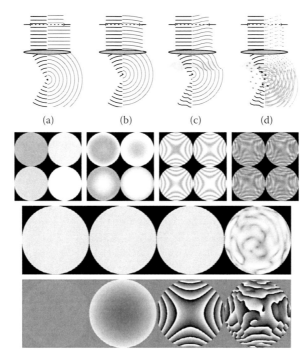

FIGURE 14.5 Comparing the light propagation and backscattering in four different samples. For clarity, incoming waves are depicted only on the left of the optical axis and outgoing waves only on the right of the optical axis. Below the schematic drawings for each sample, the four camera images taken during phase shifting are shown, as well as the logarithm of the amplitude (middle row) and the phase (bottom row) of the field reconstructed by phase shifting. Samples are (a) a single scatterer (black spot) in the focus of the objective, (b) an axially displaced scatterer, (c) an aberrating layer inside the sample, and (d) a realistic sample.

wave is transformed by the objective into a perfect plane wave (Figure 14.5a). Phase shifting will reveal a flat phase profile in the BFP (Figure 14.5, bottom row, left).

For a scatterer that is inside the CV but not exactly in the focus, the axial displacement will cause the wavefront to contain piston (absolute phase) and defocus* (Figure 14.5b), which is visible as a radial phase gradient in the lower row of Figure 14.5. A lateral displacement of the scatterer would cause the wavefront to be tilted. Since virtually all epidetecting microscopic techniques are insensitive to piston, tilt, and defocus with respect to signal strength and resolution, these modes do not have to be corrected (apart from the image distortions they induce).

If an aberrating layer exists between the objective lens and the point-like scatterer, already the incoming light is aberrated. This causes the electric field amplitude of the incoming light at the position of the scatterer to be decreased (Figure 14.5c); however, since the scatterer is assumed to be point-like, the spatial structure of the aberrated incoming field does not change the spatial structure of the scattered field, which is still spherical, until it reaches the aberrating layer. Therefore, the wavefront aberrations that are detected in the BFP correspond directly to the single-pass aberrations acquired on the return path. If the density of scatterers was low enough to have only one scatterer in the CV at the same time, the phase determined by phase shifting on the camera would correspond directly to the sample-introduced aberrations that need to be corrected by adaptive optics to provide diffraction-limited focusing in the position of the scatterer. In Figure 14.5c, this is shown for the special case of pure astigmatism. Since the

---

* It should be stressed that defocus in high-NA systems does not correspond directly to Zernike defocus—it is not simply a quadratic function of radial position in the BFP. Instead, high-NA defocus contains a certain amount of all orders of spherical aberration. See Botcherby et al. (2008).

total phase change between center and the edges of the aperture is larger than 2π in this example, phase wrapping is visible as sudden jumps in the otherwise continuous phase profile.

In reality, there will be many scatterers in the CV, each contributing its own backscattered signal to the amplitude and phase profile detected in the BFP (Figure 14.5d). As discussed for OCM, their contributions are summed coherently. If the CV is small compared to the distance to the major aberrating layers, the light from all scatterers will essentially acquire the same aberrations on the way toward the objective. However, due to their different positions with respect to the geometrical focus, each backscattered wavefront will have a different piston, tilt, and defocus. These phase changes with low spatial order lead to constructive interference in some positions in the pupil and to destructive interference in others. So while speckle in OCM was an effect that could be observed in the microscope images (produced by scanning the CV through the sample), CGWS observes speckle already in the wavefront emanating from one single CV. The speckle adds a random phase term onto the wavefront, which prohibits direct use of the measured wavefront for correction and which we call speckle error. Note how the phase (Figure 14.5d, lower row) still has local gradients resembling the speckle-free case (Figure 14.5c).

Since the random speckle phase term comes from the random distribution of scatterers in the CV, a different realization of randomly distributed scatterers will lead to a different speckle phase term. Therefore, we can hope to average away this term if we are able to measure speckled wavefronts for a multitude of different realizations of scatterers. Indeed, a mathematical analysis shows that the ensemble average of the wavefront over speckle realizations is dominated by the incoherent superposition of the signals from the individual scatterers; the coherent effects responsible for speckle formation cancel out.* This means that speckle error can be decreased by averaging over different realizations of scatterers. For example, 20 realizations reduce the speckle error by a factor of sqrt(20) = 4.5, which was found to be a suitable compromise in our experiments.

If the positions of scatterers change over time and ergodicity is satisfied, different speckle realizations can be obtained just by waiting and repeatedly measuring the wavefront. However, many biological samples are too static at the relevant scale for this to be feasible. In this case, a lateral displacement of the CV can provide a new wavefront measurement with different scatterers and therefore different speckle. Of course, the use of measurements from different lateral positions implies a certain amount of lateral averaging of the aberrations, so a compromise between reduction of speckle noise and restriction of lateral averaging needs to be found.

A second problem associated with the speckelization of individual wavefronts is the singularities in the phase. Since an interferometric phase measurement can determine the phase only up to 2π (see Figure 14.5), phase unwrapping is necessary to reconstruct a wavefront, which can easily have a peak-to-valley difference of several wavelengths. Phase singularities in the speckled wavefront can occur in points where the amplitude goes to zero. Therefore, the phase-unwrapping algorithm used needs to be stable against singularities. In existing implementations of CGWS, a virtual Shack-Hartmann sensor (vSHS; see later) was used for phase unwrapping, but other singularity stable algorithms (such as any other least-squares reconstructor) could also be used.

In summary, ensemble averaging of phase-unwrapped speckled wavefronts allows the determination of the aberrations relevant for adaptive optics in microscopy. However, it should be stressed that this statement relies heavily on the assumption of point-like scatterers and their random distribution in the CV. The assumption of point-like scatterers is supported by studies (Mourant et al. 1998; Schmitt and Kumar 1998) which have found that backscattering is dominated by particles in the size range of $\lambda/4$ to $\lambda/2$. If scatterers predominantly had sizes above the wavelength, their backscattered signal would not have a spherical wavefront but instead contain at least part of the aberrations present in the incoming field. Similarly, if point-like scatterers were arranged in a planar structure instead of randomly

---

* In principle, this is only true up to a certain maximum density of scatterers, above which coherent effects would dominate even after ensemble averaging. However, this critical density of scatterers is several tens of orders of magnitudes above physically possible densities, so that this regime is of but academic interest.

throughout the volume, they would act as a mirror, where the incoming wavefront is reflected back with all its aberrations. Because of the reflection, odd wavefront aberrations for the first and second pass through the tissue would cancel each other out, even though aberrations would double (Artal et al. 1995). So both large scatterers and nonrandom arrangements of scatterers, such as reflective surfaces inside the sample, will hinder the correct measurement of single-pass aberrations.[*]

14.4.2 Virtual Shack-Hartmann Sensor

As stated earlier, a vSHS is a convenient way to perform phase unwrapping in CGWS, which is stable against phase singularities caused by the speckelization of the coherence-gated phase measurements.

Shack-Hartmann sensors (Figure 14.3) have been commercially available for quite some time. Unfortunately, such off-the-shelf instruments are not compatible with the interferometric detection that is necessary in CGWS. Imagine one replaced the camera in our setup with an SHS to perform wave-front sensing and phase unwrapping at the same time (Figure 14.6a). The aberrated light from the CV, the background light scattered back in the other layers of the sample, and the flat reference wavefront would all pass through the lenslet array on their way to the camera. The reference light would be focused exactly on the optical axis of each lenslet, while the aberrated wavefront of the light from the CV would be more or less tilted on different lenslets. For strong enough tilts, the reference and sample spots on the camera would not overlap and could therefore not interfere. This would make it impossible to use coherence gating to separate the in-focus light from the dominant background of the light backscattered in other depths, making the whole scheme unusable.

There is a way to solve this problem and still use a physical lenslet array (Figure 14.6b): if the collimated reference beam is recombined with the sample light behind the lenslet array, interference takes place no matter where the lenslets focus the light from the CV. Although this is in principle possible (Marcus and Feierabend private communication; Tuohy and Podoleanu 2010), such a modification is not compatible with commercially available SHSs, which removes one good reason for using a real

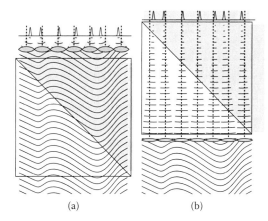

(a)   (b)

FIGURE 14.6 (a) A flat reference wavefront (depicted in gray, coupled in using a beam splitter cube) and the sample wavefront both pass through the lenslet array of an off-the-shelf SHS; (b) a custom-made lenslet array with a long focal length allows accommodation of the beam splitter cube between lenslets and detection plane.

---

[*] In a closed-loop adaptive optics system, where aberrations are measured, then corrected, and residual aberrations remeasured, an oscillatory behavior of even aberrations modes would indicate the measurement of double-pass aberrations. The solution would of course be to reduce the gain of wavefront correction to 0.5, which would solve the issue for even aberrations.

Shack-Hartmann sensor. In addition, more reference-arm light is necessary, since the reference beam remains collimated on the camera while the sample beam is focused.

In contrast, the virtual SHS relies on the fact that amplitude and phase of the electromagnetic field to be analyzed have already been determined interferometrically. By performing Fourier transforms on small rectangles of the electromagnetic field, the lenslets of a real SHS can be simulated, and the resulting intensity analyzed exactly as in a real SHS. With respect to software requirements, the only additional step a vSHS needs compared to a real SHS is the small fast Fourier transforms (FFTs) for each sublens, which can be parallelized and/or performed on a graphics processing unit (GPU) if speed is of concern.

However, the vSHS has several advantages for CGWS with respect to a real SHS. Since coherence gating inherently has to deal with a strong incoherent background from backscattering outside the coherence length, a high-dynamic-range camera is necessary. With a real SHS, the sample light from the CV would be focused into small spots on the camera, which would further increase the dynamic range necessarily. In contrast, the vSHS camera measures the electromagnetic field in a plane where the intensity is distributed as homogeneously as possible, minimizing the dynamic range needed.

Furthermore, with a vSHS, no tedious calibration and alignment between lenslet array and camera are necessary. The size and number of lenslets can be changed rapidly in software; the same data can even be analyzed several times with different parameters if necessary.

Note that the strength of aberrations that can be correctly analyzed by a real Shack-Hartmann sensor is limited by the focal length of the lenslets. If the tilt on one sublens is strong enough, the corresponding spot on the camera will be laterally displaced far enough to leave the region of the camera associated with this particular sublens. This can cause the software algorithm analyzing the data to misinterpret which spot corresponds to which sublens. Similarly, the strength of aberrations that can be analyzed by a vSHS is limited by the sampling of the camera—that is, by the pixel size of the camera. Since the phase is determined only modulo 2π, only phase slopes with a magnitude below π/pixel are determined correctly. When the tilt becomes strong enough that the phase change between two adjacent pixels is below $-\pi$ or exceeds π, aliasing occurs, and the tilt between these pixels will be over- or underestimated by a multiple of 2π/pixel.

14.4.3 Measurement Procedure

Summing up the steps described earlier, the procedure for wavefront measurement in scattering tissue is as follows. A quadruplet of phase-shifted interferograms is acquired, which, using Equation 14.1, allows the calculation of the electromagnetic field in the BFP of the objective. The aperture of the BFP is divided into quadratic subregions, and the Fourier transform of the electromagnetic field for each subregion, is calculated, simulating the effect of a real sublens perfectly aligned with this pixel region. The squared amplitude of the FFT result corresponds to the intensity of the diffraction pattern, which would have been measured on the camera of a real SHS. To quantify the local wavefront tilt, the displacement of the diffraction pattern is determined by centroid estimation. Using all local wavefront tilts, the full wavefront in terms of Zernike modes can be determined. Since this reconstructed wavefront still contains speckle, the whole procedure from phase shifting to Zernike modes is repeated in several laterally adjacent positions of the CV. All resulting Zernike coefficient vectors can be averaged, corresponding to ensemble averaging described earlier, to determine the final wavefront estimate with minimized speckle error.

Since centroid estimation, Zernike reconstruction, and averaging are all linear, a minor speed gain can be achieved by averaging the diffraction patterns directly and performing centroid estimation and Zernike reconstruction on the averaged diffraction patterns for all lenslets (as described in Figure 14.7).

Note that averaging must not be performed before the calculation of the virtual diffraction pattern, since the Fourier transform and subsequent absolute square are nonlinear operations. For example, averaging of the reconstructed complex electromagnetic field corresponds to coherently summing the signal of all scatterers, which implies that speckle contrast remains maximal. If averaging is performed even earlier on the level of image quadruplets taken for different scatterer distributions, no electric field can be reconstructed since this would average away the speckle present in the individual quadruplets.

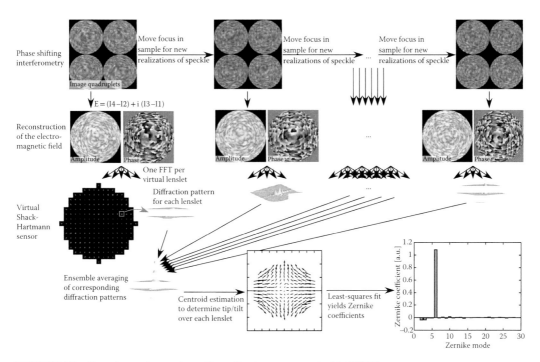

FIGURE 14.7   Experimental procedure of wavefront measurement with CGWS.

14.4.4 Implementation Considerations

No practical implementation of CGWS will be as simple as Figure 14.4, since aberration correction and other experimental constraints will invariably add additional components to the system. Back-reflections from optical components can become a limiting factor by saturating the CCD camera. One solution is to use a forward-only reference arm where the beam is not reflected back on itself and to minimize the length of the sample arm that is reflected back on itself (Feierabend et al. 2004). However, using a CCD with a large dynamic range marginalizes the problem of back-reflections. This opens up the possibility of using completely symmetric sample and reference arms (Wang et al. 2012), most notably the integration of a microscope objective in the reference arm (Linnik interferometer). The advantages are an automatic alignment of both interferometer arms because of the cat's-eye effect (objective and reference mirror reflect the beam back onto itself), as well as the perfect dispersion correction because of the symmetry between the arms.

14.5 Application

14.5.1 Experimental Implementation

In principle, CGWS is a wavefront sensing technique that works with any sufficiently scattering sample where scatterers are randomly distributed. It is independent of the imaging modality and can, in principle, be combined with anything from classical wide-field microscopy, confocal microscopy, two-photon microscopy (2PM), structured illumination microscopy, coherent anti-Stokes Raman scattering microscopy, second/third-harmonic generation microscopy, or even stimulated emission depletion microscopy.

For a first experimental demonstration of the power of CGWS, the wavefront sensing was integrated into a custom-built two-photon microscope, in combination with a deformable mirror (DM) for wavefront correction (Rueckel et al. 2006); see Figure 14.8.

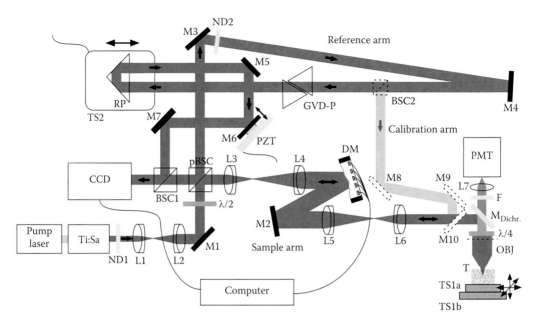

FIGURE 14.8 CGWS and DM integrated into a two-photon microscope. From Rueckel, M., J. A. Mack-Bucher, et al. (2006). Adaptive wavefront correction in two-photon microscopy using coherence-gated wavefront sensing. *Proc. Natl. Acad. Sci. USA* **103**(46):17137–17142. Copyright 2006 National Academy of Sciences, USA.

Two-photon microscopy is similar to confocal microscopy in that it is a fluorescence-based laser scanning microscopy providing three-dimensional resolution. In contrast to confocal microscopy, 2PM does not rely on the fluorescent light from the focus reaching a conjugated focal point on a ballistic trajectory, where it can be spatially filtered for z-selectivity. Instead, two excitation photons with half the energy (and hence twice the wavelength) are absorbed in one quantum event. This two-photon absorption process has a probability that depends quadratically on the excitation power, providing z-selectivity already at the level of absorption. The fact that this allows to detect and correctly assign fluorescent photons, even if they have been scattered on their way toward the detector, together with the decrease in scattering of the excitation light associated with the doubling of the wavelength, explains the superior penetration depth of 2PM, which has been shown to be feasible down to depths of 1 mm in cortical tissue (Theer and Denk 2006).

To reach the high instantaneous excitation powers necessary for two-photon excitation while keeping the average power low, femtosecond laser pulses are usually used. In our experiments, a titanium-sapphire (Ti:Sa) laser operating at 930 nm was used. Since the spectral width of the laser pulses had a full width at half maximum (FWHM) of 15 nm, the coherence length of the two-photon laser was short enough to make it a feasible source for CGWS, resulting in an FWHM of the CV of 19 µm. This has the advantage that the same source can be used for both wavefront measurement and imaging.

The laser beam from the Ti:Sa is attenuated with a neutral density filter (ND1) and its size increased with a telescope (L1/L2) before it enters the interferometer. To adapt the proportion of the light entering the two arms of the interferometer, a polarizing beam splitter cube (pBSC) is used in conjunction with a half-wave plate ($\lambda/2$) that rotates the linear polarization of the incoming laser beam.

Compared to the simplified CGWS setup discussed earlier, the sample arm (to the right of the pBSC) becomes much longer since the DM used for wavefront correction needs to be conjugated to the objective lens BFP. One telescope (L3/L4) corresponding to the telescope in Figure 14.4 conjugates the CCD camera with the DM, while a second telescope (L5/L6) is necessary to conjugate the DM to the BFP of the objective lens (OBJ).

As a result of the increased arm length dictated by the two telescopes, the reference arm (upper half of Figure 14.8) is folded using a large number of mirrors to make it fit onto the optical table. To match group velocity dispersion caused by the five lenses (including objective) in the sample arm, several prisms (GVD-P) are placed in the reference arm. A right-angle prism on a linear translation stage allows modification of the reference-arm length to match the sample arm without changing the direction of the beam. In some cases, additional attenuation of the reference arm was desired, which is realized using another neutral density filter (ND2). Phase shifting was realized using a piezo-mounted folding mirror (M6/PZT). Note that in contrast to the simple CGWS sample described earlier, the reference arm is not folded back onto itself in this implementation, but is rejoined with the light coming back from the sample by means of a second (nonpolarizing) beam splitter (BSC1).

To avoid overcomplicating the setup, no scanning mirrors (which need to be conjugated to the BFP and therefore would require another telescope in the sample arm) were integrated. Instead, x, y, and z scanning were performed by moving the sample using a piezo-driven translation stage (TS1a).

For two-photon imaging, a dichroic mirror (M_{Dichr}) separated the backscattered excitation light (for CGWS) and the fluorescence light (for imaging). The latter was collected with a lens (L7) onto a photomultiplier tube (PMT), after rejection of residual transmitted excitation light by means of an absorptive blue-green filter (F).

14.5.2 In Vivo Adaptive Correction

Using the setup described earlier, anesthetized transgenic zebra fish (*Danio rerio*) larvae with GFP interneurons were imaged with and without wavefront correction (Rueckel et al. 2006). The larvae, which had been chemically treated to inhibit pigment formation, were embedded in agarose and placed under the microscope objective to allow access to the olfactory bulb (see Figure 14.9a); imaging was

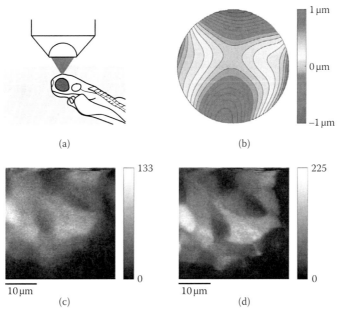

FIGURE 14.9 (a) Experimental setup: an anesthetized zebra fish embedded in agarose is placed under a two-photon microscope with adaptive optics for imaging; (b) optimal wavefront for a focus position inside the olfactory bulb (OB), 50 μm below the surface of the head; (c) two-photon image of neurons in said position in the OB, without adaptive optics; (d) two-photon image of the same region after pre-compensation of the aberrations shown in panel b. From Rueckel, M., J. A., Mack-Bucher, et al. (2006). Adaptive wavefront correction in two-photon microscopy using coherence-gated wavefront sensing. *Proc. Natl. Acad. Sci. USA* **103**(46):17137–17142. Copyright 2006 National Academy of Sciences, USA.

performed at 50 μm below surface. In this setting, dominant aberrations in this rather transparent sample can be expected to originate from the surface profile of the fish head, which has a slightly higher refractive index than the surrounding agarose gel. Without wavefront correction—that is, with the DM in a flat shape—GFP (green-fluorescent-protein)-stained neurons and their processes were visible (see Figure 14.9c) in 2PM, but resolution was notably diminished (as indicated by the lack of sharp edges in the image). Using CGWS, the single-pass aberrations represented by the lowest 28 Zernike modes could be determined (see Figure 14.9b). Applying the conjugate of this optical aberration as a wavefront bias to the DM (Figure 14.9d) resulted in an increase of 69% of peak fluorescence, along with a visible increase in resolution. Note that although CGWS is in principle capable of single-step wavefront correction, Figure 14.9d was actually a result of five iterative correction cycles involving wavefront measurement and adjustment of the DM. The fact that further improvement was possible after the first iteration indicates that the response of the DM was not perfectly incorporated into the correction algorithm.

14.5.3 Speckle Size and Multiple Scattering

As the CV is moved deeper into a strongly scattering sample such as a brain tissue, the grain size of the speckle in the BFP decreases. This effect can be seen in Figure 14.10, where CGWS measurements were performed in a phantom made up of 100 nm beads with a mean-free path (MFP) of 370 μm down to five MFPs. The speckle size decreased roughly and linearly with depth.

In principle, the speckle size corresponds to the inverse of the lateral extent of the CV* as viewed from the detection pupil. However, the lateral extent of the CV should not depend on the depth inside the scattering sample, at least not when single scattering is dominant. This implies that even though coherence gating rejects most of the multiply scattered light, its contribution and thereby the effective extent of the CV increases with focus depth. Since this multiply scattered light passes the coherence gate, it has the same optical path length as the light traveling to the focal region and exhibiting one scattering event.

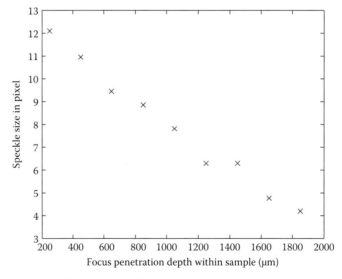

FIGURE 14.10 In a scattering phantom made of 100 nm beads in agarose with a mean-free path of 370 μm, the observed speckle size decreases with focus penetration depth.

---

* When the coherence length is longer than the Rayleigh length, the lateral extend of the CV can be much larger than the diffraction limit due to its axial extend into the geometrical part of the beam.

This implies that multiply scattered coherence-gated light has penetrated the sample less deeply than singly scattered light, which is a source for bias in the wavefront measurement.

In addition, the decrease in speckle size implies an increase in speckle error, which means more interferograms with independent scatterers need to be averaged for the same measurement precision.

One possible solution to this problem could be the use of an aperture in a plane conjugated to the objective lens focus, similar to the pinhole in a confocal microscope. Care would, however, have to be taken in choosing its size; a subdiffraction-sized aperture would completely filter all aberrations, whereas an excessively large aperture would not be effective in filtering out the out-of-focus light. Because of the high anisotropy of light scattering in tissue (most scattering events change the light angle by only a small amount), most multiply scattered light passing the coherence gate can be expected to take paths not fundamentally different from singly scattered light. This implies that the signal that is to be filtered out originates in direct spatial vicinity to the useful signal, making it doubtful if a suitable aperture size can be found.

14.6 Critical Discussion

Coherence-gated wavefront sensing, the technique reviewed in this chapter (Feierabend 2004; Feierabend et al. 2004; Rueckel and Denk 2005, 2006, 2007; Rueckel 2006; Rueckel et al. 2006; Tuohy and Podoleanu 2010), differs from most other current techniques in wavefront sensing for biological microscopy in that it does not rely on the microscope's regular images to optimize the wavefront. Instead of analyzing the images taken with different wavefront bias, simple backscattered light is used. For fluorescence microscopy, where bleaching is often a major limitation, this has the advantage that no fluorescence needs to be generated during wavefront measurement, so the full fluorescence is available for imaging after the best possible wavefront bias has been determined. For example, although 2PM needs a few milliwatts of average laser power at the focus, a decrease of the laser power by three orders of magnitude would still give sufficient backscattering to measure the wavefront using CGWS in less than a millisecond.

The most obvious downside of CGWS is the added complexity of the system, since an interferometer needs to be build around the whole microscope.

In contrast to the image-based optimization techniques, CGWS can determine all aberration modes from the same dataset, without any iteration. Since no bias wavefronts are used, sensitivity to certain modes does not depend on the ability to produce such a mode. However, it should be noted that several sets of interferograms need to be taken to average out the speckle noise that is unavoidable in interferograms because of the coherent backscattering of the large number of scatterers in the CV.

To correctly measure single-pass aberrations, CGWS depends on scatterers being small and randomly distributed, as discussed earlier, in Section 14.4. The assumptions of smallness and random distribution seem to be reasonably well fulfilled in the volume of many biological samples, except at surfaces and interfaces between regions with drastically different refractive indices, where the light reflected by the interface could cause CGWS to measure double-pass aberrations for all even aberrations modes and zero for all odd aberration modes. Variations on the order of 15% in the slope of measured Zernike defocus are seen between different samples (Rueckel and Denk 2007), which might be caused by violation of the scatterer size assumption, but more research is needed here.

The current implementation using phase shifting for phase retrieval is sensitive to sample movements, which can cause phase steps to deviate from the desired value. This issue could be solved by replacing phase shifting with a phase retrieval scheme based on off-axis holography (Leith and Upatnieks 1962), where the reference wavefront would be tilted with respect to the sample wavefront.

The issue of multiple scattering decreasing speckle size is a serious one, since it seems inevitable at higher depths, and since it increases speckle error for a given number of interferograms. However, increasing this number can be used to compensate the loss. Note that there is a minimum speckle size because of the finite NA of the transmission optics and finite wavelength; it is unclear whether a speckle pattern with this minimum speckle size can still carry any wavefront information. For CGWS at several

MFPs inside the tissue, a combination of many different interferograms with independent scatterers and a real SHS might in the end be the best solution.

Being an interferometric measurement technique, CGWS fundamentally measures differential aberrations only between the sample and reference arm. Any aberrations present in the laser beam before entering the interferometer will not be detected; similarly, aberrations in the reference arm will be added as a constant bias (with negative sign) to the measured aberrations. These two sources of a static measurement error are analogous to the noncommon path aberrations, limiting the performance of extreme adaptive optics in astronomy (Sauvage et al. 2007). A naive wavefront correction scheme would consequently insert all the aberrations present in the reference arm into the sample arm, although not correcting the aberrations already present at the entrance of the interferometer. To determine this static measurement offset, a bleaching-resistant nonbiological test sample can be used in conjunction with any image quality or signal-strength-dependent optimization technique, such as the techniques discussed in Chapters 10, 11, 12, and 13. Accounting for this static offset will allow CGWS to correct for all static system aberrations as well as any sample-induced aberrations.

14.6.1 Wavefront Shaping to Compensate Scattering?

When one thinks of adaptive optics in microscopy, one is usually concerned with optical aberrations present in the path of ballistic light, which one wants to compensate for. Any light that is being scattered is considered lost for the imaging process, so it is not considered for wavefront compensation. Although this is a natural approach when coming from the microscopy side, recent work (Popoff et al. 2010; Vellekoop and Aegerter 2010; Vellekoop et al. 2010b; Cui 2011) is showing that this is not a fundamental truth. Since scattering is a deterministic process, even multiply scattered light that produces a completely random speckled intensity pattern can be focused in one point behind a scattering sample. This requires control of the incident wavefront with a much larger number of degrees of freedom, all of which need to be chosen correctly for focusing in a given point. In general, scanning the focus over larger areas is practically impossible without reoptimizing the wavefront for each desired position. If one tries to scan the focus by adding a tilt to the wavefront, as one would do for ballistic light, the peak intensity decreases exponentially for displacements of a few wavelengths (Yaqoob et al. 2008), necessitating a new wavefront optimization. In addition, a detector needs to be placed directly in the position where the focus is to be created, so current implementations are by no means suitable for imaging inside scattering samples.

Since all contributing light paths depend on the position of all the scatterers, the procedure relies on the fact that determination of all degrees of freedom is completed before any scatterers in the sample have moved. This places another strong constraint on the usability of the method for biological imaging, since living and even fixed biological samples have rather short speckle decorrelation times (Cui et al. 2010; Vellekoop and Aegerter 2010). In summary, although the idea to use not only ballistic but also scattered light for wavefront-engineered focusing inside biological samples is very intriguing, it currently seems hard to imagine how the necessary data could in principle be acquired.

14.7 Conclusions

CGWS is a technique to measure the optical aberrations hindering diffraction-limited focusing inside scattering samples, such as biological tissue. Although true point measurements of optical aberrations are not possible in static samples because of the need for speckle averaging, an accuracy of $\lambda/50$ can be achieved by averaging over an area with a radius of only five wavelengths. In contrast to current implementations that are limited by their relatively slow data processing, the physical limit for the measurement time has been estimated to lie below 1 μs (Rueckel 2006).

In strongly scattering samples, CGWS seems to be limited by increasing dominance of multiple scattering to depths of a few scattering free path lengths, comparable to OCT, which is fundamentally based on the same physical principles as CGWS. Within its operating regime, CGWS is expected to provide a superior wavefront sensing technique for weakly stained fluorescent samples, where bleaching during image-based wavefront optimization is prohibitive. CGWS relies only on backscattering of the excitation light, so it is independent of the imaging modality it is combined with.

Further research is necessary to evaluate the possibilities to improve measurement depth in highly scattering samples. For example, polarization-sensitive detection, confocal filtering of the backscattered light, and the use of coherence lengths as short as the NA-determined optical section thickness could all contribute to an increased measurement depth.

Acknowledgments

We thank Sylvain Gigan, Jean-François Léger, Winfried Denk, and, in particular, Laurent Bourdieu for discussions and comments on the chapter.

References

Akkin, T., D. Landowne, et al. (2009). Optical coherence tomography phase measurement of transient changes in squid giant axons during activity. *J. Membr. Biol.* **231**(1):35–46.

Artal, P., S. Marcos, et al. (1995). Odd aberrations and double-pass measurements of retinal image quality. *J. Opt. Soc. Am. A Opt. Image. Sci. Vis.* **12**(2):195–201.

Booth, M. J., M. A. A. Neil, et al. (2002). New modal wave-front sensor: application to adaptive confocal fluorescence microscopy and two-photon excitation fluorescence microscopy. *J. Opt. Soc. Am. A Opt. Image. Sci. Vis.* **19**(10):2112–2120.

Botcherby, E. J., R. Juskaitis, et al. (2008). An optical technique for remote focusing in microscopy. *Opt. Commun.* **281**(4):880–887.

Cui, M. (2011). A high speed wavefront determination method based on spatial frequency modulations for focusing light through random scattering media. *Opt. Express.* **19**(4):2989–2995.

Cui, M., E. J. McDowell, et al. (2010). An in vivo study of turbidity suppression by optical phase conjugation (TSOPC) on rabbit ear. *Opt. Express.* **18**(1):25–30.

Feierabend, M. (2004). Coherence-gated wave-front sensing in strongly scattering samples. PhD thesis, Max-Planck Institute for Medical Research, Heidelberg, Germany.

Feierabend, M., M. Rückel, et al. (2004). Coherence-gated wave-front sensing in strongly scattering samples. *Opt. Lett.* **29**(19):2255–2257.

Huang, D., E. A. Swanson, et al. (1991). Optical coherence tomography. *Science* **254**(5035):1178–1181.

Leith, E. N. and J. Upatnieks (1962). Reconstructed wavefronts and communication theory. *J. Opt. Soc. Am.* **52**(10):1123–1128.

Liang, J. Z., B. Grimm, et al. (1994). Objective measurement of wave aberrations of the human eye with the use of a Hartmann-Shack wave-front sensor. *J. Opt. Soc. Am. A Opt. Image Sci. Vis.* **11**(7):1949–1957.

Malacara, D. (1992). *Optical Shop Testing.* New York, John Wiley & Sons, Inc.

Marsh, P. N., D. Burns, et al. (2003). Practical implementation of adaptive optics in multiphoton microscopy. *Opt. Exp.* **11**(10):1123–1130.

Mourant, J. R., J. P. Freyer, et al. (1998). Mechanisms of light scattering from biological cells relevant to noninvasive optical-tissue diagnostics. *Appl. Opt.* **37**(16):3586–3593.

Neil, M. A. A., M. J. Booth, et al. (2000). Closed-loop aberration correction by use of a modal Zernike wave-front sensor. *Opt. Lett.* **25**(15):1083–1085.

Popoff, S., G. Lerosey, et al. (2010). Image transmission through an opaque material. *Nat. Commun.* **1**(6):81.

Poyneer, L. A. (2003). Scene-based Shack-Hartmann wave-front sensing: analysis and simulation. *Appl. Opt.* **42**(29):5807–5815.

Rimmele, T. R. and R. R. Radick (1998). *Solar adaptive optics at the National Solar Observatory*. Adaptive Optical System Technologies, Kona, HI, USA, SPIE.

Rueckel, M. (2006). Adaptive wavefront correction in two-photon microscopy using coherence-gated wavefront sensing. PhD thesis, Ruperto-Carola University of Heidelberg, Heidelberg, Baden-Württemberg, Germany.

Rueckel, M. and W. Denk (2005). Polarization Effects in Coherence-gated Wave-front Sensing. Adaptive Optics: Analysis and Methods/Computational Optical Sensing and Imaging/Information Photonics/Signal Recovery and Synthesis Topical Meetings on CD-ROM (The Optical Society of America, Washington, DC, 2005), AThC4.

Rueckel, M. and W. Denk (2006). Coherence-gated wavefront sensing using a virtual Shack-Hartmann sensor. Advanced wavefront control: Methods, devices, and applications IV, Volume 6306, p. 63060H. *Proceedings of SPIE*, San Diego, California, USA, SPIE.

Rueckel, M. and W. Denk (2007). Properties of coherence-gated wavefront sensing. *J. Opt. Soc. Am. A Opt. Image Sci. Vis.* **24**:3517–3529.

Rueckel, M., J. A. Mack-Bucher, et al. (2006). Adaptive wavefront correction in two-photon microscopy using coherence-gated wavefront sensing. *Proc. Natl. Acad. Sci. USA* **103**(46):17137–17142.

Sauvage, J.-F., T. Fusco, et al. (2007). Calibration and precompensation of noncommon path aberrations for extreme adaptive optics. *J. Opt. Soc. Am. A* **24**(8):2334–2346.

Schmitt, J. M. (1999). Optical coherence tomography (OCT): A review. *IEEE J. Sel. Top. Quant. Electron.* **5**(4):1205–1215.

Schmitt, J. M. and G. Kumar (1998). Optical scattering properties of soft tissue: a discrete particle model. *Appl. Opt.* **37**(13):2788–2797.

Sheppard, C. and C. Cogswell (1991). Effects of aberrating layers and tube length on confocal imaging properties. *Optik (Stuttgart)* **87**(1):34–38.

Sherman, L., J. Y. Ye, et al. (2002). Adaptive correction of depth-induced aberrations in multiphoton scanning microscopy using a deformable mirror. *J. Microsc. Oxford* **206**:65–71.

Soltau, D., D. S. Acton, et al. (1997). Adaptive Optics at the German VTT on Tenerife. *1st Advances in Solar Physics Euroconference. Advances in Physics of Sunspots, ASP Conf. Ser.* Vol. 118., Eds.: B. Schmieder, J. C. del Toro Iniesta, & M. Vazquez: 351. Astronomical Society of the Pacific, San Francisco, CA.

Theer, P. and W. Denk (2006). On the fundamental imaging-depth limit in two-photon microscopy. *J. Opt. Soc. Am. A Opt. Image Sci. Vis.* **23**(12):3139–3149.

Tuohy, S. and A. G. Podoleanu (2010). Depth-resolved wavefront aberrations using a coherence-gated Shack-Hartmann wavefront sensor. *Opt. Exp.* **18**(4):3458–3476.

Vellekoop, I. M. and C. M. Aegerter (2010). *Focusing light through living tissue*. San Francisco, California, USA, SPIE.

Vellekoop, M. Ivo, et al. (2010a). Scattered light fluorescence microscopy: imaging through turbid layers. *Opt. Lett.* **35**(8):1245–1247.

Vellekoop, I. M., Lagendijk, A., et al. (2010b). Exploiting disorder for perfect focusing. *Nat. Photon* **4**(5):320–322.

von der Lühe, O. (1983). A study of a correlation tracking method to improve imaging quality of ground-based solar telescopes. *Astron. Astrophys.* **119**(1):85–94.

Wang, J., J.-F. Leger, et al. (2012). Measuring aberrations in the rat brain by a new coherence-gated wavefront sensor using a Linnik interferometer. *Proceedings of SPIE*, San Francisco, California, USA, SPIE.

Wright, A. J., D. Burns, et al. (2005). Exploration of the optimisation algorithms used in the implementation of adaptive optics in confocal and multiphoton microscopy. *Microsc. Res. Tech.* **67**(1):36–44.

Yaqoob, Z., D. Psaltis, et al. (2008). Optical phase conjugation for turbidity suppression in biological samples. *Nat. Photon* **2**(2):110–115.

15

Adaptive Optics in Wide-Field Microscopy

Peter Kner
University of Georgia

Zvi Kam
Weizmann Institute of Science

David A. Agard
University of California at San Francisco

John Sedat
University of California at San Francisco

15.1 Introduction

The history of optics may be traced back to ancient times to the realization of the magnification effect of concave gold-plated mirrors used by Egyptian queens, to the magnifying effect of morning dew droplets bringing Greeks to describe cells in leaves, and to Phoenicians condensing sunlight with water-filled spherical glass bottles and crystals (Pliny 1938). Geometrical optics books were written by Euclid, Hero, and Ptolemy from the Alexandria school, followed by later studies of the physical nature of light by Ibn al-Haytham, the medieval Arab scholar. Optics emerged into its modern form during the Renaissance with the development of telescopes and microscopes.

Tyson traces the development of adaptive optics (AO) back to Archimedes, but leaves an empty gap until our times (Tyson 1997). The Greeks identified five planets and named one of them, Mercury, as the Twinkling Star. Therefore, they obviously noticed the effect of atmospheric turbulence without explaining its source. Although the Maya were also prolific stargazers (Friedel et al. 1995), and Egyptian, Babylonian, and Greek astronomers, including Aristillus and Timocharis (~300 BC), Hipparchus (second century BC), and Ptolemy, edited catalogs of up to 1,020 stars and assigned to each brightness levels of one through five, they did not mention the twinkling effect. Twinkling could not have eased the many nights of long observations by Tycho Brahe, Johannes Kepler, Galileo Galilei, and, later, William Herschel and his sister Caroline. They all must have correlated the twinkling effects with atmospheric conditions, yet they did not seem to discuss the physics of this phenomenon and how it was combined with telescope diameter to set the limit to the number of stars they were so diligently trying to fill into their maps. Maybe this is because they could systematically improve the quality and diameter of their reflectors but could do little about the weather. Even today, with AO, builders of earth telescopes have to select faraway sites to minimize the interference of weather.

Microscopes, first developed around 1650, have been essential to the progress of biological science. Every incremental improvement in microscope resolution has been followed by a wave of findings in biology and a deeper understanding of life. In 1655, Hooke built his compound microscope leading immediately to the description of cells. In 1674, Leeuwenhoek built his simple microscopes from glass beads and described protozoa and bacteria. Improvements in transmitted light microscopy allowed Brown to describe the nucleus in orchids in 1833, and Schleiden and Schwann to state cell theory in 1838. Following Abbe's work in 1876 on increasing microscope apertures, Flemming described mitosis. Cajal used stains to see tissue anatomy in 1881 and Koch launched microbiology in 1882. Following Zeiss and Abbe's building of a diffraction-limited microscope in 1886, Golgi described the Golgi apparatus (1898). Following the discovery of radioactivity by Curie, Lacassagne developed autoradiography in 1924. Lebedeff's interference microscope, followed by Zernike's phase-contrast technique, allowed scientists to see cells migrating out of a biopsy into tissue culture dishes and to develop cell lines. Coone's development of the fluorescent microscope (1941) and Ploem's (1975) implementation of epifluorescence enabled immunofluorescence imaging of the molecular constituents of cells. Today, microscopy is the central tool for following complex spatiotemporal processes in cells, embryos, and whole animals. The development of genetic labeling with green fluorescent protein (GFP) and its variants offers a direct link between the list of genes encoding the molecular components of life and their functions.

Microscopy was originally developed to image thin, two-dimensional samples, and modern high-resolution fluorescence microscopes are optimized to create diffraction-limited images of the plane just below the coverslip. But biological samples are three dimensional (3D), and microscopy has increasingly been used over the past three decades to develop 3D fluorescence views by moving the focal plane of the microscope objective through the depth of the sample (Agard and Sedat 1983; Agard et al. 1988; Agard et al. 1989).

In 3D microscopy, as the imaging plane moves deeper into the sample, the image quality becomes degraded because of optical aberrations and scattering. Both scattering and optical aberrations are caused by refractive-index variations. Scattering is the result of submicron inhomogeneities in the tissue and can best be treated as multiple scattering events in the sample that lead to a randomized phase and amplitude distribution of the light. Aberrations are caused by larger-scale variations in the refractive index that lead to a distortion of the light wavefront (Tuchin 2007).

For imaging fixed biological samples with fluorescence microscopy, sample-induced aberrations can be reduced or eliminated by optical clearing (Sullivan et al. 1999; Tuchin 2007), and depth aberrations can be eliminated by matching the index of the fixation medium with that of the immersion oil (Staudt et al. 2007). But for live 3D imaging, the correction of aberrations is more critical because the sample index cannot be manipulated. A water-immersion objective can be used to more closely match the index of a live sample. However, this solution cannot correct for refractive index variations within the sample and only pushes the point at which depth aberrations become serious from a few microns to a few tens of microns. AO promises to correct aberrations allowing high-resolution 3D live imaging.

Today there are multiple microscopy modalities using transmitted light (phase contrast and digital image correlation [DIC]), epi-illumination (dark field and fluorescence), and nonlinear optical techniques (multiphoton fluorescence, Raman scattering, second and third harmonic imaging, and others). All these techniques would benefit, in principle, from AO. Because fluorescence microscopy is such a central tool in cell biology, much of the work on AO in microscopy to date has focused on this technique. In particular, the emphasis has been on confocal and multiphoton microscopy, techniques that are intended for deep imaging in tissue samples.

Although wide-field microscopy loses contrast more significantly with depth as compared to confocal or multiphoton microscopy, there are many conditions under which it is the preferred technique. Murray et al. (2007) demonstrated that wide-field microscopy is the most sensitive fluorescence microscopy technique under low-scattering conditions. Furthermore, live biological samples are very sensitive to the light excitation dose (Carlton et al. 2010). Thus, wide-field fluorescence microscopy is well suited to in vivo imaging in samples where scattering is not too large. Although out-of-focus light is present in wide-field imaging, it can be effectively reassigned to the location of emission by constrained deconvolution algorithms (Swedlow et al. 1997).

Under the conditions for which wide-field microscopy is best suited, aberrations can still be quite serious. As shown in Figure 15.1, imaging 20 μm into a live sample with an oil-immersion lens causes the peak intensity of the point spread function (PSF) to drop threefold and the width of the PSF in the axial direction to increase by twofold (Gibson and Lanni 1991).

15.1.1 Sources of Aberrations in Microscopy

The resolution achieved by modern objectives, which reaches the limits of what is possible with lens optics, is inevitably linked with tight constraints on the specimen geometry and the optical properties of the light reaching the front lens of the microscope objective. Typically, the design of an objective uses multiparametric optimization of the optical variables for about 15 single lenses within the objective itself, and also the optical path and refractive index within the immersion oil and coverslip (Figure 15.2). The objective accepts a given numerical aperture (NA) of the wavefront emerging from each point within the specimen and (for infinity corrected objectives) transforms it into a plane wavefront. The purpose of AO is to correct the wavefront for varying specimen geometries. It must be emphasized that this is

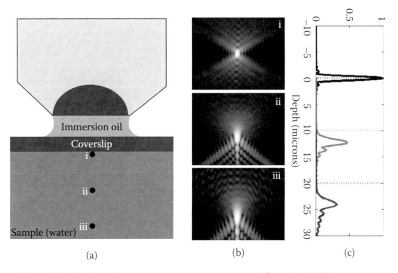

FIGURE 15.1 Effect of depth aberrations on microscope point spread function (PSF).

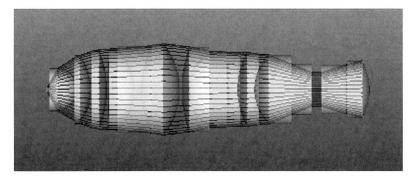

FIGURE 15.2 Zemax model of 60× oil immersion objective containing 15 separate elements (Fujimoto and Nishiwaki, 2004).

a far from trivial task and requires—ideally—optimizing the amplitude, phase, and polarization of the light field although AO can correct only the phase. In the following paragraph, we start by grossly classifying the typical aberrations occurring in biological imaging and review the methodologies used to correct them before the development of AO devices. This is of interest since the limitations of various AO technologies may require a combination of classical and new methods to tackle present biomedical challenges.

Depth aberrations arise from the refractive index mismatch between the coverslip and immersion oil (which can be considered to have the same index) and the refractive index of the sample. Figure 15.3 shows a ray diagram illustrating the optical path for a ray emerging at a given angle from a sample with refractive index lower than glass, at depth d below the coverslip, and depicts two alternative modes for correcting the phase of this ray to that of a ray at the sample angle emerging from a hypothetical sample immersed in glass. Figure 15.3a illustrates the path-length difference between a ray from a point source at depth d below the coverslip in a material with index $n_2 < n_1$ (solid line) and that from a point source in a material with $n_2 = n_1$ (dashed line). Because the path-length difference is a function only of ray angle and not field position, it can be corrected for the entire field in the back pupil plane of the objective. Since field-corrected objectives obey the sine condition (Gu, 1999), the phase correction applied in the back pupil plane is (Gibson and Lanni, 1991; Booth et al., 1998):

$$\phi = 2\pi \frac{d}{\lambda}\left(n_2 \sqrt{1 - \left(\frac{(\mathrm{NA})\rho}{n_2}\right)^2} - n_1 \sqrt{1 - \left(\frac{(\mathrm{NA})\rho}{n_1}\right)^2}\right) \tag{15.1}$$

where $\rho = n\sin\theta/\mathrm{NA}$ is the normalized radial coordinate in the back pupil plane, λ is the wavelength of the emitted light, and NA is the numerical aperture. n_1 and n_2 are the objective immersion and specimen medium refractive indices, respectively.

By adding a global curvature to the phase applied in the back pupil plane, the focal point will be moved into the sample. Figure 15.3b shows the path-length difference between a point at the coverslip and a point at depth d below the coverslip. Correcting for this additional path length difference in the back pupil plane accomplishes focusing with depth correction. In this case, the correction is

$$\phi = 2\pi \frac{n_2 d}{\lambda}\sqrt{1 - \left(\frac{(\mathrm{NA})\rho}{n_2}\right)^2} \tag{15.2}$$

The sample-induced aberrations cannot be described analytically because they will depend on the sample geometry, which must be measured. Sample-induced aberrations cannot be described by an

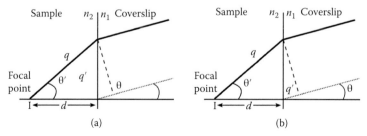

FIGURE 15.3 Ray diagram of the depth aberration. (a) Diagram of the path length difference $(n_2 q - n_1 q')$ between an unrefracted ray (dashed line) from a point a depth d below the coverslip and a ray that is refracted at the interface (solid line). Compensating for this path length difference in the back pupil plane will correct the spherical aberration. (b) The path length difference $(n_2 q - n_1 q')$ between a ray from a point just below the coverslip (dashed line) and a ray a depth d below the interface. Compensating for this path length difference in the back pupil plane will focus to a depth d and correct for spherical aberration.

aberration in the pupil plane because the source of the aberrations is close to the light source. Thus, they will vary across the field of view. The area in the field of view that can be corrected at one time will depend on the source of the aberrations and is referred to as the isoplanatic patch. To correct the entire field of view in wide-field microscopy, we need to take different images with the correction optimized for different patches. The corrected isoplanatic patches from each image would then be stitched together to correct the entire field of view. In confocal and multiphoton microscopy, there is the possibility of correcting each pixel in the image independently because these are raster scanning techniques, but in some reports this has been determined to be too time-consuming, and the wavefront correction has been set only once for each focal depth (Marsh et al. 2003). Many biological samples have a cylindrical or spherical shape, and the overall shape of the sample will also be a significant contribution to the sample aberrations (Schwertner et al. 2004a,b).

15.1.2 Earlier Correction Approaches

Many different approaches to correcting or eliminating aberrations have been developed over the years. As already mentioned, cell fixation and clearing techniques can remove sample-induced aberrations and scattering (Tuchin 2007). The refraction index of fixed cells can be matched to that of the immersion medium to minimize depth aberration. This can be carried out by changing the immersion medium or by fixing the cells. Glycerol and water-immersion lenses have slightly smaller NAs than oil-immersion objectives but reduce the refractive index mismatch with live specimens and therefore also the resulting depth aberration. After fixing cells, their sample refractive index will closely match the fixation-medium index. Using a higher refractive-index fixation medium allows the imaging system to effectively use a higher NA. Recently, a new high-refractive-index fixation medium has been developed that permits imaging with the full NA of an oil-immersion lens (Staudt et al. 2007).

Using an immersion oil with refractive index, $n = 1.518$, we can find that a point image 50 μm below the cover glass in glycerol ($n = 1.4746$, refractive index difference, $\delta n = 0.0434$) has similar distortions as a point image 12 μm inside a volume of a typical biological buffer medium ($n = 1.341$, $\delta n = 0.177$). The corresponding peak intensity is about four times lower than that would be detected for an identical point source right under the cover glass. For a water-immersion objective imaging into tissue ($n = 1.41$, $\delta n = 0.077$), comparable distortions will appear at 28 μm depth.

The depth aberrations resulting from variations in cover-glass thickness can be corrected to a great extent by adjustable collars built into some objectives (King and Cogswell 2004). Adjustment of the immersion oil refractive index has been shown to produce similar correction (Hiraoka et al. 1990). By incorporating a special telescope before the imaging detector, optimization of the correction without disturbance of the objective or the specimen can be achieved, as shown in Figure 15.4 (Kam et al. 1997). In essence, these methods shift the imaged plane above or below the designed focal plane, adding or subtracting a phase that varies approximately as second order with the angle of emergence from a point in the sample at focus. However, it should be stressed that all these methods provide only second-order corrections that may not be sufficiently good for light entering at high NAs.

Another approach to addressing depth aberrations is to deconvolve the images with a depth-dependent PSF. In deconvolution, the fluorophore density, $f(x,y,z)$, is estimated by mathematically inverting the imaging process of the microscope, which is the convolution of the instrument response, the PSF, with $f(x,y,z)$. For the case of a space invariant PSF, $h(x,y,z)$, the image, $g(x,y,z)$, can be expressed as

$$g(x,y,z) = \iiint h(x-x_0, y-y_0, z-z_0) f(x_0,y_0,z_0) \mathrm{d}x_0 \mathrm{d}y_0 \mathrm{d}z_0 \tag{15.3}$$

There are many deconvolution approaches to inverting this equation to get the best estimate of $f(x,y,z)$. Because Equation 15.3 is a convolution, it can be calculated rapidly using fast Fourier transform algorithms. But, clearly, if the PSF is not spatially invariant, as is the case when aberrations are present, Equation 15.3 is not correct, and the inversion will not result in a good estimate of the object.

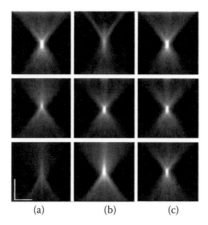

(a)   (b)   (c)

FIGURE 15.4 Point spread functions (PSFs) measured from focal series images of 0.1-μm-diameter fluorescent spheres using a 100×/1.3 objective. The bead centers were found, the three-dimensional PSFs were cylindrically averaged and presented as a vertical section through the optical axis. The intensity scale is logarithmic to depict the faint out-of-focus patterns. Images were taken in focal steps of 0.25 μm, and pixel was replicated four times along the optical axis to have the radial and axial pixel size roughly equal. The scale bars are 5 μm. The Galilean telescope version was used. The beads are in an agarose layer with an estimated refractive index of 1.4. (a) PSFs for three beads: one right under the coverslip ($F = 0$), one at $F = -5$ μm, and one at $F = -15$ μm below the coverslip. An increasingly aberrated pattern is obtained for deeper beads. (b) PSFs for the same three beads after adjusting the telescope ring to $Z = -2$ (arbitrary units) and focusing with F. The nonaberrated plane of focus is now −5 μm below the coverslip, and opposite signs of the aberrated patterns are obtained above and below this plane. Top bead: $F = -3.5$ μm; middle bead: $F = -8.5$ μm, lower bead: $F = -18.5$ μm. (c) PSFs for the same three beads now taken with both F and Z adjusted to minimize aberration at all three depths. Top bead: $F = 0$, $Z = 0$; middle bead: $F = 0-8.5$ μm, $Z = -2$; lower bead: $F = -26$ μm, $Z = -4$. From Kam et al. (1997).

When depth aberrations are present, the PSF will be a function of both the imaging plane, z, and the object plane, z_0, independently. Now,

$$g(x, y, z) = \iiint h(x - x_0, y - y_0, z, z_0) f(x_0, y_0, z_0) \mathrm{d}x_0 \mathrm{d}y_0 \mathrm{d}z_0 \qquad (15.4)$$

There are approaches to approximating and inverting Equation 15.4, but Equation 15.4 is much more computationally intensive than Equation 15.3, and these methods are correspondingly intensive, because the image must be calculated many times (Kam et al. 2001; Preza and Conchello 2004; Arigovindan et al. 2010). Equation 15.4 can in principle be extended to the case where the PSF varies with lateral as well as axial position to account for aberrations other than the depth aberration, but this will make the problem even more intractable. Because deconvolution is a post-acquisition process, it cannot restore the signal-to-noise ratio that is degraded through the lower peak intensities because of the aberrated PSF.

15.2 Microscope Design

Later we describe the design and construction of a microscope to correct for depth aberration and focus. The microscope layout used at University of California, San Francisco, is shown in Figure 15.5. The microscope uses a 60× oil-immersion objective (Olympus 1.42 NA PlanApo N) and Olympus tube lens (f_{tl}, focal length 180 mm) in a custom-built open microscope configuration assembled on an optical table. The back pupil plane of the objective is reimaged onto the deformable mirror (DM) using the tube lens and achromat f_1 chosen so that the desired NA fills the 15 mm diameter of the DM.

$$\phi_{DM} = \frac{f_1}{f_{tl}} 2NA f_{obj} = \frac{2NA f_1}{M} \qquad (15.5)$$

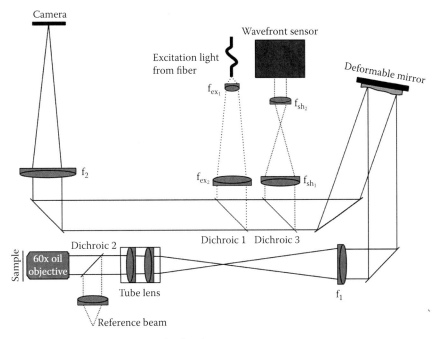

FIGURE 15.5 Microscope layout. See text for details.

where M is the objective magnification. Lens f_1 is chosen as 350 mm so that the NA of the system is 1.285. To collect light from a sample below the coverslip, the maximum NA is set by the sample refractive index, implying for aqueous media NA < 1.33, in any case. An NA of 1.285 in water corresponds to a maximum angle of 74°, greater than the 70° maximum angle for an NA of 1.42 in immersion oil (1.512); the lower NA does not correspond to a less-challenging imaging configuration. Lens $f_2 = 800$ mm was chosen so that the sample is Nyquist sampled on the CCD camera. This microscope uses a camera from Astronomical Research Cameras, Inc. (San Diego, CA) with a cooled CCD57-10 chip from e2v Technologies (Chelmsford, UK). The pixel size is 13 microns. For Nyquist sampling, we require

$$\frac{\lambda}{4\text{NA}} M \frac{f_2}{f_1} > d_{\text{pixel}} \tag{15.6}$$

For emission at 510 nm, the Nyquist pixel size is 99 nm. And for the component values chosen here, the effective pixel size at the sample plane is 94.6 nm. The additional components between the tube lens and the CCD camera reduce the light-collecting efficiency of the system by no more than 22%. This loss could be minimized by using dielectric mirrors instead of metallic mirrors.

The excitation light at 488 nm (Coherent Innova) passes through an optical shutter (UniBlitz) and is focused onto a rotating diffuser (Physical Optics) to remove spatial coherence. The light is then coupled into a multimode fiber (600 μm core, 0.22 NA; Ocean Optics). Lenses f_{ex1} and f_{ex2} magnify the fiber end onto the DM; because the DM is in the back pupil plane, this is a Kohler illumination configuration. Dichroic 1 (Semrock FF495-Di01) inserts the excitation light at 488 nm into the beam path and sends the emission light (510 nm) to the CCD camera.

The mirror shape is monitored with a Shack-Hartmann wavefront sensor (SHWFS; Haso 32; Imagine Optic, Orsay, France) that measures a light beam from a 632 nm HeNe laser inserted into the optical path after the objective lens with a notch filter, dichroic 2 (Semrock FF 579/644-Di01), and sent to the SHWFS after the DM with another notch filter, dichroic 3 (Semrock FF 579/644-Di01). Lenses f_{sh1} and f_{sh2} demagnify the beam to image the DM onto the lenslet array.

Placing the DM in a conjugate plane to the back pupil guarantees that the same wavefront correction is applied over the entire field of view.

15.2.1 Deformable Mirror Control

There are several DMs currently on the market and the best choice depends on several factors. The most important factors are likely to be the mirror throw and the number of actuators, which will determine the maximum aberration amplitude and the maximum spatial frequencies that can be corrected. Other important factors are the DM diameter and the response time. Because the back pupil plane will be magnified onto the DM, the DM diameter will largely determine the size of the microscope; the larger the DM, the longer the focal length lenses are required. The response time determines how fast the mirror shape can be changed, an important parameter in live imaging and focusing with the DM. The response time of a piezo-electric z-stage used for focusing is typically several milliseconds; the response time of a DM will range from hundreds of microseconds for micro-machined DMs to several milliseconds for membrane-based DMs.

For the DM, we chose the Mirao52D from Imagine Optic because it is capable of large displacements and thus permits the correction of aberrations deep into the sample. The 15-mm-diameter mirror has 52 actuators on a square grid with a 2.5 mm spacing. The mirror is capable of a maximum displacement of ±75 microns for the focus mode $\left(Z_2^0\right)$ and ±8 microns for first-order spherical aberration $\left(Z_4^0\right)$. The mirror can take the shape of any Zernike through order four with a root-mean-square (RMS) wavefront error of <20 nm.

We control the mirror by measuring the wavefront of the HeNe laser with the wavefront sensor. We reference the wavefront to a measurement with all actuators set to zero, and then measure the wavefront on the 32 × 32 element SHWFS for each of the 52 actuators activated individually, yielding a 1024 × 52 matrix. To set a desired mirror shape, we use the standard singular-value decomposition technique (Gavel 2003) to determine the matrix S, which will yield the actuator values for a desired wavefront. Typically, we retain the first 45 singular values.

The manufacturing process results in a small distortion of the mirror surface at each actuator, and this array of bumps ("print through") diffracts the light causing small satellite peaks around the central PSF. This can be clearly seen in Figure 15.6, which shows an in-focus PSF on a logarithmic scale.

15.2.2 Simulation of Microscope and Sample Configurations

To assess the microscope performance, we carried out ray-tracing simulations of point sources at various depths below the coverslip. While microscope optical designs can be evaluated and optimized using commercial optical software programs, the effect of the optical path within specimens cannot be readily simulated by a discrete set of optical elements. To this end, we wrote ray-tracing software that computes analytically the ray paths through a volume with continuously varying refractive index represented by a 3D grid of refractive

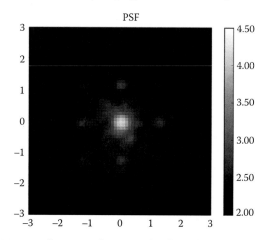

FIGURE 15.6 Image of a 200 nm yellow-green fluorescent bead on a logarithmic scale. Satellite peaks caused by the print-through of the mirror are visible. The maximum intensity of the satellite peaks is ~1% of the central peak.

index values (Kam 1998). Such refractive-index maps can be derived from microscope imaging modes, such as DIC, that record local refraction-index gradients at high spatial resolutions (Kam 1998; King and Cogswell 2004). The simulations were compared with measurements and found to be reliable. Computed PSFs from measured refractive-index maps were shown to deconvolve effectively aberrated fluorescent data (Kam et al. 2001). Figure 15.7 shows the agreement of these ray-tracing simulations with measured data.

The configuration for simulating depth aberrations is shown in Figure 15.8. A fan of rays from a point source of depth d into a sample medium is traced to the back pupil plane of an ideal microscope objective. The Fraunhoffer integral is then used to model the image on the CCD camera. By changing the focal plane of the objective in the ray tracing, the 3D PSF is created. Simulated real-space 3D PSFs were calculated by Fourier-transforming the pupil function after adding the defocus phases (Stokseth 1969). The Strehl ratio was calculated from the amplitude and phase of the pupil function (Hardy 1998) and was found to be a useful and sensitive metric for quantifying of the quality of correction. The Strehl ratio is defined later in the text and can be calculated with Equation 15.13.

To calculate the effects of adaptive optical elements, we "relayed" the rays emerging from the medium by ideal lenses, as shown in Figure 15.8, to each of the adaptive optical elements. The phase of each ray was then adjusted by the setting for the DM at that position.

With an ideal adaptive optical element placed in the back pupil plane the depth aberration can be completely corrected by applying Equation 15.1, as shown in Figure 15.9. These simulations were used to evaluate the effect of various parameters on the PSF.

Simulations can evaluate, for example, intensity changes because of increased reflection at the sample/glass interface at high incident angles. This pupil apodization effect was found to have negligible effect on PSF width and peak intensity. Simulating the effect of print-through by adding small deviations from the theoretical phase correction, matching the spacing between the actuators, indicated a small reduction in the central PSF peak intensity and the appearance of diffraction peaks around the PSF, as was indeed found experimentally. Such simulations also help us to evaluate the sensitivity of the correction to various distance and aperture geometrical parameters in the optical setup.

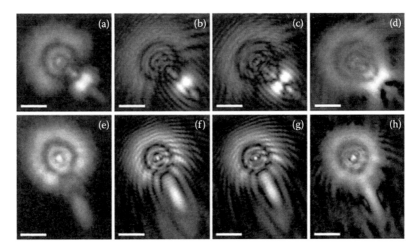

FIGURE 15.7 **(See color insert.)** Simulated and measured sample-aberrated PSFs (Kam et al. 2001). Refractive index map of the specimen was retrieved from three-dimensional digital image correlation (DIC) images using LID integration (Kam 1998). (a–d) Defocused optical sections of a bead aberrated by an oil drop below the coverslip at 3.0 μm below focus. (e–h) Optical sections of another aberrated bead at 2.75 μm below focus. (a and e) A measured image of the 0.1 μm bead. (b and f) A computed three-dimensional ray-traced PSF using a refractive-index map from the line-integrated oil drop DIC data. (c and g) A ray-traced PSF using simulation of an oil drop with uniform known refractive index. (d and h) A computed PSF in which the aberrated wavefront calculated by ray tracing through a simulated oil drop of uniform refractive index was applied to a measured, unaberrated PSF. Scale bars are 2 μm.

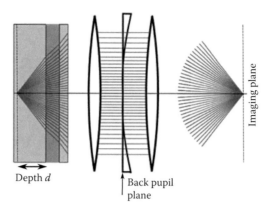

Depth *d*   Back pupil
                                                 plane

FIGURE 15.8 Schematic presentation of the simulated optical path for depth-aberration correction at the pupil plane. The phase-correction element is the thin sheet between the two lenses.

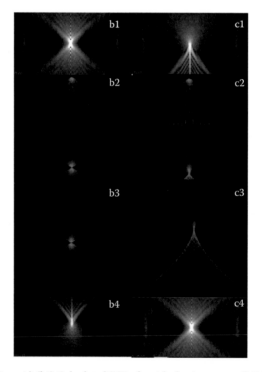

FIGURE 15.9 (**See color insert.**) (b1) Calculated PSF of an ideal microscope. (b2) Plotted ray paths emerging from a point under the microscope (bottom fan) and exiting the microscope to focus at the image plane (top fan of rays). (b3) Enlarged view of rays converging onto the focal point. (c1) The PSF computed with a layer of water in the imaging path, showing the typical spherical aberration. (c2) The geometrical optics pattern due to refraction in a layer with refractive index different from the immersion oil (depicted in the top ray fan). (c3) Enlarged view of (c2) near the focus, with the geometrical optics pattern corresponding to the aberrated PSF of (c1). (b4) The PSF computed for the ideal microscope optics in which the adaptive element introduces the phase corrections, according to Equation 15.1, for the layer of water. The PSF shows a spherical aberration "inverted" to that in (c1). (c4) The PSF for imaging into water with phases corrected by the adaptive element, showing the recovery of the nonaberrated PSF (Kam et al. 2007).

To correct sample-induced aberrations across the entire field of view, the light must travel through an "inverse sample," as shown in Figure 15.10, so that the accumulated phase distortions along each ray are reversed. The inverse sample has the same structure as the sample but with the refractive index differences reversed. This inverse sample can be calculated from DIC image stacks, and could, in principle, be constructed with photorefractive materials, but this would be difficult to say the least.

An approximation of an inverse sample can be constructed by multiconjugate adaptive optics (MCAO) where multiple adaptive elements, conjugate to different planes in the sample, are used to approximate the inverse sample (Ragazzoni et al. 2000; Ragazzoni et al. 2002). In Figure 15.11, we show the results

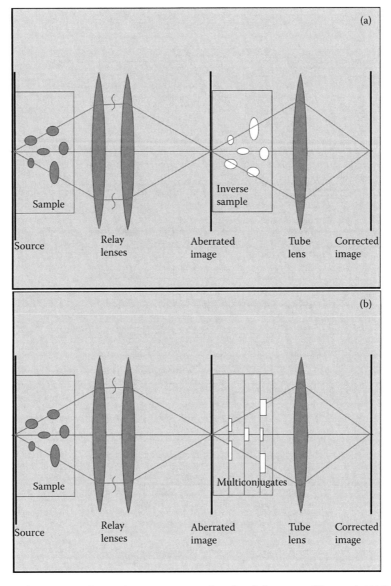

FIGURE 15.10 Schematic optical setup for correcting sample-induced aberrations (Kam et al. 2007). The sample is simulated by a set of ellipsoids with a refractive index different from the embedding medium. Ray tracing is performed through the sample volume, followed by the correcting optics. In (a), the correction is performed by a hypothetical "inverse sample" with the identical distribution of ellipsoids but with the opposite refractive-index contrast. In (b), the inverse sample is approximated by three adaptive elements, shifting the phases of rays that pass through them (Kam et al. 2007).

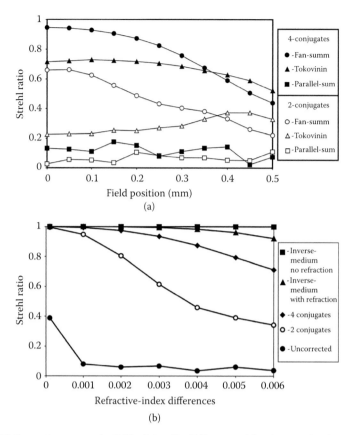

FIGURE 15.11 (a) Correction across the field of view for different multiconjugate adaptive optics schemes. The sample consists of 500 Gaussian spheres with $\delta n = 0.003$. (b) The effect of refractive-index differences on different multiconjugate adaptive optics schemes. The sample again consists of 500 Gaussian spheres with varying δn. The first data point is at a refractive-index difference of 0.0001 (Kam et al., 2007).

of simulations in which a the sample is modeled by 500 randomly positioned Gaussian fluctuations in the refractive index (Kam et al. 2007). We evaluate the use of two and four adaptive elements with three different methods for calculating the phase correction from each element. The simplest method is the "parallel sum," where the phase aberration accumulated from a finite slice within the sample volume is integrated along the optical axis and corrected by an element positioned at a plane conjugated to the middle of this slice. A more realistic method takes account of the fact that light rays forming the image traverse the sample while filling the whole NA. The "fan sum" therefore calculates the phase aberration, averaging over the NA fan of rays. The Tokovinin method is based on a proposal for an optimized integral that laterally blurs the phase contribution of each plane the farther it is from the conjugate (Tokovinin et al. 2000).We evaluate the results by calculating the Strehl ratio for each case across the field of view (Figure 15.11a) and on axis as the refractive index differences in the sample increase (Figure 15.11b).

15.3 Wavefront Measurement and Correction Approaches

15.3.1 Open-Loop Correction of Depth Aberration

The approach we follow is to correct the depth aberrations resulting from the ray path length differences given by Equation 15.1 or 15.2. The shape of the DM is then given by

$$f(\rho) = \frac{\phi(\rho)}{2k} \tag{15.7}$$

where $\phi(\rho)$ is given by either Equation 15.1 or 15.2. The variable k is the wave vector, and ρ is the normalized radial coordinate of the DM. The factor of 2 accounts for the fact that the change in path length of the light reflecting off the mirror is twice the mirror displacement.

Figure 15.12 shows the correction of a 200 nm bead 67 μm below the coverslip in a glycerol/water mixture with a refractive index of 1.42. Images taken first without correcting for the depth aberration

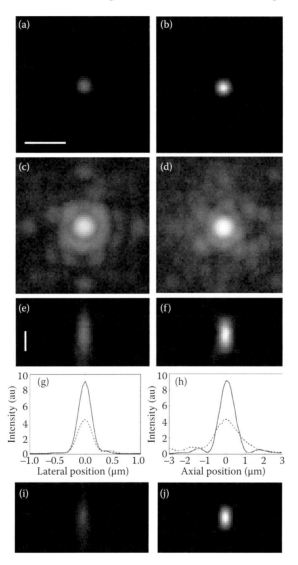

FIGURE 15.12 Images of a 200 nm bead 67 μm below the coverslip in a water/glycerol mixture with $n = 1.42$. (a) Uncorrected image of in-focus plane. (b) Corrected image of in-focus plane: same scale as (a). (c and d) The same as (a) and (b), respectively, but on a logarithmic scale. (e and f) Cross sections through the focal plane on a linear scale. The scale bars are 1 μm. (g and h) Line profiles through of the intensity through the center of the bead along a lateral and the longitudinal axis, respectively. The dashed line is from the uncorrected image and the solid line is from the corrected image. (i and j) Simulations of the PSF. (i) Uncorrected PSF 65 microns into a material with index 1.42 using a 1.2 numerical aperture (NA) objective with a 1.512 refractive-index immersion oil. (j) A simulated PSF at the coverslip. The peak intensity for (j) is 3.75 times the peak intensity for (i) (Kner et al. 2010).

(with a flat DM, Figure 15.12a, c, and e) show strong spherical aberration. Images taken with the shape of the mirror set by Equation 15.1 (Figure 15.12b, d, and f) show a more symmetric PSF, similar to that just below the coverslip. After correction, the peak intensity is a factor of two larger and the width of the peak in the axial direction is significantly smaller, meaning better axial resolution. The correction removes the low-intensity "pedestal" of light around the main peak in the uncorrected image and returns it to the central peak, as can be clearly seen in the logarithmic scale image (Figure 15.12c and d). The lateral plane full width at half maximum (FWHM) of the peak is not significantly changed by the correction, as seen in Figure 15.12g. This is because the depth aberrations do not increase significantly the width of the central peak. The aberrations create a broad pedestal on which the central peak sits (Figure 15.12c and d). After correction, the PSF is very similar to that at the coverslip, which should allow for much better reconstruction using conventional robust space-independent deconvolution algorithms. Figure 15.12i and j shows simulations of the PSF in the axial direction for parameters corresponding to the measured data. Figure 15.12i simulates a PSF 65 μm below the coverslip with the same refractive index mismatch, and Figure 15.12j simulates a well-corrected PSF at the coverslip. As can be seen in Figure 15.12, the agreement is quite good. The main difference is that in the experiment the correction increases the peak intensity by 2.05. In theory, the peak intensity of the corrected PSF is 3.75 times higher than that of the aberrated PSF. The difference is most likely due to the deviation of the objective from the ideal case and residual deviations in the mirror shape from the theoretical shape required to correct the depth-phase aberrations. The print-through intensity peaks all around the PSF support this reason for less-than ideal correction.

The wavefront before and after correction, calculated by phase retrieval (Section 15.3.5), is shown in Figure 15.13. The Strehl ratio for the corrected wavefront is 0.77, indicating quite an acceptable correction, although the edge of the wavefront in Figure 15.13b still shows approximately a wave of residual aberration. This is possibly due to imperfections in the objective. Interferometric measurements of objectives have shown that they can be severely aberrated near the edge of the pupil (Juskaitis 2007).

The results of correcting the spherical aberration in biological samples are less dramatic than those of a single bead because the biological sample introduces scattering as well as other aberrations. Figure 15.14 shows uncorrected and corrected lateral images 23 μm below the coverslip of UMUC bladder-cancer cells with GFP-TRF1-labeled telomeres. Although the intensity of the cell autofluorescence is not increased by the correction of the depth aberrations, the GFP-TRF1 signal is brighter by 60% after subtracting the broad autofluorescence signal (Figure 15.14c). In wide-field microscopy aberrations do not change the detected integrated intensity. Because aberrations spread out the energy but do not remove it, we would not expect increased intensities on correcting the depth aberration in a wide-field microscope from spatially broad sources such as cell autofluorescence and a thin fluorescent sheet (e.g., cytoplasmic membrane). However, the intensity of sharp features is increased after correction of aberrations.

Correcting the depth aberration or focusing with the DM corrects the shape of the PSF and increases the peak intensity, but as the sample becomes more complex, the increase in intensity becomes less significant. This is most likely due to both the importance of sample-induced aberrations and the fact,

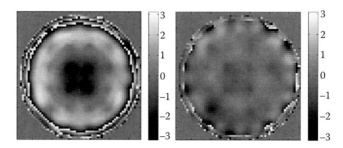

FIGURE 15.13 Phase in back pupil plane calculated by phase retrieval from the uncorrected PSF in Figure 15.12 (left) and the corrected PSF in Figure 15.12 (right).

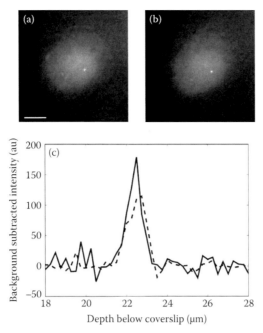

FIGURE 15.14 Images of green fluorescent protein (GFP)-TRF1-labeled telomeres in UMUC bladder cancer cells. (a) Uncorrected images 23 µm below the coverslip. (b) Depth aberration corrected 23 µm below the coverslip. The correction assumed a sample index of 1.38. (c) Background subtracted profiles through the labeled telomere (bright spot in images a and b) along the axial direction. The dashed line is from the uncorrected image (a), and the solid line is from the corrected image (b). The peak intensity in the corrected image is 60% larger over the background. The scale bar is 5 µm (Kner et al., 2010).

mentioned above, that in wide-field microscopy aberrations affect point sources and larger structures differently. For example, the intensity of a uniform distribution of fluorescent molecules will not be affected by aberrations. Because the effect of aberrations on the intensity is not linear, the intensity may not increase if aberrations remain, even if some aberrations have been corrected. Thus, the expected change in intensity from complex biological samples can be hard to predict.

Figure 15.15 shows another example: GFP-labeled centromeres in eye imaginal discs in *Drosophila* larvae. In this sample, correcting the depth aberration did not increase the intensity of the signal from the GFP centromeres, possibly because of the strong effects of scattering, but the FWHM of the centromere images in the axial direction is smaller by 27%, which can be seen in Figure 15.15.

15.3.2 Focusing

If Equation 15.2 is used instead of Equation 15.1 to set the wavefront, the DM will focus into the sample. An example of focusing through the sample is shown in Figure 15.16. DM is an attractive option for focusing through the sample quickly without perturbation. Mechanical movement of either the sample or the microscope objective could potentially disturb the sample through the acceleration and the movement of the immersion oil. DMs, depending on the technology, can be set in less than a millisecond, whereas mechanical focusing typically takes several milliseconds. But focusing with the DM requires a much larger wavefront adjustment. Imaging 20 µm into an aqueous sample at NA 1.25 requires a 16.7 µm peak-to-valley wavefront for focusing and a 4.0 µm adjustment for correcting only depth aberrations with an oil-immersion objective. The large correction for focusing means that the amount of focusing that can be achieved is limited by the DM throw. More importantly, any inability of the mirror to replicate the exact shape will lead to a residual wavefront error that will reduce the Strehl ratio of the corrected image.

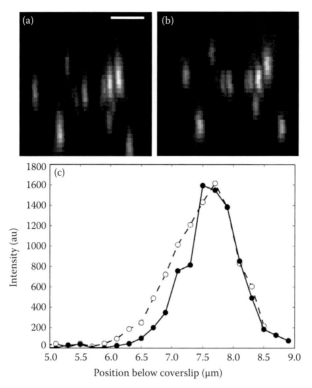

FIGURE 15.15 GFP-labeled centromeres in eye imaginal discs in *Drosophila* larvae. Scale bar is 2 μm. (a) Maximum intensity projection of uncorrected three-dimensional data stack. Vertical is the axial direction. The top of the image is 2 μm below the coverslip, and the bottom is 9 μm below the coverslip. (b) Maximum-intensity projection of a depth-aberration-corrected data stack. The correction assumed a sample index of 1.38. (c) Line profiles along the axial direction for uncorrected (dashed line, open circles) and corrected (solid line, solid circles) images of a GFP-labeled centromere (Kner et al. 2010).

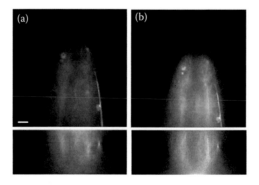

FIGURE 15.16 Comparison of images taken by focusing with the DM (a) and mechanical focusing (b). Images are of *Caenorhabditis elegans* expressing a GFP sur-5 construct; the bright feature along the right side is the ventral nerve cord. The top images are taken 6 μm below the coverslip and the bottom images are *xz*-cross sections. The difference in image intensity is due to photobleaching. Each image is scaled to its maximum intensity. A sample refractive index of 1.36 was assumed for focusing with the DM using Equation 15.2. The scale bar is 2 μm (Kner et al. 2010).

The larger the required wavefront correction, the larger the residual error. So the focus correction will in general perform less well than only the depth-aberration correction. A focus-tracking scheme using a DM has been shown in Poland et al. (2008a).

15.3.3 Deconvolution

Computational deconvolution successfully removes out-of-focus light by essentially reassigning it to its source position (Agard and Sedat 1983) but requires accurate knowledge of the PSF to work optimally. Most deconvolution algorithms assume that the PSF is uniform throughout the sample and cannot account for a spatially varying PSF because of optical aberrations. As mentioned in Section 15.1.2, deconvolution with a locally varying PSF has been implemented (Kam et al. 2001; Preza and Conchello 2004), but this approach is computationally intensive. Furthermore, aberrations degrade the signal-to-noise ratio, which will affect the results of the deconvolution even if the correct PSF is used. Thus, correcting depth aberrations improves not only the raw image but also the deconvolved results. This is shown in Figure 15.17. Figure 15.17a shows a maximum-intensity projection of the axial view through several 200 nm fluorescent beads 64 μm below the coverslip in glycerol. The depth aberration is evident from the strong axial asymmetry and the elongated PSF. Figure 15.17b shows a maximum-intensity projection through the same beads after the depth aberrations have been corrected by the DM using Equation 15.1. The PSF is now corrected and the maximum-intensity from each bead is a factor of 1.50 higher. Figure 15.17c and d show the results of deconvolution on (a) and (b) respectively, using a PSF measured at the coverslip. The deconvolution of the corrected image (Figure 15.17d) shows a significantly smaller image of each bead than that of the uncorrected image. This is true in the lateral dimension (not shown) as well as in the axial direction although the difference is greater in the axial direction. The maximum intensity from a bead in Figure 15.17d is a factor of 1.85 greater than that in Figure 15.17c; the correction of depth aberrations increases not only the intensity in the raw images but also the effectiveness of deconvolution.

Figure 15.18 shows the deconvolution of images of actin labeled with Alexa 488-Phalloidin in B16F10 cells. The deconvolved image (d) of the AO-corrected data shows significantly less background than that of the uncorrected data (b). Interestingly, the images before deconvolution (a) and (c) do not look as different. These images are *xy*-sections 4.4 μm below the coverslip and show actin labeling in cell–cell junctions. The deconvolution was performed using the AIDA software package (Hom et al. 2007).

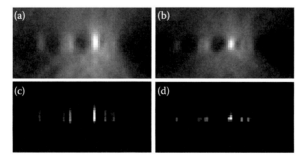

FIGURE 15.17 (a) 200 nm fluorescent beads in glycerol 64 μm below the coverslip imaged with no correction by the DM. (b) Same beads imaged with the DM set to correct the depth aberration. (c) Deconvolution of the image (a) using a PSF measured at the coverslip. (d) Deconvolution of the image (b). Each image is independently scaled to its maximum intensity. Each image is 6.0 μm in the lateral dimension and 6.4 μm in the axial dimension.

15.3.4 Optimization Approaches

Correcting only the depth aberration is attractive because it does not require extra images to optimize the DM, but there are many situations when the sample-induced aberrations are strong and must be corrected by the DM. Several groups have treated the problem as a search for the maximum of an N-dimensional surface (Albert et al. 2000; Wright et al. 2005; Booth 2006; Poland et al. 2008b). The surface height is given by the evaluation of an image quality metric such as maximum intensity or image sharpness, which can be evaluated over the N degrees of freedom of the mirror shape. This approach can be used with confocal and multiphoton as well as wide-field microscopy. An image must be taken at each point along the search path to evaluate the metric; so care must be taken to minimize the number of images taken to optimize the mirror shape. An example of an optimization approach is shown in Fig. 15.19.

15.3.5 Phase Retrieval

One approach to determining the wavefront of an optical beam that does not require additional hardware (such as a wavefront sensor) is to use phase retrieval. In phase retrieval, the phase of the electric field is calculated from multiple measurements of the intensity at different planes along the optic axis. The multiple-intensity measurements—resulting from interference of the amplitude and wavefront at known geometrical phase shifts—provide the extra information required to determine the missing phase information. There are multiple solutions for the electric field to produce an intensity distribution in one plane. Because the phase of the field determines how the field propagates, finding the field that produces the correct intensity distribution on multiple planes eliminates the ambiguity.

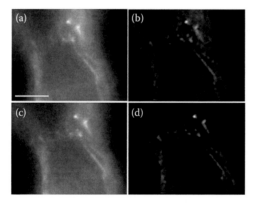

FIGURE 15.18 Deconvolved images of Alexa 488-phalloidin-labeled B16F10 mouse cells. Images are 4.4 μm below the coverslip. (a) Uncorrected image. (b) Uncorrected deconvolved image. (c) Image corrected by adaptive optics. The correction assumed a refractive index of 1.38. (d) Image corrected by adaptive optics after deconvolution. The scale bar is 5 μm (Kner et al., 2010).

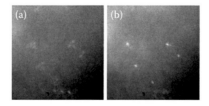

FIGURE 15.19 Image of 200 nm fluorescent beads on a microscope slide below a *C. elegans*. (a) Image taken with a flat DM. (b) Image taken with the DM shape set to optimize the bead intensity, using the algorithm described in Booth (2006).

Most phase-retrieval algorithms work by starting with an initial guess for the field. The field is then propagated to each plane where the intensity was measured. At each measurement plane, the field amplitude is replaced by the square root of the measured intensity. The process is iterated until the field is consistent with all the measurements. In the original phase-retrieval algorithm, the phase was calculated from the intensity measurements in the focal and pupil planes of a low-NA system yielding the phase in the pupil plane (Gerchberg and Saxton 1972). The algorithm of Gerchberg and Saxton was then improved upon and applied to astronomy by Fienup (1978).

Phase retrieval has been demonstrated for high-NA microscopes (Hanser et al., 2003, 2004). In this case, the phase in the back pupil plane of the microscope objective is determined from multiple images around the focal plane. Hanser et al. used a fluorescent bead with a diameter smaller than the microscope resolution to serve as an effectively coherent point source. Taking images through focus yields the 3D PSF (Hiraoka et al. 1990).

In the modified Gerchberg-Saxton algorithm used by Hanser et al., the amplitude of the field is replaced by the measured data at each measurement plane, which minimizes the difference in amplitude between the retrieved field and the measurements. This algorithm starts with an initial guess for the pupil function (the wavefront in the back pupil plane), $P(k_x, k_y) = 1$ within the back pupil plane aperture. The PSF is then calculated through focus:

$$\text{PSF}(x, y, z_i) = |A_i(x, y)|^2 \tag{15.8}$$

$$A_i(x, y) = \iint P(k_x, k_y) \exp(iz_i \gamma(k_x, k_y)) \exp(ik_x x + ik_y y) dk_x dk_y \tag{15.9}$$

$$\gamma(k_x, k_y) = \frac{2\pi n}{\lambda} \sqrt{1 - \left(\frac{\lambda k}{2\pi n}\right)^2} \tag{15.10}$$

At each measurement plane, z_i, the amplitude of the calculated light field, $A_i(x, y)$, is replaced by $\sqrt{I(x, y, z_i)}$ where I is the measured PSF. This can be expressed as

$$A_i(x, y) = \frac{\sqrt{I(x, y, z_i)}}{|A_i(x, y)|} A_i(x, y) \tag{15.11}$$

The pupil function is then calculated as

$$P(k_x, k_y) = \frac{1}{N} \sum_i \iint A_i(x, y) \exp(-iz_i \gamma(k_x, k_y)) \exp(-ik_x x - ik_y y) \mathrm{d}x \mathrm{d}y, \tag{15.12}$$

and the process is iterated.

Here we use a modification of this algorithm (Deming, 2007) that minimizes the difference in intensity between the retrieved field and the measurements. This equivalently minimizes the relative entropy between the measured 3D PSF and the PSF calculated from the retrieved field. Here we refer to Deming's method as the MRE (for minimize relative entropy) method. In effect, the only difference between these two methods is to replace Equation 15.10 by

$$A_i(x, y) = \frac{I(x, y, z_i)}{|A_i(x, y)|^2} A_i(x, y) \tag{15.13}$$

At each iteration, the difference between the calculated PSF and the measured PSF will always decrease, but the error in the wavefront will start to increase after a certain number of iterations as the wavefront is modified to fit the noise in the PSF (Deming 2007).

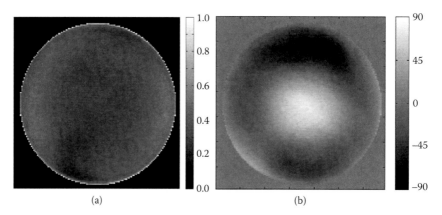

FIGURE 15.20   Wavefront in the back pupil plane of a 1.42 NA system calculated from the three-dimensional image of a 200 nm yellow-green fluorescent bead. (a) Amplitude, normalized to a maximum of 1. (b) Phase (in degrees).

Figure 15.20 shows the results of phase retrieval on a system without AO. This image was taken on a well-corrected Olympus IX71 microscope with a 1.42 NA 60× oil objective (Olympus PLAPON). The Strehl ratio calculated from the retrieved wavefront was 0.70. It was calculated as

$$S = \frac{\left| \int A(k_x, k_y) dk_x dk_y \right|^2}{\left(\int \left| A(k_x, k_y) \right|^2 dk_x dk_y \right) \left(\int dk_x dk_y \right)} \tag{15.14}$$

where the integral is over the pupil for the nominal NA of the microscope. The Strehl ratio is formulated this way to account for the possibility of a flat wavefront that does not fill the full microscope NA. Phase retrieval also works well on significantly distorted PSFs, as shown later.

The retrieved wavefront can be easily decomposed into Zernike modes. Here we use an algorithm that decomposes the gradients of the wavefront onto a set of orthonormal vector polynomials derived from the set of Zernike functions (Zhao and Burge 2007 and 2008). An important advantage of this algorithm is that the phase changes in the gradient are significantly smaller than those in the wavefront itself so that phase wrapping is not an issue. The Zernike decomposition also presents an easy way to convert from the size of the phase-retrieved back pupil plane to that of the array used to set the DM. For the following results, the phase retrieval was performed on an image stack with 256 × 256 pixel lateral size and the DM was set with the wavefront measured by an SHWFS, which is 32 × 32 pixels and not perfectly centered. A left-right inversion of the phase-retrieved pupil was performed to account for the imaging of the back pupil plane on the wavefront sensor.

To demonstrate the accuracy of phase retrieval and to test the open-loop control with the mirror, we set the DM shape to Z_2^2 (0° astigmatism) with amplitudes between –0.5 and 0.5 microns and measured the phase in the back pupil plane using phase retrieval. The phase was then decomposed into the Zernike polynomials (as described above) to test the linearity and cross-talk of the system. Figure 15.21a shows the retrieved amplitude of Z_2^2 and Z_2^{-2} (45° astigmatism) versus the amplitude of Z_2^2 set by the DM. For Z_2^2, the slope is 0.96 and the root-mean-square error (RMSE) is 20.9 nm. For Z_2^{-2}, the slope is 0.01 and the RMSE is 10 nm. Here the slope is 1% of the value for Z_2^2, indicating only weak cross talk between the Zernike modes. We tested the linearity for setting other Zernike modes on the DM with similar results. The exception was in setting tip, tilt, and focus $\left(Z_1^{-1}, Z_1^1, \text{and } Z_2^0 \right)$. With these modes, stage drift can cause substantial variations from measurement to measurement. Figure 15.21b shows the retrieved amplitudes of Z_1^{-1}, Z_1^1, and Z_2^0 versus setting Z_2^2.

Figure 15.22 shows the results of using the retrieved phase to set the DM. Column 1 shows the PSF with all the actuators on the DM set to 0 V. The PSF is strongly distorted with a calculated Strehl ratio of 0.005. (Although for such a strongly distorted PSF, the Strehl ratio is not that meaningful since the peak intensity is not at the center of the PSF.) Column 2 shows the results after correcting the DM with the retrieved phase. The PSF is dramatically improved. The Strehl ratio is 0.2, a 40-fold improvement. Another round of correction improves the Strehl ratio to 0.75. We let the loop run through two more cycles, but the Strehl ratio did not substantially improve, ultimately reaching 0.78. Because of imperfections in the microscope objective and surface roughness in the various optical elements, a perfect Strehl ratio of 1 is never achieved in practice. A Strehl ratio of 0.78 corresponds to an RMS wavefront error of roughly 40 nm and indicates a diffraction-limited, well-corrected optical system.

For Figure 15.22, each image stack was 256×256 pixels with 21 slices from 2.0 microns below focus to 2.0 microns above focus in 0.2 micron steps. The phase retrieval ran for 64 iterations, which took

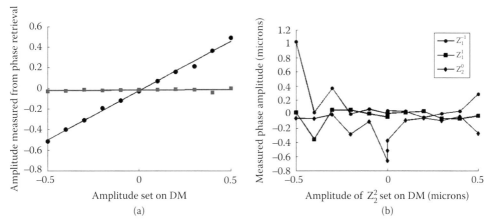

FIGURE 15.21 (a) Plot of the amplitude of Z_2^2 (circles) and Z_2^{-2}(squares) determined by phase retrieval versus the amplitude of Z_2^2 set on the DM. The solid lines are linear fits to the data. The slope of the fit to Z_2^2 is 0.96, and the root-mean-square error (RMSE) is 20.9 nm. The slope of the fit to Z_2^{-2} is 0.0097 and the RMSE is 9.9 nm. (b) Plot of the amplitude of Z_1^{-1}, Z_1^1, and Z_2^0 determined by phase retrieval versus the amplitude of set on the DM. There is substantial drift from measurement to measurement.

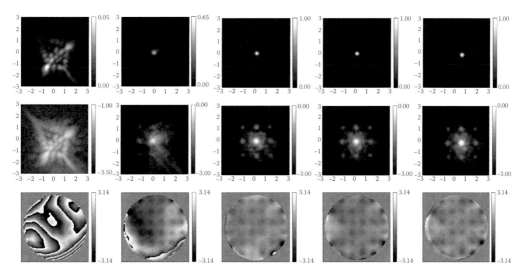

FIGURE 15.22 Iterative correction of the microscope PSF. Top row: in focus image of the PSF, linear scale. Second row: same as first row but on a log scale. Third row: phase in the back pupil plane calculated from phase retrieval with the tip, tilt, and focus terms removed. First column: before correction; DM actuators all set to 0 V. Each successive column is after a round of iteration.

46 seconds on a Core 2 Duo Intel processor. Figure 15.23 shows the improvement in peak intensity and Strehl ratio for the data of Figure 15.22. Figure 15.24 compares the width of the corrected PSF to the ideal image of a 200 nm bead. The measured PSF has an FWHM of 290 nm, about 20% greater than the ideal case. We performed the same measurement but taking only five slices per image stack with a step of 1.0 microns. Taking fewer steps speeds up the phase retrieval and also reduces the light exposure of the sample, but results in less robust phase retrieval. In this case, the Strehl ratio increased from 0.004 to 0.30, and the intensity increased by a factor of 8.2 in four cycles.

15.3.5.1 Phase Retrieval from Biological Samples

Phase retrieval can also work on point sources embedded in or below biological samples. Figure 15.25 shows phase retrieval from a 200 nm fluorescent bead embedded in a *Drosophila* embryo. The dominant Zernike mode of the retrieved phase is astigmatism (1.5 radian RMS amplitude), which is consistent with the cylindrical shape of the embryo being the major cause of the aberrations. Figure 15.25c(i) shows the image reconstructed from the calculated wavefront, confirming the accuracy of the calculation. If the phase of the wavefront is corrected, the peak intensity of the image of the bead would increase by a factor of 4.2, and the shape of the image would again resemble a point (Figure 15.25c(ii)). The amplitude

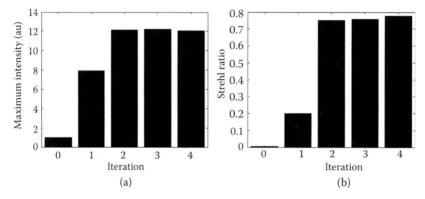

FIGURE 15.23 (a) Maximum intensity of the measured PSF for each iteration of the wavefront optimization. (b) Corresponding Strehl ratio.

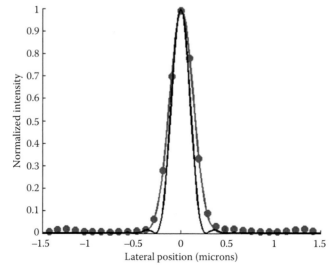

FIGURE 15.24 Lateral in-focus cross section of the final PSF from Figure 15.22 (circles). Gaussian fit to the data (gray line). Theoretically calculated image of a 200 nm bead from an ideal pupil of NA 1.285 (black line).

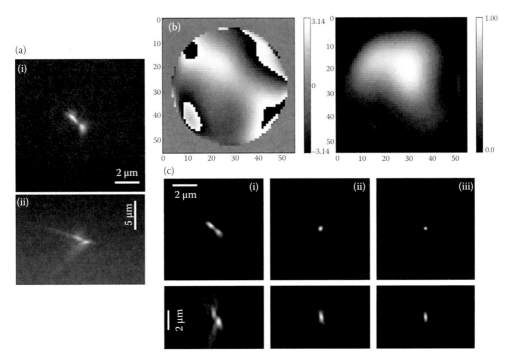

FIGURE 15.25 Amplitude and phase of the wavefront in the back pupil plane calculated using phase retrieval from a 200 nm fluorescent bead injected 40 μm below the surface of a *Drosophila* embryo. (a) Image of the bead used in phase retrieval. (i) A lateral cross section, and (ii) a longitudinal cross section. (b) Phase (radians) and amplitude (normalized) of the wavefront in the back pupil plane. (c) Point spread functions calculated from the wavefront in the back pupil plane. The top image is a lateral in-focus cross section, and the bottom image is a longitudinal cross section. (i) PSF from the wavefront in (b). (ii) PSF function from the wavefront in (b) with the phase corrected (set to 0). (iii) PSF from wavefront with uniform amplitude and flat phase.

of the wavefront is still distorted, and amplitude distortions cannot be corrected by DMs. If both the amplitude and phase could be corrected (keeping the integrated intensity the same), the peak intensity would increase by a factor of 6.8 (Figure 15.25c(iii)). (Amplitude correction could be made by attenuating the strong pupil regions to equal the weaker regions, obviously reducing the integrated intensity.) As can be seen, the image can be significantly improved by only correcting the phase.

15.3.6 Phase Diversity

An important drawback of wavefront sensing using phase retrieval is that a point-like source is required to determine the wavefront. Because there is a lack of point sources in biological samples, determining the wavefront from extended images is a significant advantage. Phase diversity is a technique that does not require a coherent source or PSF (Gonsalves 1982; Paxman et al. 1992), and thus lends itself to wavefront estimation and correction when imaging extended objects. Phase diversity has been used in used in astronomy (Paxman et al. 1996), but has not yet been used extensively in microscopy.

In phase diversity, two or more images are taken, each with a known phase aberration introduced. The aberrations and the object are estimated by finding the combination of object and PSF that best reproduces all the measured data. In general, this requires searching the entire parameter space of both the image and the PSF, but for the case of images with Gaussian noise, the image degrees of freedom can be eliminated and the cost function can be minimized by searching over only the aberration parameters (Paxman et al. 1992).

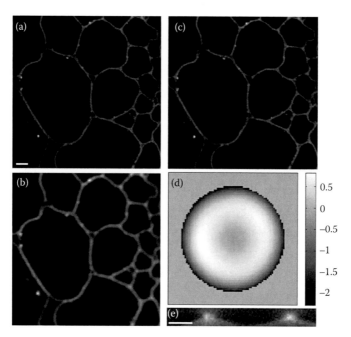

FIGURE 15.26 Example phase diversity on a fluorescence image. (a) In-focus image of GFP-labeled *Arabidopsis*. (b) Out-of-focus image. (c) Object estimated from phase diversity. (d) Wavefront estimated from phase diversity (radians). (e) Axial cross section of image stack showing the effect of the spherical aberration. Scale bars are 5 µm.

A potential advantage of phase diversity, as with phase retrieval, over other wavefront sensorless techniques is that it could potentially require fewer images to determine the wavefront aberration. Sensorless techniques typically require many images to optimize the wavefront. The number of images required typically depends on the number of degrees of freedom to be optimized. Phase diversity can in principle determine the full wavefront from only two images.

Figure 15.26 shows the use of phase diversity on a thin sample. An in-focus and an out-of-focus image of a thin sample from an *Arabidopsis thaliana* stem in which the cell walls are labeled with Alexa 488 are used as the input to the phase-diversity algorithm, and the wavefront, parameterized by the first 15 Zernike modes, that best reproduces the images is determined. These images were taken using a 40× air objective with a correction collar and show spherical aberration because of the setting of the collar. The calculated wavefront (Figure 15.26d) correctly captures the spherical aberration present in the images. Figure 15.26e shows an axial cross section of the image, showing the asymmetry in the out-of-focus light that is characteristic of spherical aberrations.

15.4 Conclusions

We have discussed our work on using AO to correct aberrations in wide-field fluorescence microscopy. To correct only depth aberrations, it is possible to use an open-loop approach, where the mirror can be set based on a priori knowledge of the depth and sample refractive index. For more complicated samples, where sample-induced aberrations are significant, this approach is not sufficient. We show that phase retrieval on a fluorescent bead can be used to correct even severe aberrations, and we discuss the possibility of using phase diversity to efficiently measure aberrations in biological samples.

AO promises to have a significant impact on wide-field fluorescence microscopy, but there is much work to do before its use is routine. The different approaches to measuring and correcting the wavefront must be evaluated and optimized before images can be reliably taken with one click of the mouse. There are also many areas of development for AO in wide-field microscopy.

Using one DM conjugate to the back pupil plane is the simplest AO configuration, but it is likely that it can be improved on. One problem with this configuration is that DMs typically have either a small number of actuators ($\lesssim$50) with a large throw (tens of microns) or a high number of actuators ($\gtrsim$128) with a small throw (a few microns). To correct large aberrations and achieve a high Strehl ratio, it will be necessary to combine two DMs in a "woofer-tweeter" combination.

A single DM conjugate to the back pupil plane can correct the depth aberration across the entire field of view but can correct only sample-induced aberrations in a small region, the isoplanatic patch, whose size will depend on the magnitude and location of the aberrations present. To achieve a well-corrected image over the entire field of view in the presence of sample-induced aberrations will require either an MCAO approach, as mentioned on Section 1.1, or an image-processing approach in which images corrected for different patches are stitched together.

Recently, different techniques for achieving super resolution (resolution beyond the diffraction limit) based on wide-field fluorescence microscopy have been developed, most notably the STORM/PALM techniques (Betzig et al. 2006; Hess et al. 2006; Rust et al. 2006) and structured illumination microscopy (Gustafsson et al. 2008; Kner et al. 2009). These techniques were all initially developed for thin, fixed samples, in part because they require well-corrected images to work. For these techniques to be extended to live imaging in thicker samples, AO will be required to provide diffraction-limited imaging deep into the sample.

Acknowledgments

We thank Kaveh Ashrafi, Xeuying Wang, Yuri Strukov, Pete Carlton, Hesper Rego, Jian Cao, William Sullivan, and Utku Avci for biological samples and help in preparing samples. ZK is the Israel Pollak Professor of Biophysics. This work was supported by NIH Grants GM25101 (JWS) and GM31627 (DAA), the National Science Foundation through the Center for Biophotonics Science and Technology under Cooperative Agreement No. PHY 0120999, and by the Keck Laboratory for Advanced Microscopy. We thank Steve Lane at the Center for Biophotonics Science and Technology, and Joel Kubby and other members of the National Science Foundation Science and Technology Center for Adaptive Optics at the University of California at Santa Cruz for helpful discussions.

References

Agard, D.A., Hiraoka, Y., and Sedat, J.W. (1988). Three-dimensional light microscopy of diploid *Drosophila* chromosomes. *Cell Motil Cytoskeleton 10*, 18–27.

Agard, D.A., Hiraoka, Y., Shaw, P., and Sedat, J.W. (1989). Fluorescence microscopy in three dimensions. *Methods Cell Biol 30*, 353–377.

Agard, D.A., and Sedat, J.W. (1983). Three-dimensional architecture of a polytene nucleus. *Nature 302*, 676–681.

Albert, O., Sherman, L., Mourou, G., Norris, T.B., and Vdovin, G. (2000). Smart microscope: an adaptive optics learning system for aberration correction in multiphoton confocal microscopy. *Opt Lett 25*, 52–54.

Arigovindan, M., Shaevitz, J., McGowan, J., Sedat, J.W., and Agard, D.A. (2010). A parallel product-convolution approach for representing the depth varying point spread functions in 3D widefield microscopy based on principal component analysis. *Opt Express 18*, 6461–6476.

Betzig, E., Patterson, G.H., Sougrat, R., Lindwasser, O.W., Olenych, S., Bonifacino, J.S., Davidson, M.W., Lippincott-Schwartz, J., and Hess, H.F. (2006). Imaging intracellular fluorescent proteins at nanometer resolution. *Science 313*, 1642–1645.

Booth, M.J. (2006). Wave front sensor-less adaptive optics: a model-based approach using sphere packings. *Opt Express 14*, 1339–1352.

Booth, M.J., Neil, M.A.A., and Wilson, T. (1998). Aberration correction for confocal imaging in refractive-index-mismatched media. *J Microsc 192*, 90–98.

Carlton, P.M., Boulanger, J., Kervrann, C., Sibarita, J.B., Salamero, J., Gordon-Messer, S., Bressan, D., Haber, J.E., Haase, S., Shao, L., et al. (2010). Fast live simultaneous multiwavelength four-dimensional optical microscopy. *Proc Natl Acad Sci USA 107*, 16016–16022.

Deming, R.W. (2007). Phase retrieval from intensity-only data by relative entropy minimization. *J Opt Soc Am A 24*, 3666–3679.

Fienup, J.R. (1978). Reconstruction of an object from the modulus of its Fourier transform. *Opt Lett 3*, 27–29.

Friedel, D., Schele, L., and Parker, J. (1995). *Maya Cosmos: Three Thousand Years on the Shaman's Path.* New York, NY:Harper Paperbacks.

Fujimoto, Y., and Nishiwaki, D. (2004). *Objective Lens System for Microscope.* Japan:Olympus Corporation.

Gavel, D. (2003). Suppressing anomalous localized waffle behavior in least squares wavefront reconstructors. Paper presented at: *Proceedings of SPIE.* Waikoloa, HI:SPIE.

Gerchberg, R.W., and Saxton, W.O. (1972). A practical algorithm for the determination of the phase from image and diffraction plane pictures. *Optik 35*, 237–246.

Gibson, S.F., and Lanni, F. (1991). Experimental test of an analytical model of aberration in an oil-immersion objective lens used in three-dimensional light microscopy. *J Opt Soc Am A 8*, 1601–1613.

Gonsalves, R.A. (1982). Phase retrieval and diversity in adaptive optics. *Opt Eng 21*, 829–832.

Gu, M. (1999). *Advanced Optical Imaging Theory.* Berlin:Springer-Verlag.

Gustafsson, M.G.L., Shao, L., Carlton, P.M., Wang, C.J.R., Golubovskaya, I.N., Cande, W.Z., Agard, D.A., and Sedat, J.W. (2008). Three-dimensional resolution doubling in wide-field fluorescence microscopy by structured illumination. *Biophys J 94*, 4957–4970.

Hanser, B.M., Gustafsson, M.G., Agard, D.A., and Sedat, J.W. (2003). Phase retrieval for high-numerical-aperture optical systems. *Opt Lett 28*, 801–803.

———. (2004). Phase-retrieved pupil functions in wide-field fluorescence microscopy. *J Microsc 216*, 32–48.

Hardy, J.W. (1998). *Adaptive Optics for Astronomical Telescopes.* New York, NY:Oxford University Press.

Hess, S.T., Girirajan, T.P.K., and Mason, M.D. (2006). Ultra-high resolution imaging by fluorescence photoactivation localization. *Microscopy Biophys J 91*, 4258–4272.

Hiraoka, Y., Sedat, J.W., and Agard, D.A. (1990). Determination of three-dimensional imaging properties of a light microscope system. Partial confocal behavior in epifluorescence microscopy. *Biophys J 57*, 325–333.

Hom, E.F., Marchis, F., Lee, T.K., Haase, S., Agard, D.A., and Sedat, J.W. (2007). AIDA: an adaptive image deconvolution algorithm with application to multi-frame and three-dimensional data. *J Opt Soc Am A Opt Image Sci Vis 24*, 1580–1600.

Juskaitis, R. (2007). Characterizing high numerical aperture microscope objective lenses. In P. Torok, and F.-J. Kao, eds. *Optical Imaging and Microscopy: Techniques and Advanced Systems*, pp. 21–43. Berlin:Springer Verlag.

Kam, Z. (1998). Microscopic differential interference contrast image processing by line integration (LID) and deconvolution. *Bioimaging 6*, 166–176.

Kam, Z., Agard, D., and Sedat, J.W. (1997). Three-dimensional microscopy in thick biological samples: a fresh approach for adjusting focus and correcting spherical aberration. *Bioimaging 5*, 40–49.

Kam, Z., Hanser, B., Gustafsson, M.G., Agard, D.A., and Sedat, J.W. (2001). Computational adaptive optics for live three-dimensional biological imaging. *Proc Natl Acad Sci USA 98*, 3790–3795.

Kam, Z., Kner, P., Agard, D., and Sedat, J.W. (2007). Modelling the application of adaptive optics to wide-field microscope live imaging. *J Microsc 226*, 33–42.

King, S.V., and Cogswell, C.J. (2004). *A Phase-Shifting Dic Technique for Measuring 3D Phase Objects: Experimental Verification.* San Jose, CA:SPIE.

Kner, P., Chhun, B.B., Griffis, E.R., Winoto, L., and Gustafsson, M.G. (2009). Super-resolution video microscopy of live cells by structured illumination. *Nat Methods 6*, 339–342.

Kner, P., Sedat, J.W., Agard, D.A., and Kam, Z. (2010). High-resolution wide-field microscopy with adaptive optics for spherical aberration correction and motionless focusing. *J Microsc 237*, 136–147.

Marsh, P.N., Burns, D., and Girkin, J.M. (2003). Practical implementation of adaptive optics in multiphoton microscopy. *Opt Express 11*, 1123–1130.

Murray, J.M., Appleton, P.L., Swedlow, J.R., and Waters, J.C. (2007). Evaluating performance in three-dimensional fluorescence microscopy. *J Microsc 228*, 390–405.

Paxman, R.G., Schulz, T.J., and Fienup, J.R. (1992). Joint estimation of object and aberrations by using phase diversity. *J Opt Soc Am A Opt Image Sci Vis 9*, 1072–1085.

Paxman, R.G., Seldin, J.H., Lofdahl, M.G., Scharmer, G.B., and Keller, C.U. (1996). Evaluation of phase-diversity techniques for solar-image restoration. *Astrophys J 466*, 1087–1096.

Pliny, G.P.S., the Elder (1938). *Natural History*. Cambridge, MA:Harvard University Press.

Poland, S.P., Wright, A.J., and Girkin, J.M. (2008a). Active focus locking in an optically sectioning microscope utilizing a deformable membrane mirror. *Opt Lett 33*, 419–421.

———. (2008b). Evaluation of fitness parameters used in an iterative approach to aberration correction in optical sectioning microscopy. *Appl Opt 47*, 731–736.

Preza, C., and Conchello, J-A. (2004). Depth-variant maximum-likelihood restoration for three-dimensional fluorescence microscopy. *J Opt Soc Am A 21*, 1593–1601.

Ragazzoni, R., Diolaiti, E., Farinato, J., Fedrigo, E., Marchetti, E., Tordi, M., and Kirkman, D. (2002). Multiple field of view layer-oriented adaptive optics. *A&A 396*, 731–744.

Ragazzoni, R., Marchetti, E., and Valente, G. (2000). Adaptive-optics corrections available for the whole sky. *Nature 403*, 54–56.

Rust, M.J., Bates, M., and Zhuang, X. (2006). Sub-diffraction-limit imaging by stochastic optical reconstruction microscopy (STORM). *Nat Meth 3*, 793–796.

Schwertner, M., Booth, M., and Wilson, T. (2004a). Characterizing specimen induced aberrations for high NA adaptive optical microscopy. *Opt Express 12*, 6540–6552.

Schwertner, M., Booth, M.J., Neil, M.A.A., and Wilson, T. (2004b). Measurement of specimen-induced aberrations of biological samples using phase stepping interferometry. *J Microsc 213*, 11–19.

Staudt, T., Lang, M.C., Medda, R., Engelhardt, J., and Hell, S.W. (2007). 2,2′-Thiodiethanol: a new water soluble mounting medium for high resolution optical microscopy. *Microsc Res Tech 70*, 1–9.

Stokseth, P.A. (1969). Properties of a Defocused Optical System. *J Opt Soc Am 59*, 1314–1321.

Sullivan, W., Ashburner, M., and Hawley, R.S., eds. (1999). *Drosophila Protocols*, 1st edn. Cold Spring Harbor, NY:Cold Spring Harbor Laboratory Press.

Swedlow, J.R., Sedat, J.W., and Agard, D.A. (1997). Deconvolution in Optical Microscopy. In P.A. Jansson, ed. *Deconvolution of Images and Spectra*, pp. 284–307. San Diego, CA:Academic Press.

Tokovinin, A., Louarn, M.L., and Sarazin, M. (2000). Isoplanatism in a multiconjugate adaptive optics system. *J Opt Soc Am A 17*, 1819–1827.

Tuchin, V. (2007). *Tissue Optics*, 2nd edn. Bellingham, WA:SPIE.

Tyson, R.K. (1997). *Principles of Adaptive Optics*, 2nd edn. New York, NY:Academic Press.

Wright, A.J., Burns, D., Patterson, B.A., Poland, S.P., Valentine, G.J., and Girkin, J.M. (2005). Exploration of the optimisation algorithms used in the implementation of adaptive optics in confocal and multiphoton microscopy. *Microsc Res Tech 67*, 36–44.

Zhao, C.Y., and Burge, J.H. (2007). Orthonormal vector polynomials in a unit circle, Part I: basis set derived from gradients of Zernike polynomials. *Opt Express 15*, 18014–18024.

———. (2008). Orthonormal vector polynomials in a unit circle, Part II: completing the basis set. *Opt Express 16*, 6586–6591.

16

Biological Imaging and Adaptive Optics in Microscopy

Elijah Y. S. Yew
*Singapore-MIT Alliance
for Research and
Technology (SMART)*

Jae Won Cha
*Massachusetts Institute
of Technology*

Jerome Ballesta
Imagine Optic

Peter T. C. So
*Massachusetts Institute
of Technology*

16.1 Introduction

In general, adaptive optics (AO) for microscopy is no different from its other applications in astronomy or ophthalmology in the sense that it relies on wavefront sensing and implementing a change in the active element to correct for the distorted wavefront. Of all the methods of microscopy, the two most successful methods are, perhaps, the laser scanning confocal fluorescence microscope and the laser scanning two-photon fluorescence microscope [1]. Because both methods build up an image by scanning the excitation beam over the object, the underlying principles of image formation are similar in both cases; the implementation of AO is likewise similar in both the methods. In this chapter, we focus on the use of a Shack-Hartmann wavefront sensor (SHWS) for the detection of reflected excitation light in the implementation of an AO two-photon microscope.

On the one hand, two-photon excitation fluorescence microscopy is a widely used and popular method of imaging biological samples as the long wavelengths used for excitation offer deeper penetration into the sample compared to single-photon fluorescence excitation. On the other hand, imaging through most biological tissue suffers from scattering and the refractive index throughout the tissue is inhomogeneous. This leads to distortions of the wavefront of the excitation light and thereby results in an aberrated focus, reduced two-photon efficiency, and reduced imaging depth. In fact, much of the aberrations that are present in imaging biological tissue are due to sample aberrations [2–4]. The causes of these aberrations are very similar to the problems experienced in astronomy and ophthalmology and, as with the previously mentioned examples in Chapter 7, an AO solution has been implemented. AO has been successfully implemented in both confocal and nonlinear microscopy techniques, such as two-photon excitation microscopy, harmonic generation microscopy, and coherent anti-Stokes Raman scattering (CARS) microscopy [5].

In an AO system, the wavefront distortions can be compensated with an active element, and the resultant image resolution and signal level can be maintained to greater imaging depths. However, the issue is not just to correct the wavefront distortions (aberrations) but also to characterize and measure the distortions, without which the correction of a single scanned position of the focused beam can take several minutes. The detection of aberrations in AO microscopy can be categorized into two main approaches—wavefront sensorless and wavefront sensors. Of the two, the former does not employ a wavefront sensor (e.g., SHWS), but instead relies on a chosen imaging metric, such as the excited fluorescence signal strength or image sharpness. With the latter, aberrations are detected from the reflected/backscattered excitation light using a wavefront sensor. In the latter method, care must be taken because the excitation light passes through the aberrating sample twice and it is difficult to distinguish between odd and even aberrations after reflection. In the following sections, we describe the main differences and various implementations of the two methods of detecting wavefront aberrations in microscopy.

16.2 Methods and Experimental Setup

A homebuilt two-photon microscope was set up following the principles laid out [6]. The implementation of AO was through a large dynamic range deformable mirror and a reflected light path with a confocal pinhole before the SHWS.

16.2.1 Large Dynamic Range Deformable Mirror

The deformable mirror (Mirao52d; Imagine Optic, Orsay, France) includes a silver-coated reflective membrane coupled with 52 miniature voice-coil-type actuators. The diameter of its pupil is 15 mm, and it has a 200 Hz bandwidth. The maximum wavefront amplitude that can be generated is 50 μm with 1-V voltage range. An important feature of this deformable mirror is its large dynamic range, and it can generate Zernike modes up to the fourth order and partially up to the sixth order.

16.2.2 Detection of Wavefront Distortion with a Shack-Hartmann Wavefront Sensor

The SHWS (HASO32; Imagine Optic) consists of a lenslet array and a charge-coupled device (CCD) as the sensor. It has a 32 × 32 lens array within a 3.6 × 3.6 mm² rectangular aperture and a bandwidth of 20 Hz. Each lenslet produces a focal spot that is detected over a region on the CCD–that is, a group of pixels (sub-detection area) corresponding to the aperture of the lenslet. If the incident wavefront is plane, the foci are all centered on the appropriate pixels. If the wavefront is aberrated, the foci shift across the sub-detection area according to the type of aberration. The slope of the wavefront is given by the ratio of the shift over the focal length of the lenslet, and the whole wavefront can be reconstructed. The advantage of the Shack-Hartmann sensor is its ability to rapidly quantify wavefront aberration with a single image.

16.2.3 Application of Confocal Detection to Select Reflected Light Signals from the Focal Plane

In the case of fluorescence microscopy, only the signal originating from the vicinity of the focus is relevant for the purpose of aberration detection and correction [7–9]. In a real biological sample, scattering and other inhomegeneities of the refractive index mean that sample-induced aberrations are the primary determinants of aberrations affecting the final image quality. Typically, the fluorescence signal is detected and changes are made to the deformable mirror till the signal is optimized. In this situation, the reflected excitation light is detected. Steps are therefore required to eliminate the reflected signals that occur from planes that are out of focus. This is accomplished through employing a confocal detection path before the wavefront sensor. As with a conventional confocal microscope, the confocal pinhole allows only the signal originating from the focal region while effectively blocking the backscattered light emanating from other planes. In theory, the depth selectivity (axial resolution) should be optimized by selecting a small pinhole [10,11]. However, this requirement is a trade-off, because the pinhole also functions as a spatial filter. Selecting a very small pinhole will therefore filter out higher spatial frequencies of the distorted wavefront, and in turn affect the accuracy of the wavefront reconstruction, which is needed for optimal adaptive correction. These two conflicting requirements must be balanced, and the optimization of the pinhole size is discussed in the following section.

16.2.4 Overall Instrument Configuration

Figure 16.1 describes the general experimental setup. The laser source was a Ti-sapphire laser (Tsunami; Spectra-Physics, Mountain View, California) pumped by a frequency-doubled Nd:YVO4 laser (Millennia V; Spectra-Physics). The excitation light was reflected by the Mirao52d deformable mirror positioned at a conjugate plane to the back aperture of the microscope objective, along the excitation optical path of a typical two-photon laser scanning microscopy. The emission fluorescence signal passing a short-pass dichroic mirror and a short-pass barrier filter (650dcxxr and E700SP; Chroma Technology, Rockingham, VT) was incident on a photomultiplier detector (R7400P; Hamamatsu,

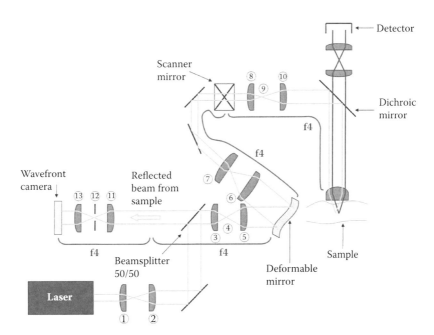

FIGURE 16.1   System configuration. The components used are a Ti-sapphire (Ti-Sa) laser, plano-convex lens L1 ($f = 50$ mm), L2 ($f = 200$ mm), L3 ($f = 100$ mm), L5 ($f = 200$ mm), L6 ($f = 400$ mm), L7 ($f = 50$ mm), L8 ($f = 20$ mm), 444232, Zeiss eyepiece, L10 ($f = 125$ mm), L11 ($f = 200$ mm), L13 ($f = 100$ mm), silver-coated mirrors, beamsplitter, deformable mirror (Mirao52d; Imagine Optics), scanning mirrors (6350; Cambridge Technology), short-pass dichroic mirror (650dcxxr; Chroma Technology), objective lens (Fluar, 1.3 NA; Zeiss), photomultiplier detector (R7400P; Hamamatsu), pinhole (P150S; Thorlabs), and wavefront camera (HASO32; Imagine Optics).

Bridgewater, NJ) and the associated single-photon counting circuitry. The scattered light signal was reflected by the dichroic scanner, descanned by passing through the scanning mirrors, and directed toward the wavefront camera (HASO32; Imagine Optic). The wavefront camera was also positioned at a conjugate plane to the back aperture of the microscope objective. A confocal pinhole (P150S; Thorlabs, Newton, NJ) was positioned before the wavefront camera, and at a plane conjugate to the focus of the objective. This completed the depth-discriminating path for the reflected light signal. As mentioned earlier, a small pinhole is necessary for good axial resolution. However, the pinhole needs to be sufficiently large to pass adequate information about the distorted wavefront contained in the spatial frequencies. In this particular case, the minimum pinhole size was set to pass the highest spatial frequency that was obtainable with the deformable mirror. The rationale behind this was that a pinhole size any larger than that would mean that the deformable mirror could not correct for those orders of aberrations while any smaller size would mean a reduced performance of the deformable mirror. The distance between any two actuators on the deformable mirror is 2.5 mm and corresponds to a minimum spatial period of 5 mm. A Fourier-transform of this spatial period gives the maximum spatial frequency that needs to pass through the wavefront-detection light path. We determined that a confocal aperture with a diameter no less than 125 μm placed at plane 12 of Figure 16.1 would transmit all the relevant aberration information. However, we chose to use a 150-μm-diameter pinhole to provide some additional margin of error. As the size of the pinhole essentially affects the spatial frequencies passed, and this in turn is inversely proportional to the depth selectivity, it is necessary to determine the axial resolution afforded by this 150-μm-diameter pinhole. Accounting for the magnification of the intermediate optics, the 150 μm pinhole at plane 12 corresponds to an effective aperture of 37.5 μm radius at plane 9 of Figure 16.1. Since the focal length of lens 10 is 125 mm, the effective magnification M is 30.3. This results in an axial resolution of about 2.0 μm. This resolution is deemed sufficient because the back aperture of the objective is underfilled, corresponding to an effective axial resolution of 4.8 μm, and is discussed later. By measuring the wavefront distortion, the wavefront camera was set such as to provide feedback to the deformable mirror. Depending on the feedback from the wavefront sensor, the deformable mirror generated a predistortion to the incident plane wave that would counter the distortion induced by the sample. This would therefore produce a PSF at the focal point closer to the ideal and improve the two-photon image.

An important issue, as described by Artal et al. [12], is now addressed. It is to be noted that the light detected by the wavefront camera reflects off the deformable mirror twice, while the actual incoming excitation light to a sample passed the deformable mirror only once. Under the assumption that the specimen is highly scattering, this is a reasonable and appropriate assumption. In the first pass, the light is aberrated. However, all phase information is lost on diffuse scattering and, as a result, the backscattered light contains information only about the aberrations on the return/reflected path, with the aberrations from the ingress being lost. The implication of this is, of course, that such an abberation-detection scheme in the transmission will pick up a different set of aberrations and is not applicable.

Figures 16.2 and 16.3 show representative wavefronts and aberration coefficients before and after AO compensation. The wavefront sensor provides the deformable mirror with a feedback signal while satisfying of the Nyquist-Shannon sampling criterion. The AO system can therefore fully utilize the maximum capability of the deformable mirror for wavefront correction, which is under 10 nm rms error, according to specifications. At the same time, our method holds only when the sample is highly scattering. This is often valid for biological tissue but is not always the case. For samples that have closer approximate specular reflection, the measured wavefront is dependent on whether the aberrations generated by the sample are odd asymmetric or even symmetric, and this has been extensively discussed [12–14]. In terms of backscattered light efficiency, the majority of the losses are due to the use of the 50:50 beamsplitter for the excitation and detection paths. These losses can be avoided by using a polarizing beamsplitter, followed by a quarter-wave plate [15]. This configuration reflects 100% of the excitation laser power instead of 50% in the absence of multiple scattering.

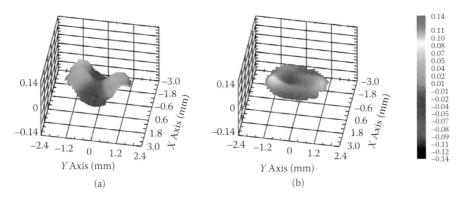

(a)   (b)

FIGURE 16.2 **(See color insert.)** Wavefront change after AO compensation. The sample is a mouse heart, as described in Section 16.3.4 and imaged at 20 μm depth. (a) Distorted wavefront without the compensation. (b) Wavefront after AO compensation.

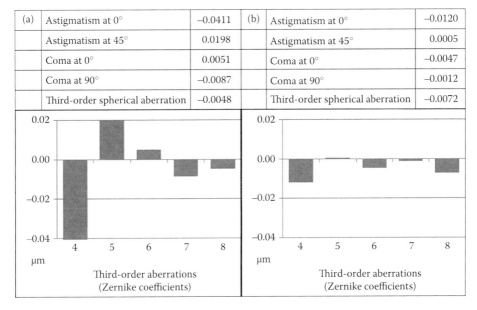

(a)			(b)		
	Astigmatism at 0°	−0.0411		Astigmatism at 0°	−0.0120
	Astigmatism at 45°	0.0198		Astigmatism at 45°	0.0005
	Coma at 0°	0.0051		Coma at 0°	−0.0047
	Coma at 90°	−0.0087		Coma at 90°	−0.0012
	Third-order spherical aberration	−0.0048		Third-order spherical aberration	−0.0072

FIGURE 16.3 Third-order aberration coefficient change after AO compensation. (a) Uncompensated coefficients corresponding to Figure 16.2a. The *x*-axis represents Zernike coefficients; 4 is $Z_{2,2}$, 5 is $Z_{2,-2}$, 6 is $Z_{3,1}$, 7 is $Z_{3,-1}$, and 8 is $Z_{4,0}$. The error of the wavefront measurement is <2 nm. (b) Compensated coefficients.

16.3 Experiment and Results

A series of experiments using artificial and biological specimens were devised to quantify the performance of our AO system. To this end, two artificial specimens with large aberrations were prepared and used to evaluate our AO system in terms of minimizing the signal loss and optimizing the image-resolution loss. Also, the performance of this AO system was evaluated in three typical biological tissue specimens: mouse tongue muscle, heart muscle, and brain slice.

16.3.1 Signal Loss Due to Aberrations as a Function of Imaging Depth

This experiment was designed to measure the signal loss due to aberration as a function of the imaging depth and the compensatory performance of the AO system. A 20x dry objective lens (Fluar, 0.75

NA; Zeiss, Thornwood, NY) was used to image into aqueous fluorescein samples with varying thicknesses: 50, 100, and 150 µm (Figure 16.4). A mirror (10D20ER.2; Newport, Irvine, CA) was placed at the bottom of the specimen to enhance the reflected light signals. A femtosecond pulsed laser was used to provide excitation light at 780 nm. The two-photon excitation focal volume was placed just above the mirror surface, within 1–2 µm. The refractive index mismatch between the air objective and the aqueous specimen resulted in the significant aberration, in particular spherical aberration [16,17]. As the focus went deeper, the signal loss became higher. The signal loss was measured with a normal two-photon excitation fluorescence microscope with and without AO compensation, and the two signal losses were compared. Figure 16.5 shows the loss of fluorescein emission signal without compensation. The uncompensated signal decreased with increasing focusing depth. However, the emission signal remained almost constant, with AO compensation. The signal improvement was measured to be 1% at 50 µm, 7% at 100 µm, and 147% at 150 µm. At 150 µm, the maximum Zernike coefficient of the uncompensated aberration among the third-order aberrations was 0.39 µm, but was 0.10 µm with AO compensation. The Zernike coefficient for spherical aberration was reduced from 0.036 to 0.004 µm. The maximum aberration was found to be astigmatism instead of the spherical aberration. This may be caused by either a slight misalignment of the intermediate optics or a surface curvature in the dichroic mirror (a common feature in some microscope dichroic-mirror holders) [18,19]. In addition, the excitation laser source was linearly polarized instead of circularly polarized. It is known that scattering depends on the polarization direction [8,12].

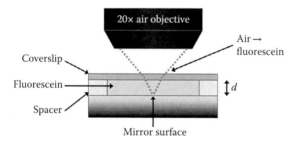

FIGURE 16.4   Signal loss experiment sample.

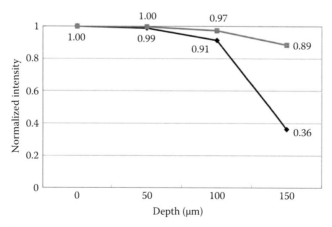

FIGURE 16.5   Signal loss improvement with AO correction (gray with squares) and without AO correction (black with diamonds). The ranges of error bars are smaller than their symbols.

16.3.2 Point Spread Function Degradation Due to Aberrations as a Function of Imaging Depth

To quantify the PSF degradation as a function of imaging depth resulting from aberrations, a 40 × oil-immersion objective lens (Fluar, 1.3 NA; Zeiss) was used to image 0.1-μm-diameter fluorescence beads (F8803; Invitrogen, Eugene, OR) deposited on a mirror through an air gap of 50 μm (Figure 16.6). The mirror was used to enhance the reflected light signal. The excitation wavelength was 780 nm. Spherical aberration was present as a result of the index mismatch between the oil objective and the air-spaced specimen. The PSF was measured with a normal two-photon excitation fluorescence microscope with and without the AO system.

Figure 16.7 shows the effect of aberration on PSF laterally and axially. The left-placed bar shows the uncompensated resolution and the right-placed bar shows the compensated resolution. Clearly, when the air-gap thickness is zero, no compensation is necessary. With a 50 μm air gap, the lateral resolution was improved by 12% and the axial resolution was improved by 38% with AO compensation. The dissimilar values between improvements to the lateral and axial resolution can be attributed to the fact that the former is proportional to, while the latter varies as, so it is reasonable that the axial resolution benefits more with AO compensation than does the lateral resolution. The Zernike coefficient of the uncompensated maximum aberration was 0.44 μm but reduced to 0.09 μm with compensation. At the same time, the Zernike coefficient for spherical aberration reduced from 0.14 to 0.02 μm. We noted that the lateral and axial resolutions were larger than the values of diffraction-limited resolutions. This was underfilling of the objective with the excitation light. As mentioned, the deformable mirror will give the same compensation to the input beam and the backscattered beam from the sample. The two beams should therefore be located at exactly the same area on the deformable mirror. In theory, their beam sizes should be the same, but for the fact that with the Gaussian characteristics of the laser beam, the objective lens was practically underfilled and the lateral and axial resolutions increased. According to

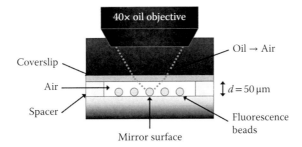

FIGURE 16.6 PSF degradation experiment sample.

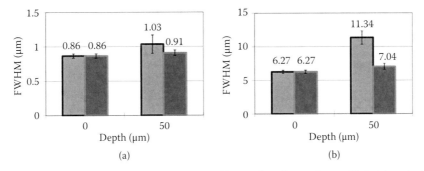

FIGURE 16.7 Resolution change after AO compensation. (a) Lateral resolution change. (b) Axial resolution change. Data with AO correction (left-placed light gray column) and without AO correction (right-placed dark gray column).

the experimental result, the effective NA of the objective was about 0.55 in our experimental setup, although its theoretical NA is 1.3, which is valid only when immersion oil is present under the objective lens till a coverslip and the sample are located immediately beneath and with a refractive index of 1.5. This current limitation can be removed by utilizing a top-hat beam profile or modifying the AO feedback algorithm to ignore information from the peripheral region of the wavefront sensor.

16.3.3 Mouse-Tongue-Muscle Imaging Using Adaptive Optics–Compensated Two-Photon Microscopy

Mouse-tongue musculature was visualized based on both endogenous fluorescence and second harmonic generation. A whole tongue excised from a female C57BL/6 mouse (10 weeks old) was fixed in phosphate-buffered saline (PBS, pH 7.4) containing 2.5% glutaraldehyde for a day. The fixed tongue tissue was then immersed in PBS for 3 h, rinsed with buffered solution repeatedly, embedded in paraffin, and mounted without a coverslip. The wavelength of the excitation light was set to 780 nm. The objective lens was a Zeiss 40× oil immersion objective. The emission signal was filtered by a green filter (535/40, Chroma Technology). The field size was 120 × 120 μm covering 256 × 256 pixels. The dwell time was 40 μs. Figure 16.8 shows representative images of the imaged mouse-tongue muscle at 80 μm depth, with and without the AO compensation. The image on the left is the uncompensated image and the image on the right is the compensated image, with both processed for background rejection. The threshold for each image was set to three times the intensity level measured in the regions outside the objects of interest (e.g., blood vessels in the mouse heart or the neurons in the mouse brain). In principle, the bandwidths of the wavefront camera and the deformable mirror are about 20 and 200 Hz, respectively. However, the current feedback algorithm takes significant CPU resources and leads to a compensation time of 4 s–5 s. The algorithm therefore forms the bottleneck for the bandwidth of the compensation process. At these time scales, it is far from practical to perform pixel-by-pixel correction over the whole image. To reduce the overly long compensated-imaging process, the assumption that the aberrations are primarily due to depth rather than field position was made [2–4], and AO compensation was performed only at the center pixel of the image. This mirror shape was kept for the rest of the image acquisition. In future studies, it would be interesting to determine the optimal number of pixels that should be corrected per image, given the trade-off between imaging speed and tissue heterogeneity. It would also be interesting to improve imaging speed by developing faster, more efficient feedback algorithm or by utilizing higher-performance computer hardware.

The compensated images show higher signal strength when compared to the uncompensated images. Figure 16.9a–d shows the histograms for the intensity distributions before and after compensation. The number of brighter pixels increased, while the number of darker pixels decreased for all imaging depths after compensation, which implies that the in-focus fluorescence signals were being optimized while

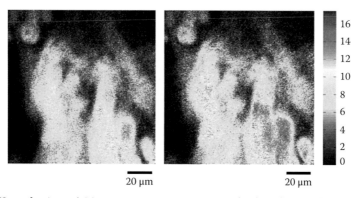

FIGURE 16.8 (**See color insert.**) Mouse-tongue images at 80 μm depth without and with AO compensation, respectively.

the out-of-focus fluorescence signals were suppressed. To find the improvement trend according to the imaging depth, mean photon counts were calculated with background rejection. Figure 16.9e shows the percentage improvement from the mean photon counts according to the imaging depth. The percentage improvement shows that at greater imaging depths, the improvement becomes larger, so the improvement for 90–100% intensity pixels was 38.1% at 80 μm depth. Figure 16.9f shows the increment to the mean photon count. In terms of photon-count increment, the result at 40 μm depth shows the greatest improvement when compared to counts at greater depths. At shallower depths, the compensation is

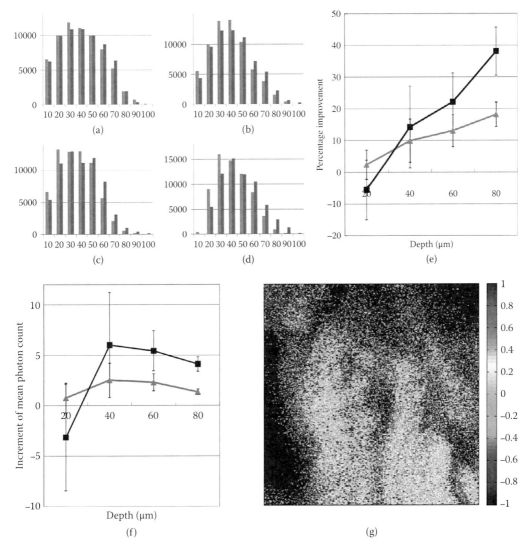

FIGURE 16.9 **(See color insert.)** Signal improvement after AO compensation. (a–d) Histograms for the number of pixels according to their intensity; the *x*-axis represents intensity of the pixels. For example, 10 means that the pixels have 0–10% intensity of the maximum in the whole image, and 100 means 90–100% intensity pixels. The *y*-axis represents the number of pixels in the intensity range. The left-placed columns show the number of pixels before the compensation, and the right-placed columns show the result after compensation. Each histogram was normalized to itself. The distribution at (a) 20 μm, (b) 40 μm, (c) 60 μm, and (d) 80 μm imaging depth. (e) Percentage improvement according to imaging depth. The curve with square markers shows the improvement with background rejection (only the fluorescent area was calculated), and the curve with triangular markers shows the improvement.

largely irrelevant, while at greater depths, the overall signal is significantly lower due to the scattering attenuation of both excitation and emission photons. Figure 16.9g shows the percentage improvement based on each pixel at 80 μm depth, as calculated from Figure 16.8. Comparing the degree of improvement on the center versus the edge, no major difference is observed. Throughout the region where fluorescence signal is generated, the improvement seems relatively uniform. This might be because the aberration from the optical system is much larger than the aberration from the sample.

16.3.4 Mouse-Heart-Muscle Imaging Using Adaptive Optics–Compensated Two-Photon Microscopy

As a second example, we imaged the muscular structures of the heart of a mouse with nuclei detection based on endogenous fluorescence, second harmonic generation, and exogenous labeling. Tail-vein injection was performed with a heavily anaesthetized mouse and nuclei were labeled with Hoechst, while the extracellular matrix surrounding the cells was labeled with Texas Red Maleimide. After staining, the mouse was sacrificed and the heart was excised, fixed in 4% paraformaldehyde, embedded in paraffin, and mounted without a coverslip after some histological processing. The wavelength of the excitation light was 780 nm. The objective lens was a 40× oil-immersion objective lens. The emission signal was filtered by a green filter (535/40, Chroma Technology). The field size was 120 × 120 μm covering 256 × 256 pixels, and the dwell time was 40 μs.

Figure 16.10 shows representative images of the mouse heart at 80 μm depth with and without AO compensation. The image on the left is the uncompensated image and the image on the right is the compensated image, with both processed for background rejection. As in the mouse-tongue experiment, adaptive correction was performed only at the center pixel of the image and the deformable mirror shape was held constant throughout the different scanned positions. To compare the signal intensity of the two images, the intensity distributions for all pixels were calculated again. Figure 16.11a–d shows the histograms for the intensity distributions before and after the compensation. The same trend was observed as in the mouse-tongue result. The number of brighter pixels increased and the number of darker pixels decreased for all the imaging depths after compensation. To find the improvement trend with imaging depth, the mean photon count was calculated with background rejection. Figure 16.11e shows the percentage improvement from the mean photon counts according to image depth. The percentage improvement shows that as the imaging plane went deeper, the improvement became larger in general. The maximum improvement of the 90–100% intensity pixels was 123% at 80 μm depth. In terms of the increment of photon count, the result at 20 μm depth showed the greatest increase due to generally higher signal levels. At greater depths, the scattering of excitation and emission photons was dominant, and the improvement due to using AO became less significant. In terms of improvement at the center compared to the edge, it seemed to be uniform, similar to the mouse-tongue result.

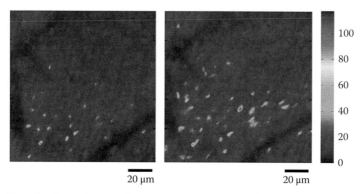

FIGURE 16.10   (**See color insert.**) Mouse-heart images at 80 μm depth without and with AO compensation, respectively.

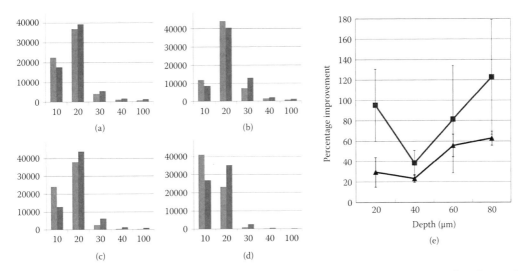

FIGURE 16.11 Signal improvement after AO compensation. (a–d) Histograms for the number of pixels according to their intensity; the *x*-axis represents the intensity of pixels the same as in Figure 16.9, except 100 (100 means 40%–100% intensity pixels in the image). The *y*-axis represents the number of pixels in the intensity range. The left-placed columns show the number of pixels before the compensation, and the right-placed columns show the result after compensation. Each histogram was normalized to itself. The distribution at (a) 20 µm, (b) 40 µm, (c) 60 µm, and (d) 80 µm imaging depth. (e) Percentage improvement according to imaging depth. The solid line with triangles shows the improvement with background rejection (only the fluorescent area was calculated), and the solid line with squares shows the improvement only with 90–100% intensity pixels.

16.3.5 Neuronal Imaging in Mouse-Brain Slices Using Adaptive Optics–Compensated Two-Photon Microscopy

Both muscle tissues have high scattering coefficients that appear to be the major limiting factor in deep tissue two-photon imaging. It is of interest to also study tissue specimens that have significantly lower scattering, and therefore allow deeper imaging. For this purpose, we chose mouse-brain slices with neurons expressing green fluorescent protein (GFP) driven by a Thy-1 promoter. Adult Thy-1-GFP-S mice were perfused transcardially with 4% paraformaldehyde in PBS. The brains were removed, and post-fixed overnight in 4% paraformaldehyde and sectioned coronally at 200 µm thickness using a vibratome. The wavelength of the excitation light was 890 nm, and a 40× water-immersion objective lens (Achroplan IR, 0.8 N; Zeiss) was used with a number 11 2 coverslip. The emission signal was filtered with a green filter. The field size was 120 × 120 µm over 256 × 256 pixels, and the dwell time was 40 µs.

Figure 16.12 shows representative mouse-brain images at 50 µm depth with and without AO compensation. The mouse-brain specimen contained sparsely distributed GFP-expressing neurons. The image on the left is the uncompensated image, and the image on the right is the compensated image, with both processed for background rejection. AO compensation was performed at the center pixel, and scanning was done with the same deformable mirror shape. The compensated image showed a higher signal than the uncompensated one. To compare the signal level of two images, the intensity distributions for all pixels are shown in Figure 16.13. Figure 16.13a and b show histograms for the intensity distributions before and after compensation. Since the structure imaged is sparse, the histogram is dominated by zero-intensity pixels. Two additional figures below Figure 16.13a and b show histogram distributions, excluding the lowest intensity range. Again, the number of brighter pixels increased, and the number of darker pixels decreased, after AO compensation. For the improvement trend according to imaging depth, mean photon counts were calculated with background rejection. Figure 16.13c shows the percentage improvement from the mean photon counts according to the imaging depth. We see a general trend of increasing percentage improvement as a function of imaging depth similar to the other two tissue cases. Since the scattering coefficient in the brain is much less than that of muscles, the achievable imaging depth is deeper with equivalent excitation power. However, the improvement at the maximum achievable depth with and without AO compensation tissues was similar in all cases, ranging up to about twofold in terms of the maximum signal strength. In a previous report, fluorescently labeled blood plasma in wild-type zebra fish was imaged [8]. The aberrations introduced by the specimens were mainly astigmatism 254 nm at 200 μm imaging depth, and the fluorescence signal at the center of a blood vessel was improved almost twofold by wavefront correction with the coherence-gated wavefront sensing method. In our measurement, the main aberration was also astigmatism 125 nm at 150 μm imaging depth, and the signal improvement was about the same, which was comparable with the previous result.

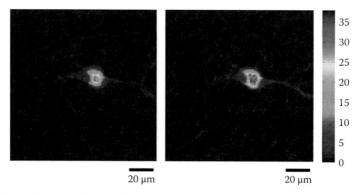

FIGURE 16.12 (**See color insert.**) Mouse-brain images at 50 μm depth without and with AO compensation, respectively.

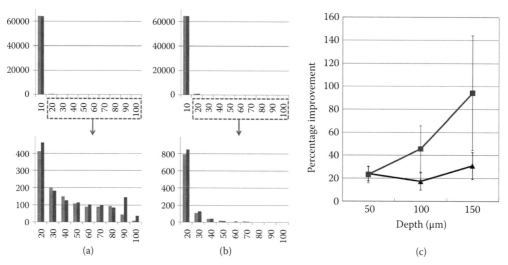

FIGURE 16.13 Signal improvement after AO compensation. (a, b) Histograms for the number of pixels according to their intensity; the *x*-axis represents intensity of the pixels, and the *y*-axis represents the number of pixels in the intensity range. The left-placed columns show the number of pixels before compensation, and the right-placed columns show the result after compensation. Each histogram was normalized to itself. The distribution at (a) 50 μm and (b) 150 μm imaging depth. The bottom figures of panels a and b are detailed distributions in the selected range. (c) Percentage improvement according to the imaging depth. The solid line with triangles shows the improvement with background rejection (only the fluorescent area was calculated), and the solid line with squares shows the improvement only with 90–100% intensity pixels.

16.4 Conclusions

Wavefront distortion by tissue specimens with inhomogeneous optical properties can be measured by an SHWS using a confocal depth selection mechanism. The measured wavefront distortion can then be corrected by a deformable mirror integrated into the microscope system. In specimens with large aberration, such as artificial specimens, significant improvement in signal strength and resolution can be achieved. In more realistic tissue specimens, the imaging depth is limited to 100–200 μm by the respective scattering coefficients of the respective tissue type. For example, in general, achievable imaging depth for the brain tissue is approximately three times that of the muscles. The measured aberration for these tissues at this depth is similar to those reported by other investigators [8,20]. The corresponding improvement in terms of signal strength is on the order of 20–70%. However, the improvement on the peak signals (90–100% strength intensity signals) is up to twofold. This means the contrast of images was increased after AO compensation; therefore, the images became more vivid, and this is one of the most essential advantages of the AO system in biological imaging. Our data appears to imply two related conclusions. First, tissue scattering is the primary factor that determines the maximum imaging depth in tissue imaging by a two-photon microscope. While the effect of scattering is exponential with depth, the accumulated aberration as a function of depth appears to remain at the same order of magnitude or varies at the most linearly. Therefore, the scattering effect is always the dominant factor. Second, AO control can improve the imaging signal at depth. However, typical improvement is relatively modest, especially for highly scattering specimens, because the very limited imaging depth implies relative little accumulated aberration, limiting the improvement that can be obtained by using AO correction. These conclusions are consistent with most of the AO-compensated nonlinear optical imaging studies. The only exceptions are the application of AO compensation in CARS microscopes. One potential explanation may be that the CARS signal is highly dependent on the overlap of the excitation volumes of the pump and probe beams, and aberration may result in spatial mismatch. The correction for aberration can optimize volume overlap and result in the observed sixfold improvement [5]. It might therefore be important to ask if AO correction is important for two-photon microscopy. Our data seem to indicate that for most tissue types with high scattering-coefficient and shallow imaging-depth, adaptive correction can improve the signal strength. However, one should note that if the excitation power is not limited and photodamage is not an issue, this level of improvement can be equivalently achieved by less than a twofold excitation power increase. In the case that higher excitation power cannot be applied, AO can provide some improvement, although at the expense of an increase in instrument complexity and cost. It may also be useful to consider in what situations AO correction would be useful for two-photon imaging. Clearly, in tissues with relatively low scattering-coefficients that allow deep imaging, AO is useful to correct for the larger aberration present. This is, of course, the well-known case with the classic examples of imaging in organs, such as the eye [21–23]. Other potential application areas are tissues that induce unusually high aberrations. Another classic example is the skin, where the stratified layers have very different indices of refraction and result in significant aberration, even in relatively shallow depths. It would be interesting to examine the possible application of AO compensation in skin imaging. In addition, there are several other potentially interesting applications of AO in nonlinear microscopy imaging of tissues. First, tissue scattering, due to a combination of Rayleigh and Mie processes, is lower for light at longer wavelength regions, which corresponds to deeper penetration depth. If the AO system is combined with a two-photon microscope operating in the 1.2–1.3 μm region and imaging near-infrared emitting fluorophores, one may achieve deeper imaging in tissues with corresponding higher aberration effect. In this case, one can expect AO compensation to be more important. Second, if the scattering effect becomes negligible, then the aberration due to the inhomogeneous refractive index of a sample becomes the major factor for image degradation. One obvious example is in specimens with extensive optical clearing that can dramatically decrease tissue scattering by dehydration of the specimen and the addition of an index matching medium [24,25]. Optical clearing reduces scattering by index-matching nanometer-scale-level inhomogeneity in the specimen. With the optical clearing, the imaging depth can clearly be increased by orders of magnitude. Given the deeper imaging depth, AO systems may play potentially a more important role, provided that this optical clearing does not similarly eliminate the index-of-refraction inhomogeneity on the optical-wavelength scale. Finally, the AO system may be important for a number of imaging applications involving microfluidic devices. In the fabrication of many microfluidic devices, the material type or thickness used in the fabrication of the device may not match the requirement of the microscope objectives that are typically designed for imaging through 0.17-mm-thick glass. In these cases, significant aberrations can be induced, and the use of an AO system may result in an improvement in image quality in these devices.

Acknowledgments

Parts of this work appeared in *Journal of Biomedical Optics* 15(4), 040622 (2010).

References

1. Denk W, Strickler JH, Webb WW. 2-Photon laser scanning fluorescence microscopy. *Science*. 248(4951), 73–6 (1990).
2. Booth MJ, Neil MAA, Wilson T. Aberration correction for confocal imaging in refractive-index-mismatched media. *J Microsc*. 192(2), 90–8 (1998).
3. Kam Z, Kner P, Agard D, Sedat JW. Modelling the application of adaptive optics to wide-field microscope live imaging. *J Microsc*. 226(Pt 1), 33–42 (2007).
4. Kner P, Sedat JW, Agard DA, Kam Z. High-resolution wide-field microscopy with adaptive optics for spherical aberration correction and motionless focusing. *J Microsc*. 237(2), 136–47 (2010).
5. Wright AJ, Poland SP, Girkin JM, Freudiger CW, Evans CL, Xie XS. Adaptive optics for enhanced signal in CARS microscopy. *Opt Express*. 15(26), 18209–19 (2007).
6. So PT, Dong CY, Masters BR, Berland KM. Two-photon excitation fluorescence microscopy. *Annu Rev Biomed Eng*. 2, 399–429 (2000).
7. Rueckel M, Denk W. Properties of coherence-gated wavefront sensing. *J Opt Soc Am A Opt Image Sci Vis*. 24(11), 3517–29 (2007).
8. Rueckel M, Mack-Bucher JA, Denk W. Adaptive wavefront correction in two-photon microscopy using coherence-gated wavefront sensing. *Proc Natl Acad Sci U S A*. 103(46), 17137–42 (2006).
9. Booth MJ, Neil MA, Wilson T. New modal wave-front sensor: Application to adaptive confocal fluorescence microscopy and two-photon excitation fluorescence microscopy. *J Opt Soc Am A Opt Image Sci Vis*. 19(10), 2112–20 (2002).
10. Wilson T, Carlini AR. Size of the detector in confocal imaging-systems. *Opt Lett*. 12(4), 227–9 (1987).
11. Sandison DR, Webb WW. Background rejection and signal-to-noise optimization in confocal and alternative fluorescence microscopes. *Appl Opt*. 33(4), 603–15 (1994).
12. Artal P, Marcos S, Navarro R, Williams DR. Odd aberrations and double-pass measurements of retinal image quality. *J Opt Soc Am A Opt Image Sci Vis*. 12(2), 195–201 (1995).
13. Booth MJ. Adaptive optics in microscopy. *Philos Transact A Math Phys Eng Sci*. 365(1861), 16 (2007).
14. Artal P, Iglesias I, Lopez-Gil N, Green DG. Double-pass measurements of the retinal-image quality with unequal entrance and exit pupil sizes and the reversibility of the eye's optical system. *J Opt Soc Am A Opt Image Sci Vis*. 12(10), 2358–66 (1995).
15. Marsh PN, Burns D, Girkin JM. Practical implementation of adaptive optics in multiphoton microscopy. *Opt Express*. 11, 8 (2003).
16. Hell S, Reiner G, Cremer C, Stelzer EHK. Aberrations in confocal fluorescence microscopy induced by mismatches in refractive-index. *J Microsc Oxford*. 169, 391–405 (1993).
17. Fwu PT, Wang PH, Tung CK, Dong CY. Effects of index-mismatch-induced spherical aberration in pump-probe microscopic image formation. *Appl Opt*. 44(20), 4220–7 (2005).

18. Kao HP, Verkman AS. Tracking of single fluorescent particles in 3 dimensions—use of cylindrical optics to encode particle position. *Biophys J*. 67(3), 1291–300 (1994).

19. Ragan T, Huang HD, So P, Gratton E. 3D particle tracking on a two-photon microscope. *J Fluoresc*. 16(3), 325–36 (2006).

20. Debarre D, Botcherby EJ, Watanabe T, Srinivas S, Booth MJ, Wilson T. Image-based adaptive optics for two-photon microscopy. *Opt Lett*. 34(16), 2495–7 (2009).

21. Liang JZ, Williams DR, Miller DT. High resolution imaging of the living human retina with adaptive optics. *Invest Ophth Vis Sci*. 38(4), 55 (1997).

22. Doble N, Yoon G, Chen L, Bierden P, Singer B, Olivier S, Williams DR. Use of a microelectromechanical mirror for adaptive optics in the human eye. *Opt Lett*. 27(17), 1537–9 (2002).

23. Roorda A, Romero-Borja F, Donnelly Iii W, Queener H, Hebert T, Campbell M. Adaptive optics scanning laser ophthalmoscopy. *Opt Express*. 10(9), 405–12 (2002).

24. Tuchin VV. *Optical Clearing of Tissues and Blood*. SPIE Press. Bellingham, WA (2005).

25. Dodt HU, Leischner U, Schierloh A, Jahrling N, Mauch CP, Deininger K, Deussing JM, Eder M, Zieglgansberger W, Becker K. Ultramicroscopy: Three-dimensional visualization of neuronal networks in the whole mouse brain. *Nat Methods*. 4(4), 331–6 (2007).

17

Adaptive Optical Microscopy Using Direct Wavefront Measurements

Oscar Azucena
University of California at Santa Cruz

Xiaodong Tao
University of California at Santa Cruz

Joel A. Kubby
University of California at Santa Cruz

17.1 Introduction

The telescope and the microscope have allowed scientists to study the universe and the world we live in (Van Helden 1977). Both microscopes and telescopes suffer from optical aberrations created by changes in the index of refraction in the optical path. Dunn and Richards-Kortum (1996) have studied the changes in the index of refraction inside biological tissues. Their results indicate that structures with large changes in the index of refraction have large contrast ratios as long as they are near the surface of the biological sample. These changes in the index of refraction degrade the contrast ratio for objects lying much deeper in the tissue. The effect is much worse for samples with a lot of fine structures, since they introduce higher-order aberrations in the images. Schwertner et al. (2004) measured the specimen-induced aberrations for a range of typical biological samples (Schwertner 2007). Their results indicate that the Zernike-mode representation is a useful tool for describing these aberrations. Their results also indicate that lower-order aberrations are more pronounced than higher-order ones and that spherical aberrations dominate overall.

Adaptive optics (AO) is a method used in the telescope for improving astronomical images. Babcock (1953) first introduced the idea of improving astronomical seeing by compensating for the atmosphere-induced aberrations. His proposal was to measure the deviations of the light rays from all parts of the mirror and feed that information back so as to locally correct for the deviations. Although the idea was scientifically sound, it had a few minor technical complications, and it was not put into action until 20 years later when the first real-time AO system was used for national-defense applications (Hardy

1998). AO might have been conceived for the purpose of improving astronomical imaging, but other scientists soon realized the importance of this technology in other areas of research. In particular, vision science was one of those fields where AO has enlightened curious researchers. The first major obstacle in adapting AO to vision science was to find a reasonable reference source for measuring the wavefront. The first Shack-Hartmann wavefront (SHWF) sensor measurements for vision science were realized by Junzhong Liang by imaging a laser spot onto the retina (Liang et al. 1994). A few years later, Liang, Williams, and Miller (1997) finally constructed the first closed-loop AO system for vision science.

The idea of using AO for microscopes is relatively new and a lot of work is still needed. Most AO microscopy systems have so far not directly measured the wavefront because of the complexity of adding a wavefront sensor in an optical system and the lack of a natural point-source reference such as the "guide star" used in astronomy. Instead, most AO microscopy systems have corrected the wavefront by optimizing a signal received at a photodetector (Booth 2007). Debarre et al. (2008) have successfully incorporated an AO system into a structured illumination microscope. In his research, Debarre used a wavefront sensorless AO technique in which each mode is corrected independently through the sequential optimization of an image-quality metric. Scientists and engineers have also been investigating ways to implement AO in two-photon microscopes (Marsh, Burns, & Girkin 2003; Rueckel, Mack-Bucher, & Denk 2006). Rueckel et al. used a coherence gate to selectively pass only the light backscattered near the focus to directly measure the wavefront using an interferometer.

Although there are a lot of important researches being done in AO microscopy, many of the AO systems are specific to each microscope, and a universal method for measuring the wavefront (or the results of the correction algorithm) is not currently available. Booth (2007) described some of the difficulties associated with the utilization of a SHWF sensor in AO microscopy. Most of these difficulties can be overcome if a suitable fluorescent point-source could be found. Beverage and others found a suitable method for measuring the wavefront of a microscope objective lens by using fluorescent microspheres as reference sources (Beverage, Shack, & Descour 2002). In his research, Beverage established that bigger beads (larger than diffraction limit) could be used, allowing for more light to measure the wavefront. The size of the beads, d_{bead}, should be smaller than the diffraction limit of the wavefront sensor when imaged through the microscope objective:

$$d_{bead} = 2.44 \frac{\lambda}{2NA_{ob}} \frac{D_o}{d_{LA}} = d_{DLO} \cdot N_{D/d} \qquad (17.1)$$

where λ is the wavelength at which the beads are emitting, NA_{ob} is the numerical aperture (NA) of the objective, D_o is the limiting aperture of the objective, and d_{LA} is the lenslet array pitch. This could also be represented as the diffraction limit of the objective (d_{DLO}) times the number of subapertures across the limiting pupil. Using this technique, we can measure the aberration introduced by a biological sample by injecting a fluorescent bead into the sample. To reduce the effect of the scattered light, a field stop can be used. The field stop can be placed at the image focal plane just before the wavefront sensor. The system must be designed so that only the wavefront sensor sees the field stop. The field stop will act as a spatial filter, so that the wavefront sensor will not see spatial frequencies above the size of the field stop. To reduce aliasing at the SHWF sensor, the size of the field stop cannot be smaller than $2d_{LA}$ as projected onto the image plane (Poyneer, Gavel, & Brase 2002).

17.1.1 Fluorescent Beads

Fluorescent microspheres, with a large variety of colors, are typically used in biology to study different biological characteristics (Invitrogen Corporation 2010; Rothwell & Sullivan 2000; Guldband et al. 2010; DeMarais, Oldis, & Quatrro, 2005; Rowning et al. 1997; Kalpin, Daily, & Sullivan 1994). The microspheres are made of polymers and are impregnated with different fluorescent dyes. The microspheres can be engineered with coatings to preserve them in different conditions and can be made to target different bio-logical tissues, organelles, cell walls, or other biological structures (Invitrogen Corporation, 2010). They can be introduced into the sample by different mechanisms such as negative pressure injection, pressure injection, matrotrophycally, diffusion, and others (Guldband et al. 2010; DeMarais, Oldis, & Quatrro 2005; Rowning et al. 1997). In particular, fluorescent microspheres have been injected previously in *Drosophila* embryos (Kalpin, Daily, & Sullivan 1994). Sufficiently small fluorescent microspheres, as described earlier, are diffraction-limited when imaged by the SHWF sensor, enabling their use as point-source reference beacons for the operation of the SHWF sensor. Azucena et al. (2010) show that multiple beads can also be used to directly measure the wavefront. The wavefront measurements from multiple beads and a single bead differ only in the higher-order aberrations (e.g., above the seventh radial order Zernike).

17.1.2 Shack-Hartmann Wavefront Sensor

A SHWF sensor samples the wavefront at different points of the pupil by using a lenslet array to measure the mean slope across each subaperture as shown in Figure 17.1. On the one hand, Figure 17.1a shows a flat wavefront impinging onto the SHWF sensor; as can be seen, there is a slope of zero measured at each of the Hartmann spots. Figure 17.1b, on the other hand, shows a distorted wavefront impinging onto the SHWF sensor. The spot displacement at each subaperture is directly proportional to the product of the mean slope at the subaperture and the focal length of the lenslet. The wavefront can be reconstructed using Equations 17.2 and 17.3 (Liang et al., 1994):

$$\frac{\partial w\left(x_i, y_i\right)}{\partial x} = \frac{\Delta x_i}{f_{LA}} \tag{17.2}$$

$$\frac{\partial w\left(x_i, y_i\right)}{\partial y} = \frac{\Delta y_i}{f_{LA}} \tag{17.3}$$

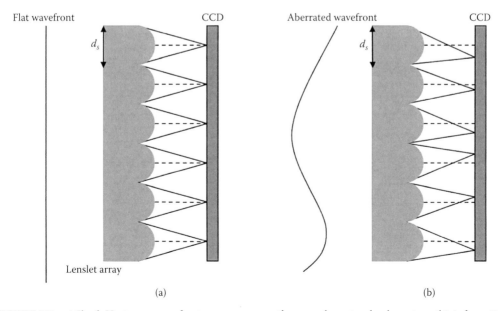

FIGURE 17.1 A Shack-Hartmann wavefront sensor measures the mean slope at each subaperture; this information is used to reconstruct the wavefront over the whole aperture. (a) Flat wavefront; (b) distorted wavefront.

where f_{LA} is the focal length of the lenslet array, Δx and Δy are the slope measurements at the subaperture i in the x and y directions, respectively, and $w(x_i, y_i)$ is the wavefront at the point x_i and y_i.

17.1.3 Cross-Correlation Centroiding

Many astronomy AO systems use a quad-cell detector in the SHWF sensor to determine the mean slope at each subaperture (Hardy 1998). Although the quad cell may be very easy to implement, it images the subaperture into four pixels, so it can be used to detect only approximately half a wavelength of tilt at each subaperture. To overcome the limitations of the quad cell, more pixels per subaperture are needed to better sample the Hartmann spots. Introducing more pixels does have the negative effect of increasing the effects of noise, which current detectors have not managed to overcome. On the other hand, there are many algorithms that help overcome the effects of noise (Thomas et al. 2006; Adkin, Azucena, & Nelson 2006; Poyneer 2003).

The cross-correlation algorithm requires a reference Hartmann spot with which all the Hartmann spots will be compared to. The reference image shown in Figure 17.2 is first analyzed to acquire the image of a Hartmann spot. Figure 17.3 shows the average Hartmann spot of the reference image in Figure 17.12. In a real time AO system, the reference spot should not be oversampled too much, since a lot of resources are required to analyze the data. Thomas et al. (2006) suggested that the Hartmann spot

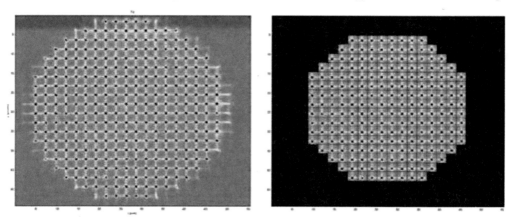

FIGURE 17.2   SHWF sensor image with a microsphere as the reference source. *Left*: The raw WFS data; *Right*: The same image with the subapertures separated.

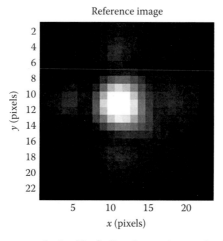

FIGURE 17.3   Average Hartmann spot, obtained by finding the maximum of each subaperture in Figure 17.2 and adding all the images around their corresponding maxima.

should be imaged into an area of 2 × 2 pixels to acquire the most information without losing precision in centroiding. One of the advantages of using the cross-correlation algorithm is that the computation can be done in the Fourier domain, taking advantage of the Fast Fourier Transform (FFT) algorithm available to improve the speed of the AO system. Although the reference function can be modeled as a Gaussian spot, it can be shown that using a real-time Hartmann-spot measurement from the SHWF sensor images can improve the accuracy of the centroiding algorithm (Thomas et al. 2006). This is mainly due to the amount of information afforded by the real-time spot measurement compared to that of a Gaussian image. The final step needed to measure the amount of slope at each subaperture using the cross-correlation algorithm is to find the maximum for each subaperture (Thomas et al. 2006).

17.1.4 Reconstruction

There are various ways of estimating a wavefront from the Hartmann slopes (Hardy 1998; Gavel 2003). Two essential pieces of information are needed for this: (1) the phase difference (slope measurements times subaperture size) from each subaperture and (2) the geometrical layout of the subapertures. The wavefront can then be calculated by relating the slope measurement to the phases at the edge of the subaperture in the correct geometrical order. A method for directly obtaining the deformable mirror (DM) commands from wavefront sensor measurements is described by Tyson (1998). First a mask with the subapertures must be created; this will generate the geometric layout of the subapertures in the aperture. The next step is to measure and record the response of all the subaperture slope changes while actuating each actuator. The results obtained are set of linear equations that show the response of the wavefront sensor for each actuator command known as the poke matrix (also known as the actuator influence matrix). The DM commands can then be obtained by solving the equation

$$\mathbf{s} = \mathbf{Av} \tag{17.4}$$

where $\mathbf{s}$ is an n-size vector obtained from the SHWF sensor slope measurements, $\mathbf{v}$ is an m-size vector with the DM actuator commands, and $\mathbf{A}$ is an $n \times m$–sized poke matrix. In the linear approximation, Equation 17.4 can be pseudo-inverted to obtain an estimate of the DM command matrix. Note that the DMs are nonlinear devices, but the matrix given in Equation 17.4 performs well in a closed-loop system, as only very small voltage changes occur, thus reducing the nonlinear effects. There are various methods for inverting the matrix $\mathbf{A}$, including singular value decomposition (SVD). The advantage of using SVD is that the mode space can be directly calculated. The noisier modes, and all the null space modes by default, can then be removed by setting a threshold on the singular value space (Gavel 2003).

Figure 17.4a shows a poke matrix obtained by using the method described here. The process begins by poking an actuator to a predetermined voltage (V); each of the subapertures' slope changes is then recorded. This process is repeated 100 times for each actuator, and the recorded data are then averaged to reduce the effect of noise. The slope changes are determined using the cross-correlation centroiding algorithm. Further conditioning of the poke matrix is performed by thresholding the data to 20% of the maximum slope changes measured for all actuators. For each actuator poke, there will be an area in the SHWF sensor that will show stronger slope changes (i.e., an influence function). In AO, a 15–20% influence function between actuators is usually considered good as this allows for high spatial deformations to be well reproduced by the DM. The thresholding step mentioned earlier essentially windows the slope measurements to an area near the center of the actuator poke with a 20% slope influence matrix. Considering a larger area can introduce higher spatial frequencies that are dominated by noise, thus introducing noisier modes into our singular-value space.

Figure 17.4b shows the singular-value pseudo inverse of the poke matrix shown in Figure 17.4a. The pseudo inverse has the singular-value space shown in Figure 17.4c, which has been regularized to remove the singular-value modes that are lower than 15% of the maximum mode as described by Gavel (2003). By multiplying the poke matrix in Figure 17.4a and its pseudo inverse in Figure 17.4b, the actuator space

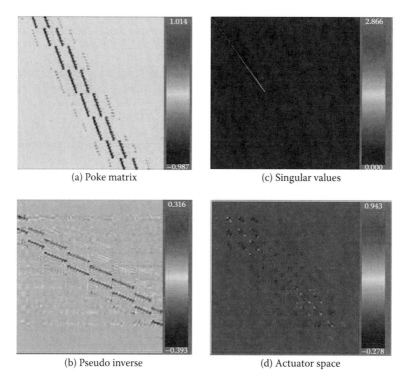

(a) Poke matrix

(c) Singular values

(b) Pseudo inverse

(d) Actuator space

FIGURE 17.4 (**See color insert.**) (a) Poke matrix obtained by the method described in the main text. (b) SVD inverse of the matrix in (a). (c) Singular-value modes with modes lower than 15% of maximum mode have been removed. (d) Actuator space obtained by multiplying (a) and (b).

can be obtained as shown in Figure 17.4d. By analyzing the diagonal of the product matrix, each actuator response can be determined. In particular, the actuators that have little or no effect on the system can be detected, and the actuators that are too far from the aperture can be removed as they have no registration on the poke matrix.

17.2 AO Wide-Field Microscope

Figure 17.5 shows the design of an AO wide-field microscope. An AO system was added to the back port of an Olympus IX71 inverted microscope (Olympus Microscope, Center Valley, PA). This allowed use of the side image port for point spread function (PSF) measurements, which were compared with the PSF viewed after the AO system to ensure that the AO system did not add aberrations. Using a camera with very small pixels (flea2 with 4.65 μm pixels, Point Grey, NY), we were able to verify a very close match between the PSF before and after the AO system. The AO system was designed around an Olympus 60X oil-immersion objective (Ob) with an NA of 1.42 and a working distance of 0.15 mm. Lenses L1 and L2 have 180 and 85 mm focal lengths, respectively, and are used to image the back pupil of the 60× objective onto the DM (Boston Micromachines, Boston, MA). The DM has 140 actuators on a square array with a pitch of 400 μm, a stroke of 3.5 μm, and an aperture of 4.4 mm. Note that 0.5 μm of stroke was lost to AO path compensation and flattening of the DM. L3 and L4 are 275- and 225-mm focal length lenses, respectively, and are used to reimage the back pupil of the objective onto the SHWF sensor. The system has two illumination and imaging arms: the first is a science arm in which we used a set of filters F1 and F2 (Semrock, Rochester, NY) to redirect a beam from an argon 488 nm laser (blue laser) to the objective for excitation of a green fluorescent sample. The light emitted from the green fluorescent sample is imaged by the green science camera (green SC). Filter F3 (Semrock) is used to redirect the HeNe 632.8 nm laser (red

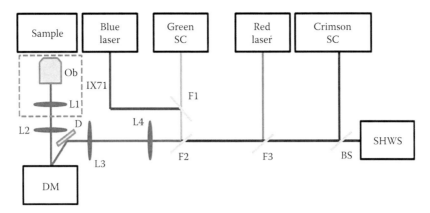

FIGURE 17.5 AO wide-field microscope setup. DM, SHWS, 488 nm laser (blue laser), green flourescent science camera (green SC), HeNe 632.8 nm laser (red laser), crimson flourescent science camera (Crimson SC). L1, L2, L3, and L4 are 180, 85, 275, and 225 mm focal length lenses, respectively. Fold mirror D helps to bring the optical path into alignment for the SHWS.

FIGURE 17.6 Adaptive optics wide-field microscope set up with Olympus IX71 inverted microscope and Boston Micromachines DM.

laser) through a confocal illuminator (not shown) onto the optical path for excitation of the crimson reference beads (Azucena et al. 2010, 2011a). This confocal illuminator allows us to illuminate a single crimson reference bead to create a single diffraction-limited spot. The beam splitter (BS) lets 90% of the emitted light coming from the crimson reference beads go to the SHWF sensor for wavefront measurement and 10% for imaging in the Crimson Science Camera (Crimson SC). The SHWF sensor is composed of a 44 × 44 element lenslet array each with a focal length of 24 mm and a diameter $d_{LA} = 328$ μm (AOA Inc., Cambridge, MA), and a cooled CCD camera (Roper Scientific, Acton, NJ). Figure 17.6 shows the setup on the optical table, highlighting a few of the components like the DM, science cameras, and SHWF sensor.

17.2.1 Fluorescent Microsphere Reference Beacons (Guide Stars)

We have developed a microsphere injection process, which works very well for introducing reference beacons, or guide stars, into live embryos. This process does not harm the embryos as they have the ability to heal the wound around the injection site. Using a 1:1000 concentration microsphere solution assures that there will be a microsphere within 10 μm of the center of the injection discharge. The microspheres spread in a random manner around the discharge site and can usually be found to spread

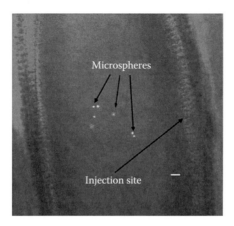

FIGURE 17.7   Combination of a differential interference contrast image and a confocal image of injected microspheres in fruit-fly embryo 40 µm below the surface of the embryo. The white bar in the figure is 10 µm long.

throughout the embryo. Figure 17.7, a combination of a differential interference contrast (DIC) image and a confocal image, both taken with a Leitz inverted photoscope equipped with laser confocal imaging system (Leica Microsystems, Bannockburn, IL), illustrates that relative to the injection site and the embryo walls, the microspheres have spread randomly inside the embryo. A higher microsphere concentration could also be used as the optical setup shown in Figure 17.5 allows for the laser to illuminate one bead at the time. A bead concentration as high as 1:100 has been used, and the results show many more beads available in the field of view (FOV).

17.2.2 Wavefront Measurements

A reference Hartmann sensor image was obtained to cancel the aberrations introduced by the optical setup and coverslip. The reference image was taken by imaging a single fluorescent bead onto the Hartmann sensor. The bead was dried onto a glass slide and imaged with the coverslip and mounting media. The image was processed to obtain the location of the Hartmann spots using the cross-correlation centroiding algorithm described earlier. For each wavefront measurement, a new Hartmann sensor image was acquired with the sample prepared as described in the Section 17.2.1. The new measurement was then processed to determine the displacement of the Hartmann spots (slope measurements) relative to the reference image described earlier. The slope measurements were finally processed to obtain the wavefront by using an FFT reconstruction algorithm (Poyneer, Gavel, & Brase 2002). For each measurement, the peak-to-valley (PV) and the root-mean-square (RMS) wavefront errors were collected. The wavefront function was also expanded into the Zernike's circle polynomials to determine the relative strength of the different modes (Porter et al. 2006; Poyneer 2003). The Zernike polynomials are normalized and indexed as described by Porter et al. (2006). The wavefront measurements were also used to analyze the PSF by taking the Fourier transform of the complex pupil function:

$$
\mathrm{PSF}(x,y) = \frac{\left| \mathrm{FT}\left\{ P(x',y') \cdot \exp\left(i\frac{2\pi}{\lambda} w(x',y') \right) \right\} \right|^2 \Big|_{\xi=\frac{x}{\lambda f},\, \eta=\frac{y}{\lambda f}}}{\left| \mathrm{FT}\left\{ P(x',y') \right\} \right|^2 \Big|_{\xi=0,\, \eta=0}}
\tag{17.5}
$$

where P is one inside the pupil and zero everywhere else, x' and y' are the coordinates at the pupil plane, ξ and η are the spatial frequencies in the transform domain, x and y are the coordinates at the image plane, w is the wavefront measurement, and λ is the wavelength at which the measurement was taken.

A measurement of the wavefront from a 1 μm crimson fluorescent microsphere embedded 45 μm below the surface of a *Drosophila* embryo using a dry 20× (0.40 NA) objective lens is shown in Figure 17.8. The distance between points is equal to the subaperture diameter, d_{LA}, for a total of 216 apertures on the circular pupil. As can be seen from Figure 17.8, the PV wavefront error is ~0.56 μm and the RMS wavefront error for this measurement is 0.09 μm. Figure 17.9 shows the Zernike coefficients for the wavefront shown in Figure 17.8. As can

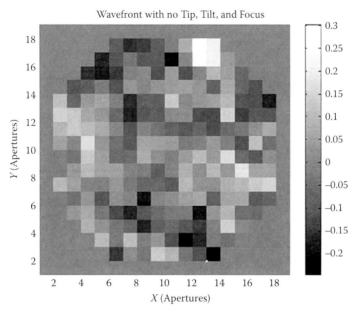

FIGURE 17.8 A wavefront measurement from a 1 μm fluorescent microsphere embedded 45 μm below the surface of a *Drosophila* embryo using a 20× (0.40 NA) objective lens with tip, tilt, and focus subtracted. The x and y axes are scaled to the subaperture diameter. The grayscale is labeled in micrometers.

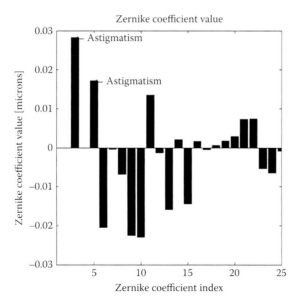

FIGURE 17.9 Zernike coefficient values for the wavefront shown in Figure 17.8. Astigmatism is labeled.

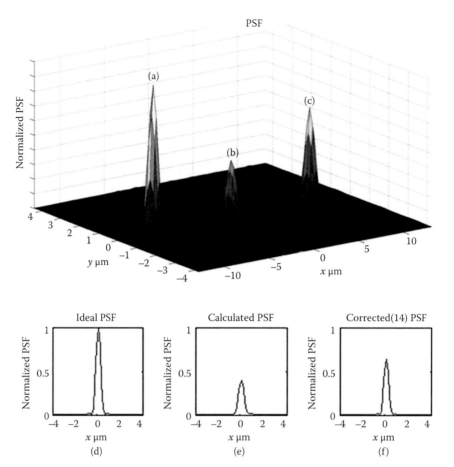

FIGURE 17.10 PSF analysis. (a) Calculated using a flat wavefront. (b) Calculated using the wavefront shown in Figure 17.8. (c) Calculated by removing the first 14 Zernike modes of Figure 17.8. (d) Cross-sectional view of (a). (e) Cross-sectional view of (b). (f) Cross-sectional view of (c).

be seen from Figures 17.8 and 17.9, astigmatism and spherical aberrations dominate the wavefront error. This is mainly due to the index mismatches in the optical path and the curved body of the embryo, which mostly introduced the lower-order aberrations. Note that the optical aberrations due to the coverslip and air-glass interface, including tip, tilt, and focus, have been removed by the reference image. A reassuring sign shown in Figure 17.9 is that the amplitude of the higher-order aberrations are decreasing, and therefore, by correcting a finite number of Zernike modes, the imaging qualities of the optical system can be expected to improve. Note that the spatial resolution for the wavefront measurement setup for the 20× objective with a limiting aperture $D = 5.9$ mm is the Zernike mode 200 so that the decreasing amplitudes shown in Figure 17.9 are not an artifact of the sensor. The spatial resolution of the SHWF sensor is directly proportional to the number of degrees of freedom (i.e., the number of subapertures inside the pupil) (Porter et al. 2006).

Figure 17.10a shows the PSF for an optical system with no aberrations. Figure 17.10b displays the PSF calculated by using Equation 17.5 and the wavefront shown in Figure 17.8. The Strehl ratio is defined as the ratio of the peak intensity of the PSF relative to the peak intensity of the diffraction-limited PSF (Porter et al. 2006). Figure 17.10e shows that the Strehl ratio is approximately 0.37. The effect of removing the first 14 Zernike modes can be seen in Figure 17.10c and f. Using this simulation, we can estimate that correcting the first 14 Zernike modes will improve the Strehl ratio to 0.70.

Table 17.1 shows the statistical data gathered from the measurements taken. Each measurement comes from different fluorescent microspheres ranging in depth from 40 to100 μm below the surface of the

TABLE 17.1 Statistical Data for Dry 20× (0.40 NA) and Dry 40× (0.75 NA) Objectives

No.	PV (μm)	RMS (μm)	S	S(14)
20× 1	0.595	0.092	0.267	0.595
20× 2	0.496	0.076	0.417	0.672
20× 3	0.876	0.127	0.11	0.593
20× 4	0.504	0.063	0.568	0.722
20× 5	0.853	0.097	0.314	0.627
20× 6	0.568	0.076	0.485	0.734
20× 7	0.878	0.127	0.18	0.627
20× 8	0.565	0.081	0.374	0.695
20× 9	0.506	0.089	0.286	0.726
20× Mean	**0.649**	**0.092**	**0.334**	**0.665**
40× 1	0.624	0.089	0.325	0.539
40× 2	0.627	0.076	0.457	0.664
40× 3	0.568	0.089	0.29	0.675
40× 4	1.371	0.161	0.051	0.285
40× 5	1.15	0.189	0.132	0.389
40× 6	1.299	0.132	0.169	0.415
40× Mean	**0.940**	**0.123**	**0.237**	**0.494**

PV, Peak-to-valley; RMS, root-mean-square; S, Strelh; S(14), Strehl after correcting first 14 Zernike modes

embryo. The measurement error for the SHWF sensor was measured to be less than 5% of the wavelength at 647 nm. This was measured by repeating a single measurement 10 times and measuring the RMS error for that one data point. Measurements 20× 1–9 were taken with a 20× objective; measurements 40× 1–6 were taken with a dry 40× objective. The measurements show a maximum PV wavefront error of 0.88 and 1.37 μm for the 20× and 40× lenses, respectively. The maximum RMS wavefront error was 0.13 and 0.19 μm for the 20× and 40× objective lenses, respectively. This demonstrates only some of the typical aberrations that can be encountered for the *Drosophila melanogaster* sample. For a similar study on some of the early phases of this work, please see Azucena et al. (2010). The higher PV and RMS measurement in the 40× objective are mainly due to the spherical aberrations introduced by the higher NA lens. Table 17.1 also shows the Strehl ratio (column four) obtained by finding the global maximum of the PSF image for each measurement using a search algorithm in MATLAB®. By removing different Zernike modes, we can also approximate the effect of removing different amounts of wavefront error. Column 5 in Table 17.1 demonstrates the effect of removing the first 14 Zernike modes from each measurement. The data show that correcting a small number of modes improves the imaging capabilities of the system.

Figure 17.11 shows the statistical data for each Zernike mode for the measurements shown in Table 17.1. The data show a gradual decrease in value with increasing Zernike mode. From this we can verify that low-order aberrations are the main source of wavefront error and that the aberration values are higher in the 40× objective. The gradual decrease in the strength of each Zernike value for higher Zernike modes shows that there is little wavefront aberration introduced for modes higher than 25, which is well within the range of our sensing capabilities. Note that the spatial resolution for the wavefront measurement setup for the 40× objective with a limiting aperture $D = 3$ mm is the Zernike mode 100. This helps to verify the simulation results obtained in Figure 17.10 that correcting only a few low-order Zernike modes helps to improve the Strehl ratio by at least a factor of two. This point will also be shown again in our correction of the wavefront that follows.

17.2.3 Wavefront Corrections

Validation of the wavefront measurements can be obtained by correcting the wavefront and thus closing the loop in our system. Figure 17.12 shows the results of the correction steps, where each correction step

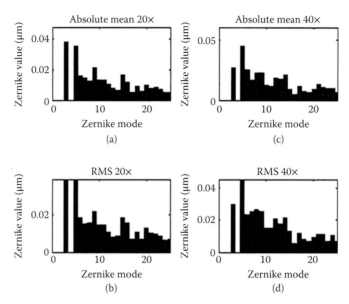

FIGURE 17.11 Zernike statistical data for the measurements in Table 17.1. (a) and (b) The mean of the absolute value and the RMS values for each Zernike mode for the 20× objective (0.40 NA), respectively. (c) and (d) The mean of the absolute value and the RMS values for each Zernike mode for the 40× objective (0.75 NA), respectively.

was 10 ms apart. Each correction was done using the light coming from a single bead to directly measure the wavefront. The measurement was then fed back to the DM by using a proportional gain of 0.4, which was the highest possible gain for this sample before the onset of oscillations (Lyapunov stability criteria) (Slotine & Li 1991). The AO loop gain can be described by the feedback equation

$$E(s) = W(s) - D(s) = W(s)H(s) = \frac{W(s)}{1 + KD(s)G(s)} \tag{17.6}$$

where $E(s)$ is the wavefront error measured by the wavefront sensor, $W(s)$ is the Fourier transform of the input wavefront coming from the sample, $H(s)$ is the transfer function of the AO system, $G(s)$ is the transfer function of wavefront sensor, $D(s)$ is the transfer function of the DM, and K is the gain on the feedback loop (0.4). The goal of the AO system is to reduce the difference between the applied phase on the mirror and the incoming wavefront, the error $E(s)$, thus flattening the wavefront (Hardy 1998; Poyneer, Gavel, & Brase 2002). In AO, DM correction usually requires a gradual change in shape to account for the nonlinearity of the wavefront sensor and the DM. This comes about mainly due to the nonlinear effects of the DM and secondly (usually much smaller) due to the nonlinear effects of the SHWF sensor. The nonlinear effects of the DM come from the nonlinear dependence of the electrostatic actuation force on the applied voltage and plate separation for a parallel plate actuator and the nonlinear restoring force from stretching of both the mechanical spring layer and the mirror surface. Figure 17.12a shows the original PSF of the microsphere before correction taken with the science camera. Figure 17.12b shows the result of correcting for 40% of the measured wavefront error in Figure 17.12a. These steps were repeated until there was no additional significant reduction in wavefront error (i.e., <7 nm). Figure 17.12e demonstrates the results of correcting the wavefront after four steps in the AO loop. Each image has been normalized to its own maximum to clearly show the details of the PSF. The bar in Figure 17.12c is approximately equal to the diffraction limit of the 40× objective, 0.45 μm. The improvement in Strehl was approximately 10×. The relative Strehl ratio S was obtained by measuring the peak intensity in Figure 17.12a divided by the peak intensity in Figure 17.12e using the same integration time Δt for each, as shown in Equation 17.7:

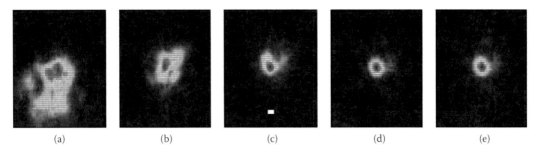

FIGURE 17.12 AO microscope loop correction steps. (a) An uncorrected image of the fluorescent microsphere. (b)–(e) Result of closing the loop by using a loop gain factor of 0.4. The length of the bar in (c) is equal to the diffraction limit of the 40× (0.75 NA) objective lens, 0.45 μm. The bead was located 100 μm beneath the surface of the embryo.

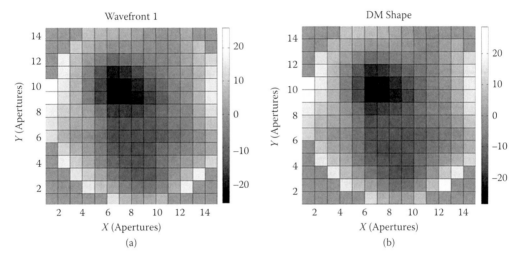

FIGURE 17.13 (a) Initial wavefront measurement. (b) Closed-loop DM wavefront. The legend is scaled in percent wavelength at 650 nm.

$$S_{\text{relative}} = \frac{I_{\text{peak,e}}}{I_{\text{peak,a}}} \tag{17.7}$$

Figure 17.13 shows the initial wavefront measurement (wavefront before correction loops started) and the final shape of the DM for the adaptive loop corrections seen in Figure 17.12. The DM shape was obtained by summing the shape commands that were sent to the DM for each loop step. The small steps in voltage reduce the nonlinear effects from the DM since only small changes in the mirror surface are produced for each time step. There was a 30 nm RMS error difference between the final DM shape and the original wavefront measurement. The final error between these measurements can be partly attributed to the wavefront reconstructor. The effect comes from the lack of measurements outside the edge of the aperture (Poyneer, Gavel, & Brase 2002). The DM wavefront is inverted to help comparison of the wavefront error and the mirror shape.

17.2.4 The Isoplanatic Angle and Half-Width

The isoplanatic angle is a relative measure of the FOV over which the AO system can operate and is defined as (Hardy 1998)

$$\sigma_\theta^2 = \left\langle \left(\varphi(\mathbf{X}, 0) - \varphi(\mathbf{X}, \theta_0) \right)^2 \right\rangle = 1 \text{ rad}^2 \tag{17.8}$$

TABLE 17.2 Isoplanatic Angle Measurements for the 40× Magnification, 0.75 NA Objective Lens

1	14	10.7	1.60	1.90	0.73
2	18	13.8	0.98	1.24	0.77
3	25	19.1	1.69	1.10	1.30
Mean	**19 ± 5.57**	**14.5 ± 4.25**	**1.42 ± 0.39**	**1.41 ± 0.43**	**0.93 ± 0.32**

where φ is the wavefront in radians, X is a vector representing the two-dimensional (2D) coordinates, θ_0 is the isoplanatic angle, and σ_θ^2 is the mean-square error between the measured and observed wavefront. We can determine the isoplanatic half-width by multiplying the isoplanatic angle by the focal length of the objective.

To determine the isoplanatic angle, we took wavefront measurements from two microspheres separated by a distance d. A microsphere was excited by shining a laser on it. Each microsphere was excited individually. Each wavefront sensor measurement was collected over a period of 500 ms, much longer than the typical AO loop bandwidth. This ensures that there is little noise on the data. The standard deviation for each individual wavefront was measured to better than 1% of the wavelength at 647 nm. Table 17.2 shows three different measurements taken with a 40× (0.75 NA) objective lens. The first measurement shows that the wavefront error for the bead located at the center of the FOV RMS(1) is 1.60 radians, the wavefront error for the bead located 14 μm from the center RMS(2) is 1.90 radians, and the wavefront difference between the two measurements RMS(1 − 2) is 0.73 radians. Taking the average of three measurements shows that the isoplanatic half width is 19 ± 5.6 μm. These results show that a reference microsphere together with an AO system can help improve the quality of the images taken, not just at the location of the microsphere but also within a circle 20 μm in radius.

17.2.5 Wavefront Corrections and Fluorescent Imaging at Different Wavelengths

We have taken wavefront measurements at one wavelength (red) that have been used to make corrections to fluorescently labeled beads at a different wavelength (green). Green beads were used to emulate biological structures that are labeled with green fluorescent protein (GFP) using well-known structures that have a size (1 μm) that is near the diffraction limit of the optical system.

The images in Figure 17.14 are of green fluorescent beads that were excited using the 488 nm laser and imaged with the green science camera (Azucena et al. 2011). In Figure 17.14a, the AO system is off, and the mirror was put on a flat position that had been calibrated using an interferometer. The flat position is kept on regardless of correction on or off to reduce systematic aberrations. We can see some details about structures 20 μm below the surface of the embryo, but we are not able to resolve the individual beads that make up the clumps of material shown in the image. In Figure 17.14b, the AO system had been turned on, and we can clearly resolve the individual 1 μm fluorescent beads. Figure 17.14c and d show cross-sectional profiles along the gray lines in Figure 17.14a and b, respectively. These figures show that with the AO system on, we can clearly resolve the individual beads and thus are able to obtain higher resolution structural information. Even though the wavefront aberrations were measured using the crimson beads, the corrections applied to the mirror still improve the image of the green fluorescent beads, which are more than 100 nm apart in wavelength.

17.3 AO Confocal Fluorescence Microscope

The principle of the confocal microscope is shown in Figure 17.15. A patent on this invention was filed in 1957 by Marvin Minsky and was issued in 1961 (Minsky 1961). The focal plane of a coherent light from a laser source is conjugate with the pinhole located in front of the detector. The light is reflected by a dichromatic mirror and focused on the sample through an objective lens. The emission light from the

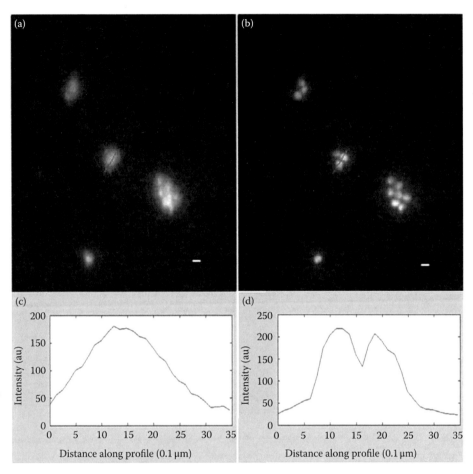

FIGURE 17.14   (**See color insert.**) Real-time AO correction of 1 μm green fluorescent microspheres 20 μm beneath the surface of a fruit-fly embryo, using a 1 μm crimson fluorescent microsphere guide-star located at the center of the image. The size of the scale bar in the figures is 1 μm.

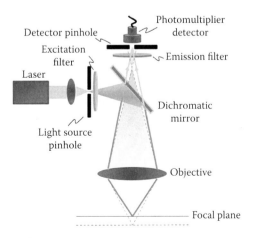

FIGURE 17.15   Diagram of a confocal microscope.

sample passes through the objective lens and dichromatic mirror and focuses on the detector pinhole. Because the focal plane and the detector pinhole all share the same conjugate plane, only the light from the focal plane can pass through the detector pinhole. This minimizes the background fluorescence and improves the contrast of the final image. Compared to the light from the focal plane, the effect of the background light on the final image is negligible. The excitation filter is used for the selection of the excitation wavelength of the light source, which is especially important when a multichannel excitation laser source is used. The dichromatic mirror is used to separate the excitation and emission light paths. The emission filter selects the emission wavelength of the light more narrowly and removes any traces of excitation light. To obtain a 2D image, a confocal microscope is often equipped with a 2D scanner and a Z scanning stage, which can provide lateral sections (x-y plane) and vertical sections (x-z and y-z planes). With a fast resonant scanner or a spinning disk, the confocal microscope can provide real-time imaging.

17.3.1 Optical Setup

Figure 17.16 shows the layout of the AO confocal fluorescence microscope (AOCFM). The whole system was designed and optimized using the optical design software (CODE V). A 60× water-immersion objective with a NA of 1.2 was used (Olympus Microscope, Center Valley, PA) for imaging of both fixed and living cells. The optical system includes three telescope relay subsystems. Lenses L1 and L2 image the exit pupil of the objective onto the Y scanner. Lenses L3 and L4 relay the X scanner conjugate onto the Y scanner. This design minimizes the movement of the scanning beam at the exit pupil of the objective and the emission light at the DM, which is important for accurate wavefront measurement and correction. The lenses L2 and L3 also serve as scanning lenses. The current design is optimized for an optical scanning angle of 4.4°, which provides a FOV of 128 μm on the sample with the 60× objective. By changing the control signal to the scanners, the FOV can be easily adjusted. Lens L5 and L6 image the pupil of the X scanner onto the DM. Lenses L7 and L8 image the pupil of the DM onto the wavefront sensor. The DM (Boston Micromachines) has 140 actuators and 3.5 μm of stroke. The diameter of the effective aperture on the DM used in this design is 4 mm, which is slightly smaller than the 4.4 mm of aperture of the mirror, to decrease edge effect. The exit pupil of the objective is 7.2 mm. To match the aperture of the DM with the objective, the telescope formed by L1 and L2 demagnifies the pupil from 7.2 to 4 mm.

A HeNe laser emits light at 633 nm for excitation of the crimson fluorescent reference beacon. A solid-state laser emits light at 515 nm that excites yellow fluorescent protein (YFP) bred into the sample. F2 and F4 are excitation filters. Light emitted from the reference beacon is passed through filter F1 to the

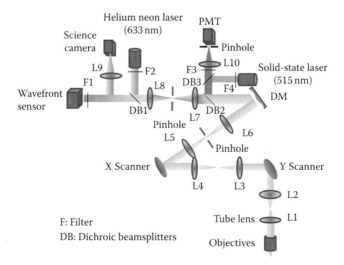

FIGURE 17.16   System layout of the adaptive optics confocal microscope.

wavefront sensor. The SHWF sensor is composed of a 44×44 element lenslet array with a lenslet diameter of 400 μm and focal length of 24 mm (AOA Inc., Cambridge, MA) and an electron multiplying CCD (EMCCD) camera (Cascade II, Photometrics). The dichroic BS DB2 separates the light from the crimson reference beacon and the YFP-labeled sample. The dichroic BSs DB1 and DB3 are used to separate the excitation light and its associated back reflection. The fluorescent light emitted by the YFP is filtered by F3 and detected by the photomultiplier tube (PMT). The wavefront aberration is corrected by the DM.

Two laser channels are included in the system: a solid-state laser ($\lambda = 515$ nm) and a helium-neon laser ($\lambda = 633$ nm) used for confocal fluorescence imaging and wavefront sensing, respectively. The two channels share the same light path through the DM, relay lenses, scanner, and scanning lens and feed into an Olympus IX71 inverted microscope (Olympus Microscope, Center Valley, PA) through its side optical port. The focused beam is scanned on the sample in a raster pattern with a resonant scanner (SC-30, 16 kHz, 5°, Electro-Optics Products Corp.) and a vertical scanner (GVS001, Thorlabs). The control signals for the scanners are generated by a data-acquisition board (PCIe-6363, National Instruments Corporation). The emission light from both the crimson microsphere guide-star and the YFP-labeled sample are collected by an objective lens (Olympus 60×, NA1.2 water immersion) and separated by a dichroic mirror (DB2). The light from the sample was focused onto a pinhole by an achromatic lens L10. A GaAs PMT (H422-50, Hamamatsu) was used as a photon detector. The signal from the PMT is then fed to a frame-grabber board (Helios eA/XA, Matrox Imaging). With the vertical and horizontal synchronized signal from the DAQ, the frame-grabber board generates a raw image of 512×512 pixels at 30 frames/second for live imaging. When the signal levels are low, more frames can be averaged to increase the signal-to-noise ratio (SNR). We have also used two galvo scanners (6215H, Cambridge Technologies) at lower frame rates (3 Hz) to increase the SNR. The experimental system was setup on an optical table with vibration isolation as shown in Figure 17.17. The beam from the scanning system was fed into the Olympus IX71 inverted microscope by a periscope.

17.3.2 Results for Fixed Mouse-Brain Tissue

To investigate the feasibility of the proposed system, a fixed brain slice from a YFP-H line transgenic mouse was prepared. YFP is labeled on the cell body and protrusions of the neurons. One micron

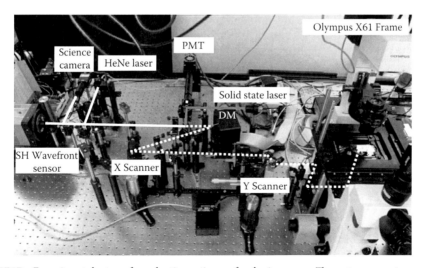

FIGURE 17.17 Experimental setup of an adaptive optics confocal microscope. The system was setup on an optical table with vibration isolation. The optical path for the crimson reference beacon (guide star) is shown as a white solid line. A HeNe laser is used for excitation of the guide star and its emission wavefront is measured with a Shack-Hartman wavefront sensor. The optical path from the microscope to the photomultiplier tube (PMT) for detecting the sample fluorescent emission (YFP) is shown as a dashed line. The excitation path for the sample (515 nm solid-state laser) is shown as a grey solid line.

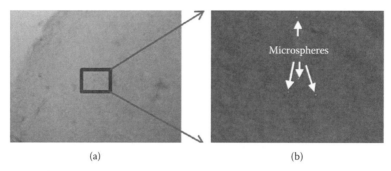

(a)   (b)

FIGURE 17.18   Wide-field microscopy images of mouse-brain tissue (thickness = 100 μm) with (a) 10× objective and (b) 60× water immersion objective. The focal plane is at the bottom of the tissue. The microsphere is deposited on the glass slide.

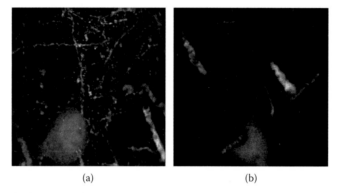

(a)   (b)

FIGURE 17.19   Confocal imaging of mouse-brain tissue without AO. The thickness of tissue is 100 μm. (a) The image on the top surface of the sample. (b) The confocal image at the bottom surface of the sample.

diameter crimson fluorescent microspheres (Invitrogen, Carlsbad, CA) were deposited onto a glass slide and below a cover plate for use as laser guide-stars. Sample brain coronal sections of different thicknesses (15, 50, and 100 μm) were cut with a microtome. The tissues are mounted with antifade mounting medium (Invitrogen) on glass slides coated with fluorescent microspheres. Figure 17.18a shows the image of the brain tissue with a thickness of 100 μm. The microspheres below the tissue are shown in Figure 17.18b.

17.3.2.1 Confocal Imaging without Wavefront Compensation

The spherical aberration induced by the cover glass is initially compensated by adjustment of a correction collar on the objective lens. The system aberration is further corrected by the DM by measuring the wavefront aberration from a microsphere at the bottom of the cover plate using the SHWF sensor. The thickness of the brain tissue is 100 μm. The confocal imaging system scans from the top surface to the bottom surface. The FOV is about 50 μm. The frame rate is 30 frames/second for fast scanning. To improve the SNR, 300 frames are averaged to generate the final image. Figure 17.19a shows the confocal image at the top surface, where the cell body, dendrite, and spine can be observed. Figure 17.19b shows the image at the bottom surface. Because of the aberrations induced by the tissue, it is very hard to see the dendrites and spines around the cell body.

17.3.2.2 Confocal Imaging with Wavefront Compensation

Brain tissues with thickness of 15, 50, and 100 μm are imaged in the AO system. A motorized Z stage under the sample focuses the HeNe laser on the microsphere at the bottom of the tissue. The wavefront

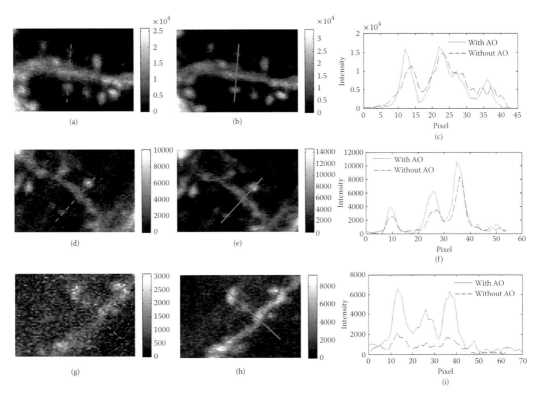

FIGURE 17.20 (**See color insert.**) Confocal images with and without wavefront error correction, respectively, and the intensity profiles along the line indicated in the confocal images. (a) and (b) The images before and after correction for brain tissue with 15 μm thickness. (c) The intensity profile along the lines indicated in (a) and (b). (d)–(f) The images before and after correction and intensity profile for brain tissues with 50 μm thickness. (g)–(i) The results for brain tissue with 100 μm thickness.

error induced by the sample is measured by the SHWF sensor. A stack of 20 confocal images with a flattened mirror surface are collected by scanning along the z-axis with a range of 3 μm and step size of 0.15 μm between each image plane. The final image is obtained by using the maximum-intensity projection applied on these images. After turning on the wavefront correction loop, the wavefront error converges after around 10 iterations, which takes about 0.35 seconds. Then the confocal images are collected with the same settings as were collected before correction.

The confocal images of the brain tissue with 15 μm thickness before and after correction are shown in Figure 17.20a and b, respectively (Tao et al. 2011b). The images of the dendrite and spines are clearer after correction. The intensity profile along the line crossing the dendrite and the spine is shown in Figure 17.20c. Both the signal intensity and the image contrast are improved. The signal intensity increases by 32%. The RMS wavefront error is 0.11λ (λ = 633nm) before correction. After correction, RMS wavefront error was reduced to 0.01λ. Figure 17.20d and e show the confocal images of brain tissue with 50 μm thickness before and after correction. Because the system suffers more aberration, the image of the spines becomes dimmer. The intensity profile along the dashed line and solid line is shown in Figure 17.20f. The signal intensity increases by 43%. The RMS wavefront error was reduced from 0.19λ to 0.03λ. Although the diffraction limit image is achieved, the image is still suffering from the scattering effect of the brain tissue. The increase in the signal intensity could be smaller than the theoretical calculation without consideration of the scattering effect. For the brain tissue with 100 μm thickness, it is very hard to observe any feature as shown in Figure 17.20g. After wavefront correction, the dendritic spine can be clearly observed as shown in Figure 17.20h. The intensity profile is shown in Figure 17.20i. The signal

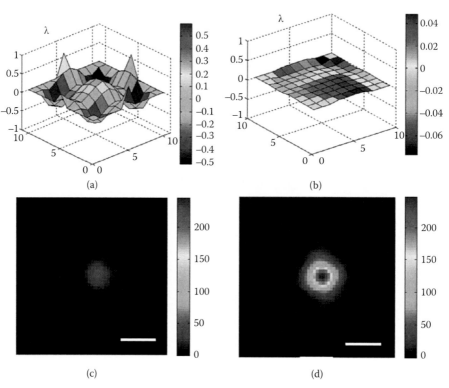

FIGURE 17.21 Wavefront measurements from a fluorescent microsphere guide-star through 100-μm-thick brain tissue. Wavefront error before correction (a) and after correction (b). The RMS errors for (a) and (b) are 0.24λ and 0.028λ (λ = 633 nm), respectively. Images of the microsphere guide-star before correction (c) and after correction (d). The scale bar is 1 μm.

TABLE 17.3   Statistical Properties of the Strehl Ratio for Brain Tissues

		Strehl ratio			
		Before correction		After correction	
No.	Thickness (μm)	Mean	Std Dev	Mean	Std Dev
1	15	0.52	0.025	0.99	0.003
2	50	0.27	0.027	0.96	0.021
3	100	0.05	0.017	0.92	0.034

intensity increases by 240%. The wavefront error for the crimson guide-star before correction is shown in Figure 17.21a, which suffers a large amount of spherical aberration (Tao et al. 2011). The RMS error is 0.24λ. After correction, the RMS wavefront error decreased to 0.028λ as shown in Figure 17.21b. The images of the microsphere before and after correction are shown in Figure 17.21c and d. The Strehl ratio was measured using the method described in Azucena et al. (2010) at 10 different positions on each sample (Table 17.3). The improvement in the Strehl ratio was approximately 4.3× for a 100-μm-thick sample.

17.3.2.3 Comparison between a Commercial Confocal Microscope and the AO Confocal Microscope

To compare our system's performance with a standard commercial confocal system (Leica TCS SP5 II), the resolution was investigated for a 50-μm-thick brain tissue sample on both systems using objective lenses with the same NA (1.2) and detectors with the same-size pinholes (0.9 Airy units). The transverse/axial resolutions were improved from 0.32/1.42 μm for the commercial system to 0.23/0.8 μm for our AO system after correction. Although reducing the pinhole diameter can increase the resolution, the commercial system still suffers from aberration. The lateral/axial resolution for the commercial system with 0.5 Airy units is 0.29/1.11 μm.

17.4 Using Fluorescent Proteins as Guide Stars

To make a stable and fast wavefront measurement, a direct wavefront measurement method has been demonstrated, where a fluorescent microsphere injected into the sample was used as a reference source for a SHWF sensor (Azucena et al. 2011; Tao et al. 2011a, b). The exposure time during wavefront sensing, typically 35 ms, is much shorter than the indirect wavefront methods (Debarre et al. 2009; Ji, Milkie, & Betzig 2010), which reduces the possibility of phototoxicity and photobleaching. It also enables higher-speed imaging for dynamic live samples. However, this method requires the injection of fluorescent microspheres into the sample, which complicates the sample preparation procedure. The microspheres are required to cover the whole region of interest, and the distance between two microspheres should be less than the isoplanatic width (Azucena et al. 2010). As an invasive method, the side effects of injection to the functionality of the biological tissue need to be considered for live imaging.

To overcome these disadvantages and generalize the direct wavefront sensing method, we use fluorescent proteins as laser guide-stars (Tao et al. 2011c). Here we use fluorescent proteins that label particular cellular structures rather than specific layers such as the retina in the eye (Biss et al. 2007; Diaz Santana Haro & Dainty 1999). An example is green fluorescent protein (GFP), from the bioluminescent jellyfish *Aequorea victoria* (Lakowicz 2006). The fluorophore unit is located within a barrel of β-sheet protein, which shields the chromophore from the local environment. Because of this special structure, GFP has good photostability and high quantum yields (Lakowicz, 2006). This feature makes GFP a good candidate as a laser guide-star in AO microscopy. As a noninvasive fluorescent marker, GFP has been extensively used in live cell imaging. AO microscopy can be easily applied to those studies without special preparation of the samples. The possible damage to the sample from the injection of a guide-star reference bead can be avoided. Here the genetic mutant of GFP, YFP yellow fluorescent protein, was used for wavefront measurement (Patterson et al., 1997).

17.4.1 Method

Figure 17.22 shows the system setup, which integrates an AO system into a confocal microscope. A solid-state laser ($\lambda = 515$nm) provides the excitation light for both fluorescence imaging and wavefront sensing instead of separate lasers (515 and 633 nm) for the system, shown above in Figure 17.16. The light is fed into an objective lens (×60 water objective, NA 1.2, Olympus) and scanned on the sample in a raster pattern with two galvo scanners (6215H, Cambridge Technology). The emission light from the fluorescent protein is divided by a BS with a 50/50 splitting ratio. Half of the light was collected by the PMT (H422-50, Hamamatsu), which generates the confocal image. The other half of the light is used for wavefront sensing. We are investigating the use of a flip-mirror to guide 100% of the fluorescent light to the SHWF sensor for wavefront sensing or to the PMT for imaging to avoid splitting the light 50/50 with the BS. Four relay lens groups make the exit pupil of the objective lens conjugate with the SHWF sensor, DM, and scanners. To minimize the amount of out-of-focus light that enters the SHWF sensor, irises I1 and I2 were placed in the light path. These irises also block stray light from the DM, scanner, and lenses. The position of the spots in the image from the wavefront sensor was detected using a cross-correlation centroiding algorithm (Thomas et al., 2006). Then an FFT reconstruction algorithm was implemented to obtain the wavefront (Poyneer, 2003). To make an accurate wavefront measurement using a SHWF sensor, the diameter of the guide star should be smaller than the diffraction limit of the wavefront sensor, which is defined in Equation 17.1. In our current setup, $d_{\text{diffraction\_limit}} = 5.36$ μm.

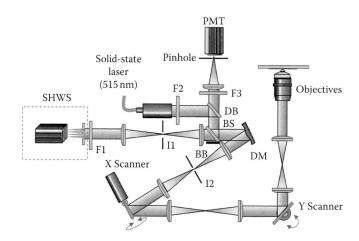

FIGURE 17.22 AO confocal microscope setup for using fluorescent protein guide-stars. A solid-state laser (515 nm) is used as the excitation light source for YFP. A dichroic beam splitter (DB) separates the emission and excitation light. Light emitted from a YFP-labeled cellular structure is collected by an objective lens and split for wavefront measurement and confocal imaging by a 50/50 BS. The wavefront error is measured by a SHWS and then corrected by a DM. Defocused and stray light is blocked by irises I1 and I2. F2 is an excitation filter. F3 and F1 are emission filters. BB is a beam blocker. A PMT located behind a pinhole collects a confocal image of the sample's fluorescence.

Because the fluorescent light from a given point is proportional to the light intensity illuminating that point, the size of the guide star is determined by the illumination PSF. The PSF is calculated from the wavefront measurement using the method described in Beverage, Shack, and Descour (2002). For GFP at a depth of 70 μm, the diameter of the PSF with 80% of the encircled energy is 1.4 μm, which is small enough to be used as a guide star.

To investigate the feasibility of the proposed system, a fixed brain-tissue slice from a YFP-H line transgenic mouse was prepared. Sample brain coronal sections (100 μm) were cut with a microtome. The tissues were mounted on the slide with antifade reagents (Invitrogen). The spherical aberration induced by the cover plate was compensated by adjustment of a correction collar on the objective lens. The system aberration was further corrected by the DM by measuring the wavefront aberration from the top surface of the tissue. During the experiment, the confocal images were initially captured without wavefront correction. The approximate location of the GFP was identified from this initial image. Then the scanning mirror was used to steer the laser beam to the region of the GFP, and a wavefront measurement was performed. The DM corrected the aberration in a closed loop using the direct-slope algorithm (Porter et al. 2006). Then the corrected confocal image of the isoplanatic region (Azucena et al. 2010) around the GFP guide star is captured with the optimal shape of DM as determined from the wavefront measurement.

During the wavefront measurement, the laser is focused on a stationary point, which may cause photobleaching of GFP. By decreasing the laser power and increasing the exposure time of the SHWF sensor, sufficient signal can be captured by the SHWF sensor during the wavefront measurement before photobleaching. In this system, the laser power is controlled by a software and is turned down to 23 nW at the back aperture of the objective lens during wavefront sensing and turned up to 20 μW during confocal scanning. The exposure time of the SHWF sensor could be decreased before photobleaching by increasing the laser power and using a high-speed shutter to limit the exposure time of the laser on the sample, which will be considered in the future. GFP on the dendrite and cell body of the neuron were tested as laser guide-stars at depths of 25 and 70 μm, respectively, as shown in Figure 17.23a and b. The crossed lines show the location of the laser focus. The exposure time of the wavefront measurement for the dendrite and cell body is 500 and 30 ms, respectively. Because the dendrites have less GFP than the cell bodies, longer exposure times are required to collect enough signal for accurate wavefront measurement.

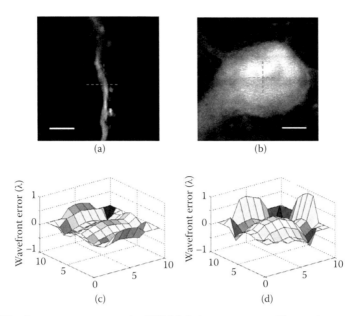

FIGURE 17.23 Wavefront measurements using YFP-labeled structures as guide stars in mouse brain tissue. The excitation light source is focused on a dendrite in (a) and on a cell body of the neurons in (b). The crossed lines indicate the location of the focus point for the wavefront measurements. The wavefront measurements using a YFP-labeled structure is shown for a dendrite in (c) and a cell body in (d). The scale bar is 5 μm.

The measurement errors for different numbers of photons per subaperture can be measured using the method described in Morzinski et al. (2010). The EM gain setting for the EMCCD in SHWF sensor was 200×. The photons per subaperture for the wavefront measurements shown in Figure 17.23c and d are 1950 and 2047, which produce 0.03λ in measurement error. This amount of measurement error is small and improvements at this level will also depend on the errors caused by limitations of the DM and wavefront temporal variations. The RMS wavefront errors shown in Figure 17.23c and d are 0.1742λ and 0.3471λ (λ = 527 nm), respectively. Photobleaching is tested for the YFPs in a dendrite, because it has a lower concentration of fluorophores and thus photobleaches more quickly. Confocal images were obtained every 30 seconds during the exposure of the laser with 23 nW at the back aperture of the objective lens for wavefront sensing. The intensity change in the focus area is shown in Figure 17.24. After the first three minutes, the intensity drops less than 10%. Considering the exposure time (500 ms) for the wavefront measurement, photobleaching caused by the wavefront measurement is very limited. For live samples, higher laser power can be applied because the fluorescence recovers after photobleaching. Therefore, shorter exposure times can be achieved. The line profile along two subapertures on the SHWF sensor image is shown.

The confocal images for the sample are collected by scanning along the *z*-axis with a 3 μm range and a 0.15 μm step size. The final images, as shown in Figure 17.25, are achieved using the maximum intensity projection applied to the images. After turning on the wavefront correction loop, the wavefront error converges after 10 iterations, which takes 0.30 seconds. The YFP on the cell body was used as a guide-star, which is located at a depth of 70 μm, as shown in Figure 17.23b. Before correction, the RMS wavefront error is 0.3471λ. After correction, the measured wavefront error is 0.03λ RMS, as shown in Figure 17.25e. The confocal images before correction and after correction are shown in Figure 17.25a and b. The intensity profile along the dashed lines across a dendrite and a spine is shown in Figure 17.25f. The intensity increased by 3×. The image of the dendrite and the spines are much clearer after correction with improved image contrast.

We also tested the fruit-fly embryos labeled with GFP-polo and EGFP-Cnn (Tao et al., 2012). To test the ability of the wavefront correction at a deep depth, the confocal images with and without corrections

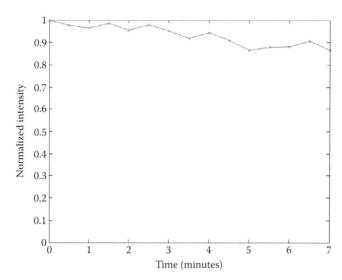

FIGURE 17.24 Photobleaching analysis. Normalized intensity at the focal point on the dendrite.

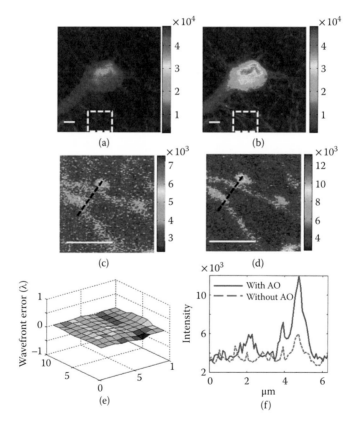

FIGURE 17.25 Confocal fluorescence imaging of mouse-brain tissue using an YFP-labeled cell body of a neuron as a guide star. The maximum-intensity projection image within the isoplanatic region of the guide star before correction (a) and after correction (b). The dashed boxes indicate enlarged images before correction (c) and after correction (d). (e) The wavefront error after correction. (f) Intensity profiles along the dashed line in the uncorrected image (c) and along the solid line in the corrected image (d). The scale bar is 5 µm.

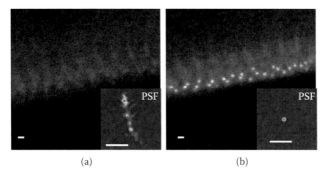

FIGURE 17.26 The images and PSF without (a) and with (b) correction for a cycle 14 fruit fly embryo with green fluorescent protein-polo at the depth of 83 μm. The scale bar is 2 μm.

are captured at the depth of 83 μm as shown in Figure 17.26. The GFP-polo-labeled centrosomes can be observed clearly after correction but cannot be observed before correction. The size of the PSF decreases from 1.7 to 0.21 μm. The Strehl ratio calculated based on the PSF shows an increase from 3.3×10^{-3} to 0.7. The penetration depth for live imaging of a fruit-fly embryo is tested by performing AO correction during Z scanning from the top surface to depth of 100 μm. The maximum-intensity projection of the scan series shows the GFP at the edge of embryo at different depths as shown in Figure 17.27a and b. Before correction, the EGFP-Cnn-labeled centrosomes can be observed only up to 60 μm depth. After correction, they can be observed below a depth of 80 μm. Using the 3D view function in ImageJ with the resampling factor of two, the 3D images of the fruit-fly embryo show that the penetration depth increases from 60 to 95 μm, with more than a 50% increase in imaging depth as shown in Figure 17.27c and d. Without correction, the RMS wavefront error reaches to 0.8λ when the imaging depth goes to 90 μm. The decrease in the Strehl ratio shows the degradation of the optical performance with the imaging depth as shown in Figure 17.27e. After correction, even at the depth of 90 μm, the system can still achieve a Strehl ratio of 0.6 with an RMS wavefront error of 0.1λ. Aside from improving the penetration depth, the system also improves the optical resolution. Although the EGFP-Cnn-labeled centrosomes can be observed at a depth of 60 μm without AO, the resolution is still poor because of the aberrations. Before correction, the size of the PSF is 1.67 μm at a depth of 60 μm. After correction, it decreases to 0.2 μm as shown in Figure 17.27f. At a depth of 90 μm, it shows a significant improvement of the PSF by a factor of nine.

17.5 Discussion and Conclusion

One of the challenges in designing an SHWF sensor is imposed by the amount of light the reference source can provide. Polystyrene microspheres are loaded with fluorescent dye, and the light emitted is proportional to the radius cubed; thus, smaller beads provide less light. The size of the beads should be smaller than the diffraction limit of one subaperture of the Hartmann wavefront sensor. Note that this is larger than the diffraction limit of the microscope aperture by the ratio D (size of the aperture)/d_{LA}, as shown in Equation 17.1. Since the diffraction limit of the microscope is inversely proportional to the NA, smaller beads are needed for higher-NA systems. Fortunately the light gathered by the objective also increases with increasing NA (light-gathering power ~ NA^2). Increasing the wavefront sampling by a factor of four increases the size of the microsphere radius by a factor of two and the amount of light emitted by a factor of eight. Current results show that for a 40× objective with an NA of 0.75, a 1 μm fluorescent microsphere provides enough light to run the AO system loop with a 10 ms period. If the size of the bead was reduced by a factor of 10 to 100 nm in radius, we could still obtain a good correction by using the AO system but with the disadvantage of a lower SNR. This could be compensated by increasing the AO loop period to 100 ms or by reducing the number of pixels used for each subaperture. Both these approaches come from the fact that the SNR of the SHWF sensor can be improved by increasing the integration time of the CCD camera or by using fewer pixels to detect the movement of the centroids.

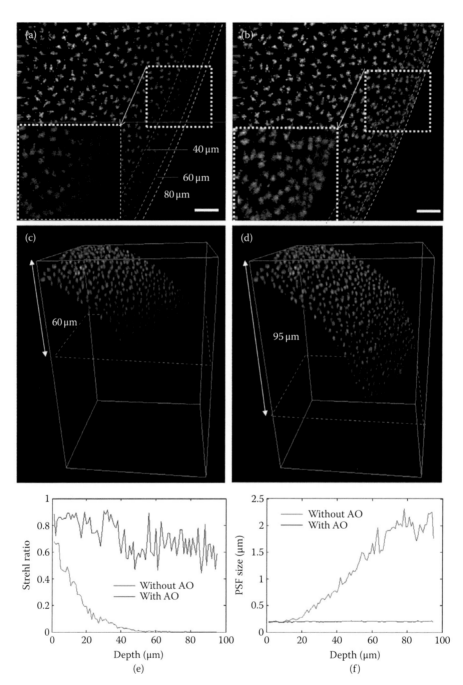

FIGURE 17.27 (**See color insert.**) Comparison of penetration depths between without and with correction for imaging of cycle 13 fly embryos with EGFP-Cnn label. (a)–(b) The maximum intensity projection of the scan series from the top surface to 100 μm with and without adaptive optics. (c)–(d) The three-dimensional reconstructions with and without AO. (e)–(f) The Strehl ratio and PSF size change for different depth. The black and gray lines indicate without and with AO, respectively. The scale bar is 10 μm.

When using fluorescent microspheres for guide stars, the microsphere solution concentration injected into the embryo ensures that there is at least 1 microsphere within 10 μm of the injection site. This can be used to accurately target different embryo locations for imaging. Higher microsphere concentrations have also been tested and the results showed that beads can spread much more densely without impacting embryo development. Having multiple microspheres relatively close to each other does not present a problem since the confocal illumination setup used in this experiment can accurately target one bead at a time. Experimental results showed that even if two or more microspheres are in the same focal plane and are "relatively" close to each other (i.e., within 5 PSF's FWHM), the wavefront sensor would see an extended object and the resulting measurement would be the average wavefront seen from each microsphere (Azucena et al. 2010). Microspheres that are in different focal planes present even less of a problem since the finite focus of the confocal illumination prevents them from being fully illuminated (because of the high NA), and hence, they are not the brightest object in the FOV. Microspheres that are in different focal planes do show up as a background light in the wavefront sensor camera. It has been shown by Thomas et al. (2006) that a robust centroiding algorithm, like the cross-correlation technique used here, does not sense the background and detects only the brightest object in the FOV, be it an extended or a point source. This still holds true even for a very small SNR (i.e., the background would add to the peak of the Hartmann spot centroid but it does not move it). The same can be said for light that is being scattered inside the embryo. It adds to the background but it does not change the wavefront sensor measurement.

The accuracy and stability of direct wavefront sensing depends on the amount of light from the fluorescent microsphere. The lateral and axial positions of the microsphere affect the accuracy of the wavefront measurement. In astronomy, the isoplanatic angle was defined as a relative measure of the FOV over which the AO system can operate. In AO microscopy, the isoplanatic half-width was used as the size of the FOV that the AO system can correct. It has been tested for the *Drosophila* embryo using a 40× objective (0.75 NA). This value varies for different types of samples and objectives. It will be evaluated for mouse-brain tissue in future research. In the confocal microscope, the images are acquired point-by-point and reconstructed by the computer. If there are multiple microspheres in the FOV, AOCFM has the ability to correct the different aberrations for different parts of the FOV in one image. The variation of wavefront error along with the depth is also being investigated. Although the wavefront measurement requires microspheres in the focal plane, the wavefront can also be estimated by linear interpolation if measurements of the wavefront are available in at least two planes close to the focal plane.

For a live sample, real-time wavefront measurement and correction is a more challenging task. The wavefront measurement is based on the emission light from the microsphere. However, the internal flow or motion of the liquid inside the cell or tissues can cause an unpredictable motion of the microsphere. Focusing the illumination laser onto the microsphere in three dimensions can also be difficult. One possible solution is to steer the beam to compensate for the motion of the microsphere with an appropriate image stabilization algorithm.

An emerging field in AO is tomography AO, where multiple light sources together with multiple SHWF sensors are used. The information from each wavefront sensor is then processed using a reconstructor to acquire a tomographic image of the changes in the refraction index in the optical path (Hardy, 1998). One of the advantages of using tomography AO is that it can provide information on the depth dependence of variations in the refraction index in the tissue, thus allowing the AO system to correct for the wavefront aberrations only in the optical path. This technique can also extend the isoplanatic angle by correcting wavefront aberrations that are common to a larger FOV. By depositing multiple fluorescent beads into the biological sample and using multiple wavefront sensors, we can also apply the tomographic techniques that have been developed for astronomical AO.

AO wide-field and confocal microscopes were designed using an SHWF sensor and Micro-Electro-Mechanical Systems DM to directly measure and correct the wavefront error induced by the biological samples. For the wide-field microscope, the wavefront measurements were taken by using a new method of seeding an embryo with fluorescent microspheres that are used as "artificial guide stars." The experiment for *Drosophila* embryo tissue samples for the adaptive optical wide-field microscope has demonstrated a Strehl-ratio improvement as high as 10× when imaging through 100 μm of tissue. For the confocal microscope, the wavefront measurements were taken using fluorescent microsphere guide-stars and also using fluorescent protein guide-stars that are incorporated in the cellular structure of the tissue being studied. The experiment for the fixed mouse-brain tissue sample for the AO confocal microscope demonstrated a 4.3× improvement in the Strehl ratio and 2.4× increase in intensity in the final image after wavefront error correction. These AO microscopes with direct wavefront sensing can effectively measure and correct aberrations at a high speed, which is important for live in vivo imaging.

References

Adkin, S., Azucena, O., & Nelson, J. (2006). The design and optimization of detectors for adaptive optics wavefront sensing. *SPIE.* **6272**, 62721E.

Albert, O., Sherman, L., Mourou, G., Norris, T., & Vdovin, G. (2000). Smart microscope: an adaptive optics learning system for aberration correction in multiphoton confocal microscopy. *Opt. Lett.* **25**, 52–54.

Azucena, O., Cao, J., Crest, J., Sullivan, W., Kner, P., Gavel, D., Dillon, D., Olivier, S., & Kubby, J. (2010). Implementation of adaptive optics in fluorescent microscopy using wavefront sensing and correction. *Proc. SPIE.* **7595**, 75901–75950I-9.

Azucena, O., Crest, J., Cao, J., Sullivan, W., Kner, P., Gavel, D., Dillon, D., Olivier, S., Kubby, J. (2011a). Wavefront aberration measurements and corrections through thick tissue using fluorescent microsphere reference beacons. *Opt. Express.* **18**, 17521–17532.

Azucena, O., Crest, J., Kotadia, S., Sullivan, W., Tao, X., Reinig, M., Gavel, D., Olivier, S., & Kubby, J. (2011b). Adaptive optics wide-field microscopy using direct wavefront sensing. *Opt. Lett.* **36**, 825–827.

Babcock, H. W. (1953). The possibility of compensating astronomical seeing. *Pub. Astron. Soc. Pac.* **65**, 229–236.

Beverage, J., Shack, R., & Descour, M. (2002). Measurements of the three-dimensional microscope point spread function using a Shack-Hartmann wavefront sensor. *J. Microscopy.* **205**, 61–75.

Biss, D. P., Sumorok, D., Burns, S. A., Webb, R. H., Zhou, Y., Bifano, T. G., Côté, D., Veilleux, I., Zamiri, P., & Lin, C. P. (2007). In vivo fluorescent imaging of the mouse retina using adaptive optics. *Opt. Lett.* **32**, 659–661.

Booth, M. (2007). Adaptive optics in microscopy. *Phil. Trans. A. Math Phys Eng. Sci.* **365**, 2829–2843.

Debarre, D., Botcherby, E., Booth, M., & Wilson, T. (2008). Adaptive optics for structured illumination microscopy. *Opt. Express.* **16**, No. 13, 17137–17142.

Debarre, D., Botcherby, E. J., Watanabe, T., Srinivas, S., Booth, M. J., & Wilson, T. (2009). Image-based adaptive optics for two-photon microscopy. *Opt. Lett.* **34**, 2495–2497.

DeMarais, A., Oldis, D., & Quatrro, J. M. (2005). Matrotrophic transfer of fluorescent microspheres in Poeciliid fishes. *Copeia.* **2005**, 632–636.

Diaz Santana Haro, L., & Dainty, J. C. (1999). Single-pass measurements of the wave-front aberrations of the human eye by use of retinal lipofuscin autofluorescence. *Opt. Lett.* **24**, 61–63.

Dunn, A., & Richards-Kortum, R. (1996). Three-dimensional computation of light scattering from cells. *IEEEE J. Sel. Top. Quantum Electron.* **2**, 898–905.

Gavel, D. (2003). Suppressing anomalous localized waffle behavior in least-squares wavefront reconstructors. *Proc. SPIE.* **4839**, 972–980.

Guldband, S., Simonsson, C., Goksor, M., Smedh, M., & Ericson, M. (2010). Two-photon fluorescent correlation microscopy combined with measurements of point spread function: investigations made in human skin. *Opt. Express.* **18**, 15289–15302.

Hardy, J. W. (1998). *Adaptive Optics for Astronomical Telescopes.* New York: Oxford University Press.

Invitrogen Corporation. (2010). *Fluorecent SpectraViewer*. Retrieved 13 January 2011, from http://www .invitrogen.com/site/us/en/home/support/Research-Tools/Fluorescence-SpectraViewer.html

Ji, N., Milkie, D. E., & Betzig, E. (2010). Adaptive optics via pupil segmentation for high-resolution imaging in biological tissues. *Nat. Methods. 7*, 141–147.

Kalpin, R., Daily, D., & Sullivan, W. (1994). Use of detran beads for live analysis of the nuclear division and nuclear envelope breakdown/reformation cycles in the *Drosophila* embryo. *Biotechniques. 17*, 730–733.

Lakowicz, J. R. (2006). *Principles of Fluorescence Spectroscopy*. New York: Springer.

Liang, J., Grim, S., Goelz, S., & Bille, J. F. (1994). Objective measurements of wave aberrations of the human eye with the use of a Hartmann-Shack wavefront sensor. *J. Opt. Soc. Am. A11*, 1949–1957.

Liang, J., Williams, D. R., & Miller, D. T. (1997). Supernormal vision and high-resolution retinal imaging through adaptive optics. *J. Opt. Soc. Am. A14*, 2884–2892.

Marsh, P., Burns, D., & Girkin, J. (2003). Practical implementaion of adaptive optics in multiphoton microscopy. *Opt. Express. 11, No. 10*, 1123–1130.

Minsky, M. (1961). Microscopy Apparatus, *Patent No. 3,013,467*. USA.

Morzinski, K., Johnson, L. C., Gavel, D. T., Grigsby, B., Dillon, D., Reinig, M., & Macintosh, B. A. (2010). Performance of MEMS-based visible-light adaptive optics at Lick Observatory: Closed- and open-loop control, *Proc. SPIE. 7736*, 773659–773659-16.

Patterson, G. H., Knobel, S. M., Sharif, W. D., Kain, S. R., & Piston, D.W. (1997). Use of the green-fluorescent protein (GFP) and its mutants in quantitative fluorescence microscopy. *Biophys. J. 73*, 2782–2790.

Porter, J., Queener, H., Lin, J., Thorn, K., & Awwal, A. (2006). *Adaptive Optics for Vision Science*. Hoboken, NJ: Wiley-Interscience.

Poyneer, L. A., Gavel, D. T., & Brase, J. M. (2002). Fast wave-front reconstruction in large adaptive optics systems with use of the Fourier transform. *J. Opt. Soc. Am. A. 19*, 2100–2111.

Poyneer, L. (2003). Scene-based Shack-Hartmann wave-front sensing: analysis and simulation. *Appl. Opt. 42, No 29*, 5807–5815.

Rothwell, W., & Sullivan, W. (2000). Fluorescent analysis of *Drosophila* embryos. In W. Sullivan, M. Ashburne, & R. Hawley, *Drosophila Protocols* (pp. 141–157). Cold Spring Harbord, NY: Cold Spring Harbor Laboratory Press.

Rowning, B., Well, J., Wu, M., Gerhart, J., Moon, R., & Larabell, C. (1997). Microtubule-mediated transport of organelles and localization of bacatenin to the future dorsal side of Xenopus eggs. *Proc. Natl. Acad. Sci. USA 94*, 1224–1229.

Rueckel, A., Mack-Bucher, J. A., & Denk, W. (2006). Adaptive wavefront correction in two-photon microscopy using coherence-gated wavefront sensing. *Proc. Natl. Acad. Sci. USA 103*, 17137–17142.

Slotine, J., & Li, W. (1991). *Applied Nonlinear Control*. Englewood Cliffs, NJ: Prentice Hall.

Schwertner, M. (2007), Specimen-induced distortions in light microscopy, *J. Microscopy. 228*, 97–102.

Schwertner, M., Booth, M., Neil, M., & Wilson, T. (2004). Measurement of specimen-induced aberrations of biological samples using phase stepping interferometry. *J. Microscopy. 213*, 11–19.

Tao, X., Fernandez, B., Azucena, O., Fu, M., Garcia, D., Zuo, Y., et al. (2011a). Adaptive Optics confocal microscopy using direct wavefront sensing. *Opt. Lett. 36*, 1062–1064.

Tao, X., Fernandez, B., Chen, C., Azucena, O., Fu, M., Zuo, Y., & Kubby, J. (2011b). Adaptive optics confocal fluorescence microscopy with direct wavefront sensing for brain tissue imaging. *Proc. SPIE. 7931*, 79310L.

Tao, X., Azucena, O., Fu, M., Zuo, Y., Chen, D., & Kubby, J. (2011c). Adaptive optics microscopy with direct wavefront sensing using fluorescent protein guide stars. *Opt. Lett. 36*, 3389–3391.

Tao, X., Crest, J., Kotadia, S., Azucena, O., Chen, D.C., Sullivan, W., & Kubby, J. (2012). Live imaging using adaptive optics with fluorescent protein guide-stars. *Opt. Express. 20*, 15969–15982.

Thomas, S., Fusco, T., Tokovinin, A., Nicolle, M., Michau, V., & Rousset, G. (2006). Comparison of centroid computation algorithms in a Shack-Hartmann sensor. *MNRAS. 371*, 323–336.

Tyson, R. (1998). Reconstruction and controls. In R. Tyson, *Principles of Adaptive Optics*, 2nd ed (252–275). San Diego, CA: Academic Press.

Van Helden, A. (1977), The invention of the telescope. *Tran. Am. Phil. Soc. 67, no. 4*.

Index

Printed and bound by CPI Group (UK) Ltd, Croydon, CR0 4YY

24/10/2024

01778309-0006